J. Vad · T. Lajos · R. Schilling

Modelling Fluid Flow

János Vad · Tamás Lajos
Rudolf Schilling (Eds.)

Modelling Fluid Flow

The State of the Art

With 281 Figures

Springer

János Vad
Department of Fluid Mechanics
Budapest University of Technology and Economics
Bertalan Lajos u. 4-6
H-1111 Budapest
Hungary

Tamás Lajos
Department of Fluid Mechanics
Budapest University of Technology and Economics
Bertalan Lajos u. 4-6
H-1111 Budapest
Hungary

Rudolf Schilling
Lehrstuhl für Fluidmechanik,
Abteilung Hydraulische Maschinen
Technische Universität München
Boltzmannstr. 15
D-85748 Garching
Germany

Library of Congress Control Number: 2004106672

ISBN 978-3-642-06034-2 ISBN 978-3-662-08797-8 (eBook)
DOI 10.1007/978-3-662-08797-8

springeronline.com

Originally published by Springer-Verlag Berlin Heidelberg New York in 2004
Softcover reprint of the hardcover 1st edition 2004

Typesetting: Digital data supplied by editors
Cover-Design: medio Technologies AG, Berlin
Printed on acid-free paper 62/3020 Rw 5 4 3 2 1 0

Preface

This volume, „Modelling Fluid Flow – State of the Art", presents the invited lectures, summaries of workshops and a selection of papers from the international conference CMFF '03 on fluid technology, organised at the Budapest University of Technology and Economics (BUTE), Hungary, 3 to 6 September 2003.

CMFF'03 was the twelfth event of international conference series on fluid flow technologies held in Budapest since 1959. The conference has been organised by the Department of Fluid Mechanics, BUTE, in collaboration with Department of Hydraulic Machines, BUTE; Department of Heat and Fluid Engineering, University of Miskolc; Scientific Society of Mechanical Engineers, Hungary (Flow Technology Section); Committee of Fluid Mechanics and Thermodynamics of the Hungarian Academy of Sciences; The Japan Society of Mechanical Engineers; and Visualization Society of Japan. The conference attracted almost two-hundred participants from about thirty countries worldwide.

The invited lectures follow the evolution and challenges of computational methods and measuring techniques related to fluid flow up to the present days. The workshop summaries reflect the recent trends, remaining questions and problems to be solved in the fields of experimental and computational fluid mechanics being in symbiosis. The selected papers cover a wide range of fluid engineering, including flows with chemical reactions, chemical and process engineering, environment, external fluid dynamics, turbulence modelling and numerical methods, and fluid machinery.

The Editors express their gratitude to the Journal of Computational and Applied Mechanics (JCAM), http://www.uni-miskolc.hu/home/web/pumns/mechanics/ for permission to include certain JCAM papers in the present volume.

The Editors Budapest, March 2004

Table of Contents

Invited Lectures

Workshop Summaries

Flows with Chemical Reactions; Chemical and Process Engineering; Environment; External Fluid Dynamics

Turbulence Modelling and Numerical Methods

Fluid Machinery

INVITED LECTURES

Application of CFD-Techniques in Fluid Machinery

R. SCHILLING*
Institute of Fluid Mechanics Hydraulic Machinery Dept.
Munich University of Technology

* Boltzmannstr. 15, 85748 Garching, Germany
Tel. (+49) 89/289 16296, Fax: (+49) 89/289 16297
Email: schilling@lhm.mw.tu-muenchen.de

Abstract The CFD-techniques being applied for the simulation of the operating behaviour and performance prediction as well as the design optimisation of fluid machinery are described and briefly discussed. Then, the developed CFD-codes are applied to predict the performances of axial and diagonal fans as well as of a centrifugal pump for which also the head drop due to cavitation has been simulated. Furthermore, the predicted performance curves are compared with measurements. Finally, the capability of the real-time design system is demonstrated with respect to the optimum design of the inlet and outlet blade angels as well as to the blade length of a centrifugal pump impeller.

Key Words: Cavitation, CFD, Fluid Machinery, Performance Prediction, Real-Time Design

Nomenclature

b	[m]	blade height
g	[m/s²]	acceleration due to gravity
$i = \beta s1 - \beta 1$	[°]	incidence angle
k	[m²/s²]	turbulent kinetic energy
m, n	[m]	meridional coordinates
nq	[1/min]	specific speed
p	[Pa]	pressure
r, φ, z	[m], [rad]	cylindrical coordinates
rref	[m]	reference radius
s	[m]	developed arc length
sk	[m]	tip clearance
t	[s]	time
u	[m/s]	circumferential velocity

D	[m]	impeller diameter
H	[m]	total head
NPSH	[m]	Net Positive Suction Head
NPSHi	[m]	incipient NPSH-value
NPSH3%	[m]	NPSH-value for 3% head drop
Q	[m^3/s]	flow rate
R	[m]	radius of bubbles
T	[°]	Temperature
V	[m^3]	total cell volume
α	[-]	vapour void fraction
β	[°]	relative flow angle
βs	[°]	blade angle
ε	[m^2/s^3]	turbulent dissipation rate
$\Delta\varphi w$	[°]	wrapping angle
ν	[-], [m^2/s]	hub to tip ratio, viscosity
ρ	[kg/m^3]	density
ω	[1/s]	angular velocity

Subscripts and Superscripts

1, 2	Blade inlet, outlet
l	liquid
t	turbulent
th	theoretical
va	vapour

Definitions

$c_p = \dfrac{2(p - p_{va})}{\rho u_2^2}$	[-]	pressure coefficient
$\varphi = \dfrac{8Q}{\pi D^3 \omega}$	[-]	flow coefficient
$\varphi_\nu = \dfrac{\varphi}{1-\nu^2}$	[-]	axial flow coefficient
$\varphi_r = \dfrac{\varphi}{\left(4\dfrac{b_2}{D}\right)}$	[-]	radial flow coefficient
$\Psi_t = \dfrac{8gH}{\omega^2 D^2}$	[-]	total energy coefficient
$\lambda = \varphi \cdot \Psi_t$	[-]	power density

$$\eta_h = \frac{H}{H_{th}}$$ [-] hydraulic efficiency

Abbreviations

3D	three dimensional
ANN	Artificial Neural Networks
B-B	Blade to Blade
DNS	Direct Numerical Simulation
H-S	Hub to Shroud
LES	Large Eddy Simulation
NEL	National Engineering Laboratory
RANS	Reynold-averaged Navier-Stokes equations
RTD	Real Time Design
TFA	Fachgebiet Turbomaschinen und Antriebstechnik, Darmstadt University of Technology
TUBS	Technical University of Braunschweig
URANS	unsteady Reynolds-averaged Navier-Stokes equations
VDMA	Verband Deutscher Maschinen- und Anlagenbau e.V.

1. Introduction

In the field of fluid machinery the techniques of CFD (Computational Fluid Dynamics) are widely used today for the computation of the fluid flow, the performance prediction and the design optimisation. The methods being in use to solve the governing set of partial differential equations, i.e. the conservation equations of mass, momentum and energy, are mainly the Finite Volume and the Finite Element approaches, s. [1].

Since the application of the Direct Numerical Simulation (DNS) and the Large Eddy Simulation (LES) approaches still are much too expensive to be applied to fluid machinery the Reynolds-averaged Navier-Stokes equations (RANS) are used to predict the quasi steady state operating behaviour. In the case of time-dependent flow simulations the unsteady Reynolds-averaged Navier-Stokes equations (URANS) are solved by a suitable time marching scheme. Despite the fact that the numerical techniques to solve the RANS or the URANS have reached a considerable level of maturity some sources of uncertainty are still existing.

Firstly, the discretization of the convective terms of the momentum equations is still a problem. Considering complex flows, such as the time-dependent and three-dimensional (3D) flow in turbomachinery components, numerical solutions can often be gained only by means of upwind approximation schemes of an order less than 2. Secondly, the turbulent structure of the flow in the various components of fluid machinery has to be modelled in a statistical way and verified by comparing with precise experimental data. Finally, the prescription of the inflow and outflow boundary conditions as well as the modelling of the rotor-stator interaction in the case of reverse flows may induce considerable numerical uncertainties. This kind of flow occurs at part load operating points where the real flow is strongly time-dependent and even separated. However, to save CPU-time the performance curves usually are computed by assuming steady flow conditions. But with decreasing flow rates the convergence behaviour of Navier-Stokes solvers is becoming worse and the order of the discretisation scheme has to be reduced towards a first order upwind scheme.

Thus, CFD-codes being in use to solve the RANS or the URANS still have to be validated and calibrated by comparing the numerically predicted results with highly precise measurements. In the following, the incompressible flow in hydraulic turbomachinery is considered.

2. CFD-Techniques

For the numerical simulation of the flow through the various components as well as for the CFD-aided design of bladings of fluid machinery four different CFD-codes have been developed following a multi-level CFD-strategy.

The low level codes, i.e. the quasi-3D Euler – and Navier-Stokes code as well as the 3D Euler code are only used in the real-time design process whereas the 3D Navier-Stokes code is applied for the steady and unsteady flow computation, the performance prediction and on the final design optimisation level.

2.1. Quasi-3D Euler Code (EQ3D)

Applying the theory of WU [2] the three dimensional inviscid flow through the bladings of hydraulic machinery may be approximated by iterating between a number N of blade-to-blade (B-B) flows on S1 - surfaces and a number M of hub-to-shroud (H-S) flows on S2 – surfaces as well as a number of flows in S3 – planes being perpendicular to the flow direction. A simplified but often used and very efficient flow model may be obtained by combining $N \times \mathrm{S1}$ axisymmetric flow surfaces with one representative $\mathrm{S2_m}$ flow surface, s. SCHILLING [3]. The resulting elliptical partial differential equations are solved by means of a node-centered finite volume scheme.

2.2. 3D Euler Code (E3D)

Since the real flow in hydraulic machinery bladings is three-dimensional the blading designed and optimised by the RTD-system has to be recalculated by a 3D code. Considering only the design point, the effects of friction may be neglected and a 3D Euler code can be applied to predict the operating behaviour sufficiently accurate.

To solve the 3D Euler equations a cell-centered finite volume scheme based on Chorin's approach, s. [4] has been developed [5]. For the time integration an implicit procedure and for the damping of occurring oscillations due to the central discretization Roe's flux difference splitting scheme is used.

2.3. Quasi-3D Navier-Stokes (NSQ3D)

Assuming a quasi-3D flow field of a viscous fluid consisting of a given number of $S1_i$-flows on (B-B) surfaces of revolution and of one $S2_m$-flow in a representative (H-S) plane, the governing equations in cylindrical coordinates (r, φ, z) have been transformed into two sets of 2D equations. First, the (H-S)-flow is described in the (r-z)-plane neglecting the derivates in φ-direction but taking into account the r-and z-components of the blade forces. Second, the (B-B)-flow is defined in (m, φ)-coordinates, where m means the meridional coordinate on an arbitrary surface of revolution [6]. In a first approach these two sets of equations have been solved by means of the artificial compressibility concept first applied by RIEDEL to solve the incompressible 3D Euler equations. However, this solution procedure was so time-consuming that the pressure correction scheme SIMPLE [7] is now being implemented [8].

This NSQ3D-concept will allow to optimise efficiently the blade shape near the shroud with respect to minimum losses or optimum cavitation behaviour based on prescribed flow surfaces as results of the (H-S)-computation.

2.4. 3D Navier-Stokes Code (NS3D)

To investigate the full 3D viscous flow in hydraulic machinery a Navier-Stokes 3D code has to be used. The developed program is based on a cell-centered finite volume discretization described by RITZINGER [6]. Mass conservation is achieved by using a SIMPLE pressure correction scheme [7]. A standard k-ε-model with log-law wall functions as well as various linear and non-linear Low Reynolds number models are used to calculate the turbulent viscosity ν_t [9].

Both the convective and the viscous fluxes are discretized using deferred correction schemes. The equation system is using a SIP algorithm of Stone. For the calculation of stages of hydraulic machinery a coupling algorithm is used that is both momentum and mass conservative. With this powerful tool it is possible to calculate the hydraulic efficiencies and the characteristic curves. A comparison of the numerical and experimental results was performed many times proving the accuracy of the CFD code, s. [10], [11], [12] and [13].

2.5. Turbulence Modelling

Despite the fact that an enormous progress can be observed in the development of LES methods for turbomachinery relevant test cases, s. TEMMERMAN and

LESCHZINER [14] and MAHESH et al. [15], the geometries are relatively simple and the Reynolds-numbers are rather low compared to applications of practical relevance. Therefore, the forecast for the substitution of the RANS approach by LES is pessimistic, s. SPALART [16].

The RANS methods are expected to dominate the calculation of turbulent flows with practical relevance for the next decades. In spite of the forced development of 2nd moment closure models these models are seldom applied due to the increased cost and mainly due to their numerical difficulties. On the other hand, eddy-viscosity models are commonly considered an industrial standard, because they are numerically robust, simple and computationally cheap. Many corrections to these models exist with regard to stagnation point and rotating flows, s. DURBIN, [17], PETTERSSON REIF et al. [18], KATO and LAUNDER [19].

A compromise between 2nd moment closure and linear eddy-viscosity models are non-linear or explicit algebraic stress models, which combine the simplicity and robustness of eddy-viscosity models with the higher order Ansatz of Reynolds stress transport models. Examples of non-linear eddy-viscosity models are the models of LIEN et al. [20] and CRAFT *et al.* [21].

The V2F model developed by DURBIN [22] and co-workers takes non-local effects into account by the use of a scalar stress component v'^2 and an elliptic operator f.

At present it seems that non-linear eddy-viscosity models or linear eddy-viscosity models with suitable additives are the favourite choice for the steady and unsteady computation of real technical flows.

2.6. Cavitation Modelling

The cavitation model which has been implemented in the NS3D code is based on bubble dynamics and describes the complex and highly transient bubble growth and collapse, s. SAUER [23]. The bubbles originate from very small nuclei, i.e. particles or air bubbles, usually existing in the fluid. When the nuclei reach the low pressure region at the leading edge of the blades they grow to vapour bubbles while they are convected downstream. Reaching the region of increasing pressure the bubbles collapse inducing the erosion of material.

Considering spherical bubbles and neglecting the bubble-bubble interaction and coalescence as well as the effects of viscosity and surface tension the bubble dynamics may be evaluated by means of the Rayleigh-equation which describes the time derivative of the bubble radius R as a function of the difference between the liquid and vapour pressure and the density of the liquid.

$$\frac{dR}{dt} = \sqrt{\frac{2}{3}\frac{p_{va} - p}{\rho_l}} \tag{1}$$

Due to the change of the bubble volume the velocity field is no more divergence-free and the continuity equation has to be extended by a source term on the right hand side of this equation. This term also appears on the right hand side of the transport equation for the void fraction α which has to be solved additionally:

$$\frac{\partial\alpha}{\partial t}+\frac{\partial(\alpha u_i)}{\partial x_j}=\frac{4\pi\cdot n_0\cdot R^2}{1+n_0\cdot\frac{4}{3}\pi R^3}\cdot\dot{R} \tag{2}$$

where n_0 means the number of nuclei per m^3 fluid.

Thus, the liquid-vapour flow is described by a one-fluid model assuming a homogeneous liquid-bubble mixture for which the density and viscosity are defined by simple constitutive relations.

3. Performance Prediction

3.1. Fans

Three types of axial fans designed and experimentally investigated at the PFLEIDERER-Institute of the Technical University of Braunschweig (TUBS), Germany, s. BRODERSEN [24], as well as two types of diagonal fans which have been designed and tested at the National Engineering Laboratory (NEL) in Glasgow, UK, s. CAREY [25], have been numerically analysed.

3.1.1. Axial Fans

At first, the three types of axial TUBS fans 7165, 7160 and 7161 were analysed with respect to their performances assuming a vanishing tip clearance between the rotor and the housing, s. BADER [26]. In Table 1 the design data as well as the computed performance data considering the optimum point of operation are listed.

Table 1. Comparison of measured and predicted performances of the axial TUBS fans

	Measurements						**Simulation**		
	n_q	φ_v	ψ_t	$\psi_{t,th}$	λ	η [%]	ψ_t	$\psi_{t,th}$	η [%]
7165	142	0,35	0,481	0,573	0,17	83,9	0,466	0,550	84,7
7160	130	0,45	0,601	0,706	0,27	85,1	0,601	0,713	84,3
7161	133	0,60	0,701	0,798	0,42	87,9	0,711	0,811	87,7

Fig. 1 shows the performance curves of the fan 7165 having the lowest power density λ and of the fan 7161 with the highest λ - value.

Obviously, the predicted performance curves coincide well with the measured data.

Secondly the influence of the tip clearance s_K on the inner efficiency was studied numerically. Fig. 2 shows the decrease of the relative efficiency $\Delta\eta^*$ versus the relative tip clearance s_K/D for the three axial fans considered. The weakly loaded axial fan 7165 shows the biggest efficiency decrease due to the smaller stagger angle compared with the highly loaded 7161 fan.

Considering the 7165 fan the axial component of the pressure forces acting against the flow related to the axial momentum forces is bigger than that of the 7161 fan, s. Fig. 3.

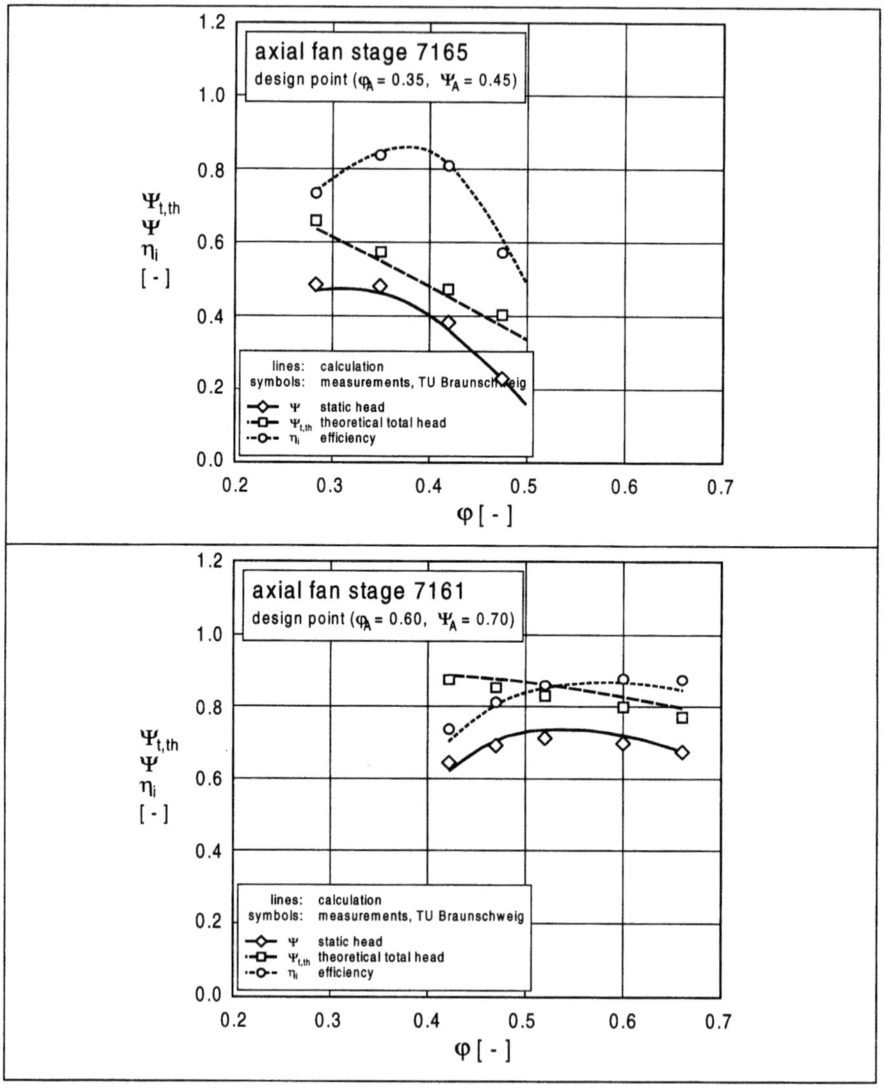

Fig. 1. Comparison of the measured and predicted performance curves of the axial TUBS fans 7165 and 7161

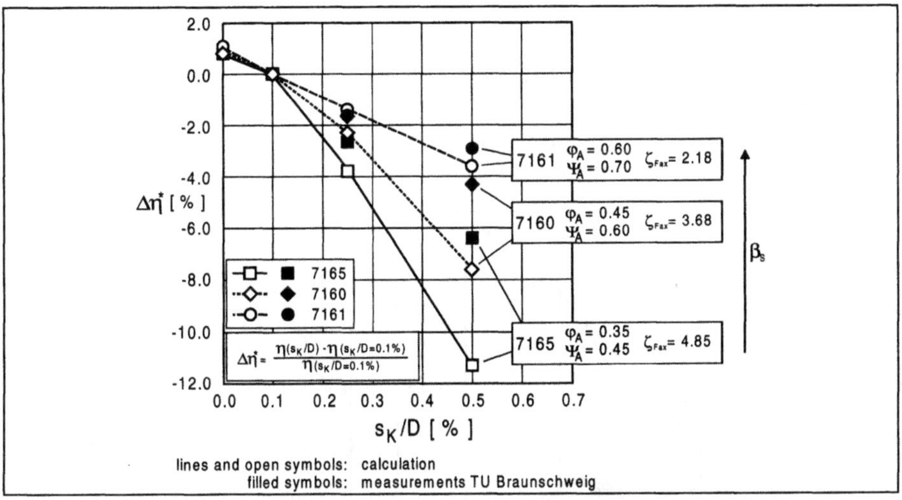

Fig. 2. Influence of the relative tip clearance s_K/D on the relative hydraulic efficiency decrease $\Delta\eta^*$

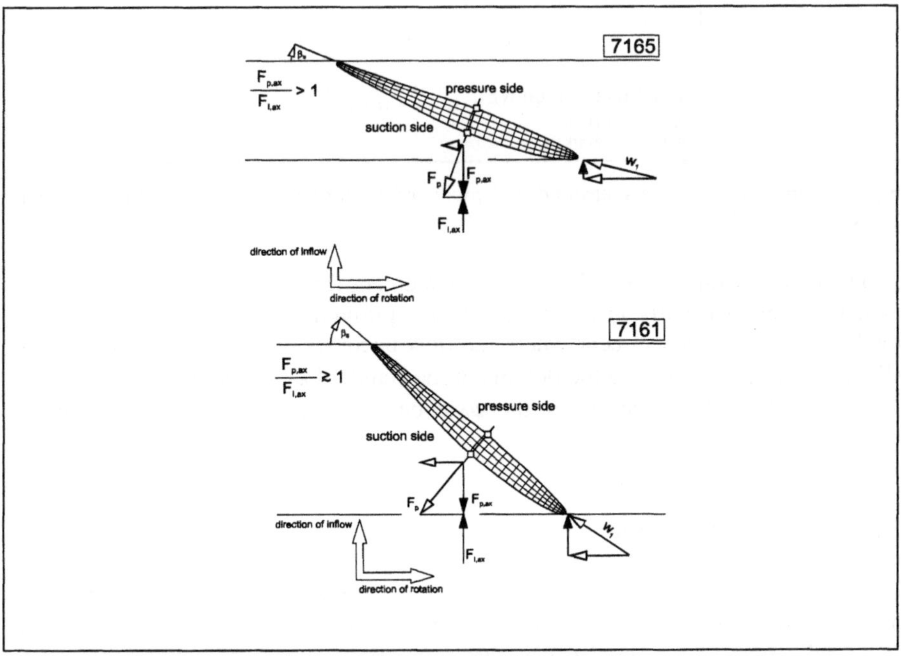

Fig. 3. Plausibility of the influence of the stagger angle βs on the axial component of the pressure forces related to the axial momentum forces

3.1.2. Diagonal Fans

The NEL in Glasgow designed two types of diagonal fans, one with an unstable characteristic, Mark1, and another with a stable one, Mark2. Both were analysed numerically, s. BADER [26] and the performance curves have been evaluated and compared with the measurements, s. Fig. 4.

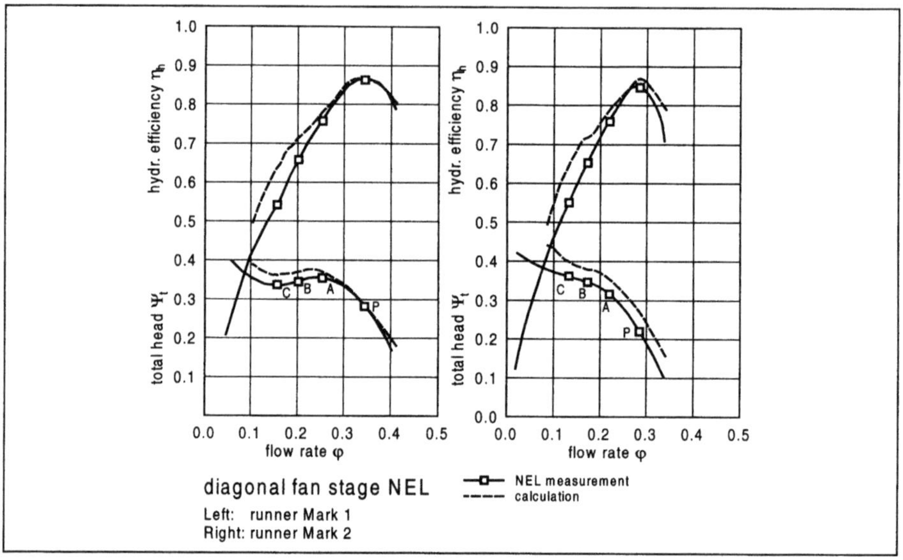

Fig. 4. Comparison of the measured and predicted performance curves for the NEL-fans Mark1 and Mark2

The comparison shows a fairly good agreement for both the head and the efficiency curves especially in the range of the optimum point of operation. At part load clear discrepancies occur where the flow becomes highly 3D, fluctuating and slightly separated reflecting the deficits of the standard *k*-ε-turbulence model.

Nevertheless, the instability in the $\psi_t(\varphi)$-curve of the Mark1 fan has predicted well.

3.2. Centrifugal Pumps

On behalf of the "FV Pumpen" in the VDMA, Germany, a software system is being developed with the aim to use it in the near future as a validated "Numerical Test Rig". Basically, it consists of a fast computing, robust and calibrated 3D Navier-Stokes code, enabling the user to simulate the steady and unsteady performance behaviour of centrifugal pumps including the effect of flow cavitation. For calibrating the 3D code the numerical simulation results are compared with measurements which are being carried out at the TFA of the Darmstadt University of Technology.

3.2.1. Single-Phase Performance Prediction

At a first step a performance prediction has been carried out for a radial centrifugal pump with a spiral casing having a specific speed $n_q = 26\ 1/min$ excluding the effect of flow cavitation by assuming a sufficiently high pressure level. In Fig. 5 the measured and predicted performance curves are compared to each other.

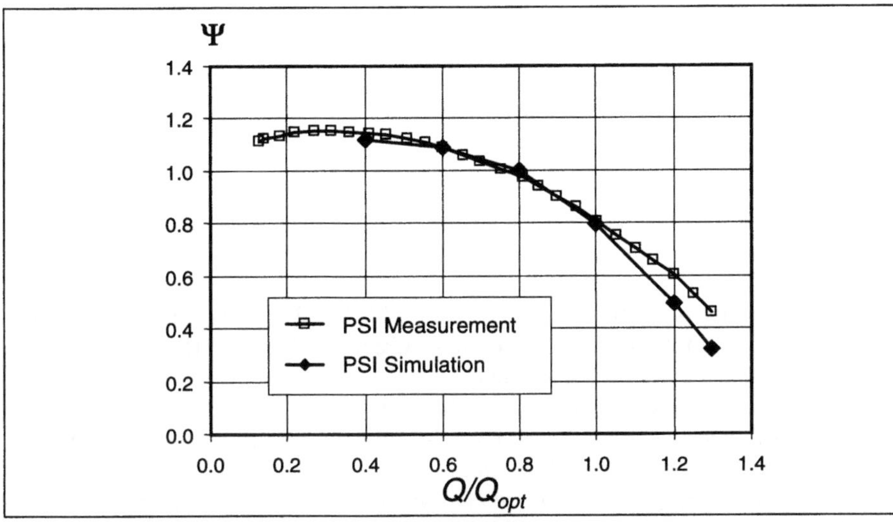

Fig. 5. Comparison of the measured and predicted performance curve of the radial centrifugal pump RP26

Within the considered range of operation the calculated and measured performance curves coincide well.

3.2.2. Simulation of the Head Drop of Centrifugal Pump

After having got a fairly good agreement between the measured and simulated operating behaviour of the centrifugal pump RP26 the described cavitation model has been implemented, validated and calibrated by comparing again with measurements performed at the TFA at the Darmstadt University of Technology, s. FROBENIUS et al. [27].

The validation and calibration of the cavitation model has been carried out by comparing with the measurements of the cavitating flow around two inclined profiles, one being a 2D profile and the other a 3D profile due to the swept leading edge.

However, the main point of interest is to simulate the head drop due to cavitation induced by lowering the pressure level, i.e. the NPSH-value, successively. Fig. 6 shows a comparison of the measured and predicted head drop curve considering the optimum point of operation, s. [27].

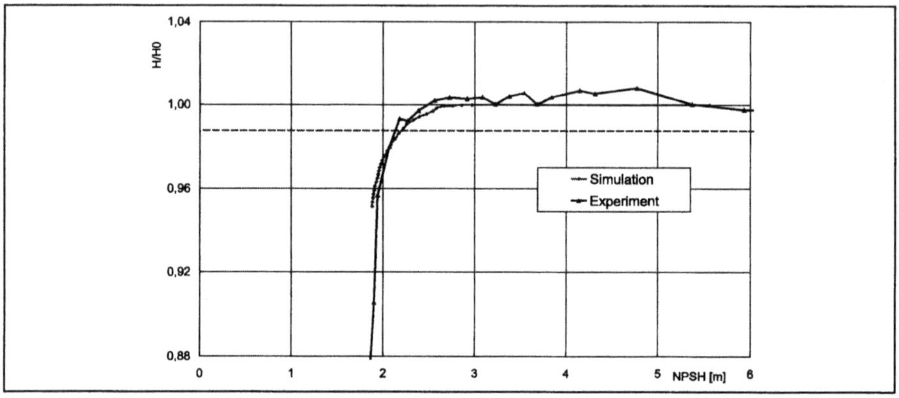

Fig. 6. Comparison of the measured and predicted head drop curve of the centrifugal pump RP26 considering the optimum point of operation

A second point of interest is to know the *NPSH*-value where the first bubbles occur, i.e. the point of incipient cavitation, $NPSH_i$, and the value where the head drop reaches 3%, i.e. the $NPSH_{3\%}$-value.

For this purpose the cavitating flow through a radial pump with a 2D blading and a specific speed $n_q = 27.5\ 1/min$ was simulated. The experiments have been carried out at the PFLEIDERER-Institute, Braunschweig. In Fig. 7 the predicted $NPSH_i$ – and $NPSH_{3\%}$ -values are plotted versus the relative flow rate and compared with the measurements of FRIEDRICHS [28], s. also FROBENIUS et al. [29]. Within the corresponding range of uncertainty of the numerical and experimental data the comparison shows a reasonably good agreement.

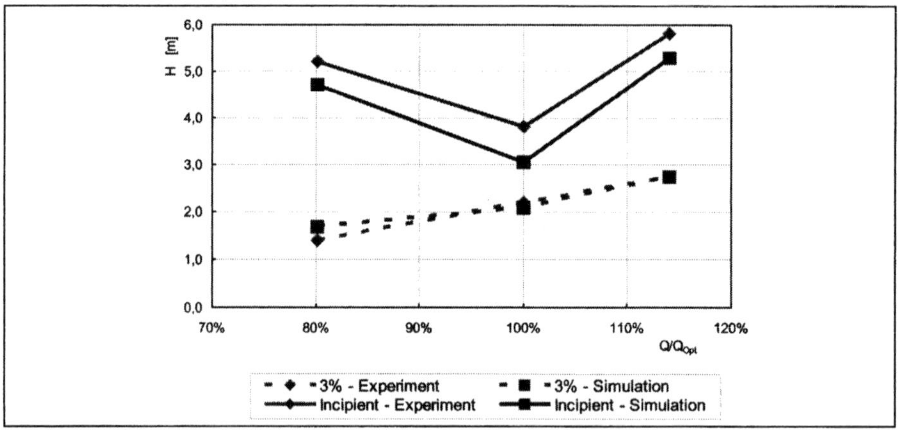

Fig. 7. Comparison of the measured and pre-dicted NPSHi – and NPSH3% -values versus the relative flow rate for the centrifugal pump RP 27.5

4. Real-Time Design of Impellers (RTD)

A real-time design system which may be run on a PC-cluster has been developed on the basis of a multi-level CFD-strategy, s. LEPACH, SCHILLING et al. [30]. The design and optimisation process is started with an initial impeller design generated by an artificial neural network (ANN) yielding the meridional section of the impeller with the leading and trailing edges and the blade angle distributions on prescribed areas of revolution. Fig. 8 shows the meridional sections of four impellers with different specific speed generated by ANN, s. KRÄMER [31].

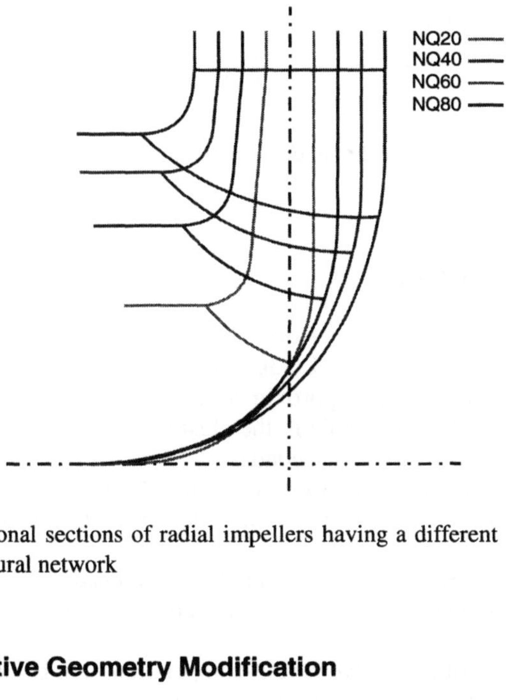

Fig. 8. Meridional sections of radial impellers having a different specific speed generated by artificial neural network

4.1. Interactive Geometry Modification

On the basis of the initial design a first calculation of the flow through the impeller is carried out by means of the developed NS3D-code. Then, the numerical results are analysed with respect to the potential of possible improvements.

In the next step the geometry of the impeller may be modified by interactively moving the describer points of the B-spline representations which characterise the hub and shroud contours, the leading and trailing edges as well as the blade angle distribution of the camber surface and the thickness distribution. The modification of an arbitrary contour or curve is realised by defining a modification spline which is superimposed to the original spline curve. Setting the describers of the modification spline to zero the original curve is again obtained.

In Fig. 9 the basic principle of the interactive modification of the shroud contour and leading edge of a pump impeller is shown, where the modification spline is scaled one magnitude smaller than the original spline.

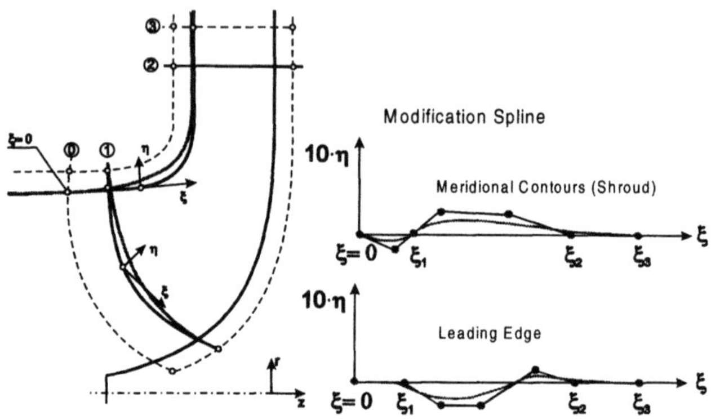

Fig. 9. Interactive modification of the shroud contour and the leading edge on an impeller

4.2. Multi-Level CFD-Strategy

The real-time design process may be performed on different CFD-levels described before. Starting the RTD-process the EQ3D-code may be used in combination with a relatively coarse geometrical approximation of the impeller geometry. Depending on the number of PCs available in the cluster a fluid flow answer may be obtained in real-time, i.e. in a very few seconds.

After having got the first approach of an optimum design the level may be increased to an E3D-simulation and then to a first Navier-Stokes simulation assuming a Q3D flow structure.

On the highest level a full NS3D-code is applied combined with a highly accurate geometrical approximation using a sufficiently big number of describer points.

Thus, the total amount of expensive NS3D-computations can be decreased considerably and the whole design and optimisation process to develop a new impeller may be reduced to one day only.

4.3. Demonstration of the RP39-Design

The developed RTD-System has been applied to optimise different geometry parameters of a radial impeller having a specific speed $n_q = 39\ 1/min$ shown in Fig. 10.

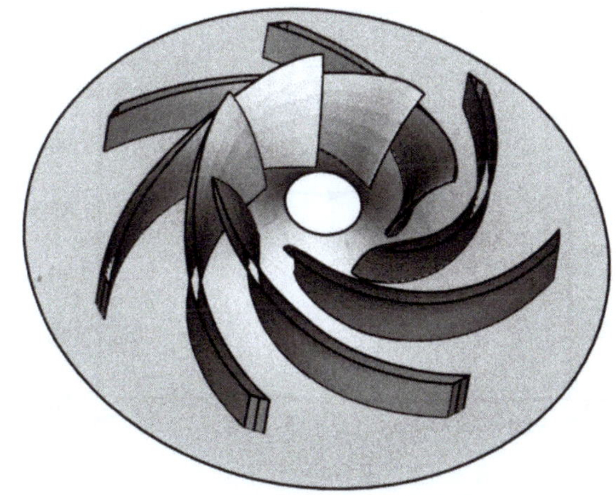

Fig. 10. Impeller of a radial centrifugal pump with a specific speed nq = 39 1/min

The main design parameters to be optimised are the inlet angle distribution β_{s1}, the outlet angle β_{s2} and the wrapping angle $\Delta\varphi_w$ which mainly characterise the performance curves of centrifugal pump impellers. The optimisation of the inlet and outlet blade angles $\beta_{s1,2}$ as well as the blade length are described in detail in the following chapters.

4.3.1. Optimum Design of the Inlet Blade Angles

Based on the initial impeller design the blade angles at the impeller inlet β_{s1} have been varied over a wide range keeping the inlet flow angle β_1 corresponding to the prescribed flow coefficient φ_r constant. The numerical results were analysed by plotting the total energy coefficient Ψ_t and the hydraulic losses (1-η_h) versus the incidence angle i related to the optimal value i_{opt} which has found to be approximately zero, s. Fig. 11. Furthermore, the development of the pressure coefficient c_p near shroud, i.e. streamline 2 (SL2), along the leading edge of the blade for three different incidence angles $i < i_{opt}$, $i \cong i_{opt}$ and $i > i_{opt}$ is shown, s. Fig. 12.

Fig. 11 clearly shows that the minimal losses occur at $i \cong 0$, meaning that the blade angle β_{s1} has to be adapted almost exactly to the inflow direction. Consequently, the development of the total energy coefficient $\Psi_t(i)$ shows the opposite trend.

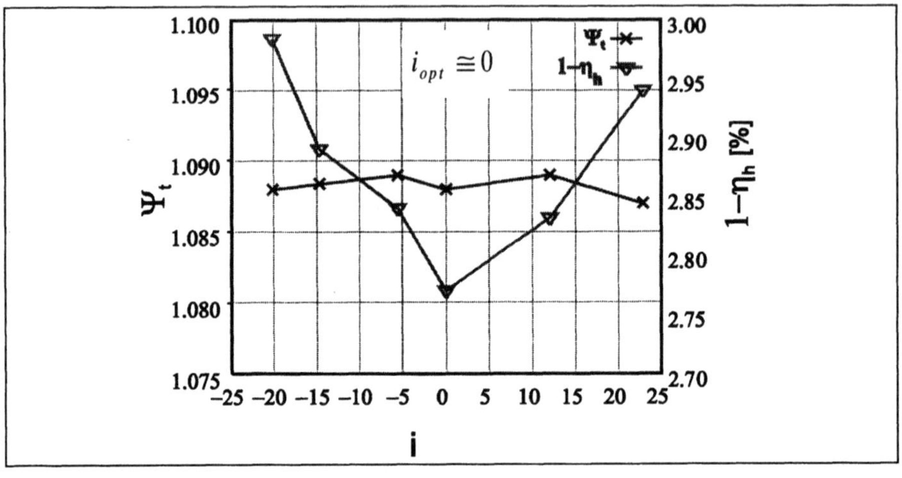

Fig. 11. Evaluation of the optimal incidence angle

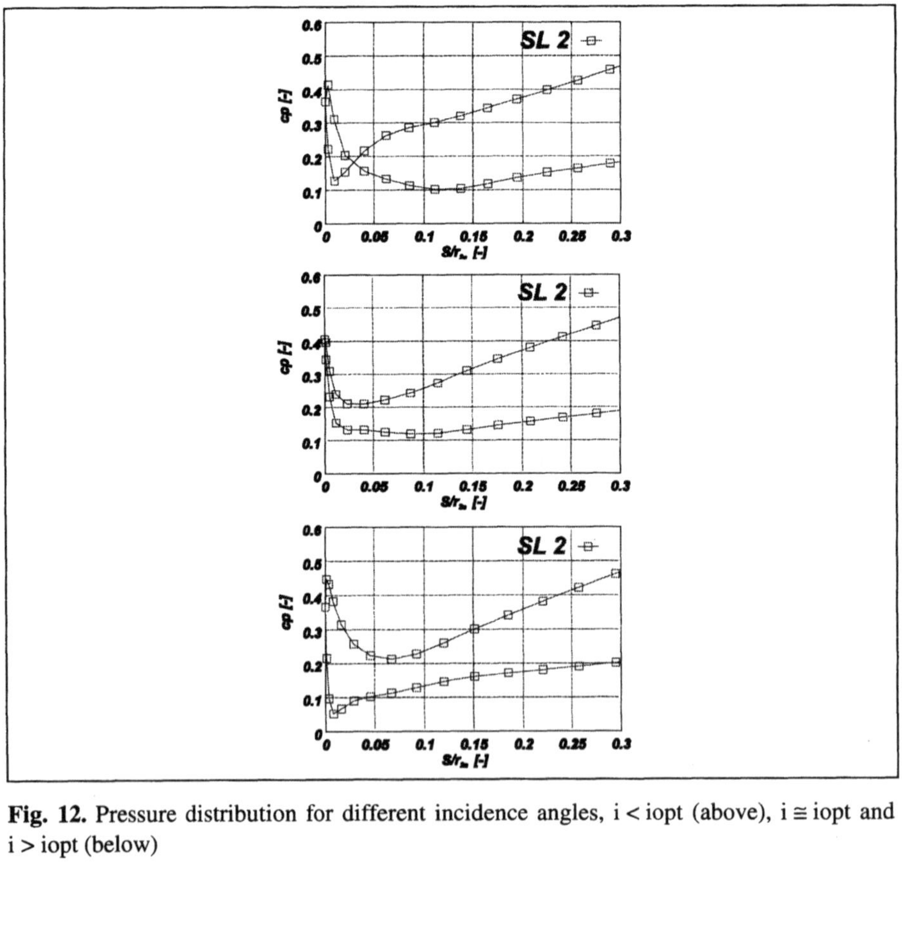

Fig. 12. Pressure distribution for different incidence angles, i < iopt (above), i ≅ iopt and i > iopt (below)

4.3.2. Optimum Design of the Outlet Blade Angle

In the next step, the outlet blade angle β_{s2} has been varied to check for which value the design point may be reached. In Fig. 13 the total energy coefficient Ψ_t and the hydraulic losses $(1\text{-}\eta_h)$ are plotted versus the outlet blade angle β_{s2} showing that for $\beta_{s2} \cong 22.5°$ the impeller design satisfies the required criterion $\Psi_t = \Psi_{t\,requ.}$. With further increasing the blade angle β_{s2} the blade loading and consequently the hydraulic losses are increasing progressively.

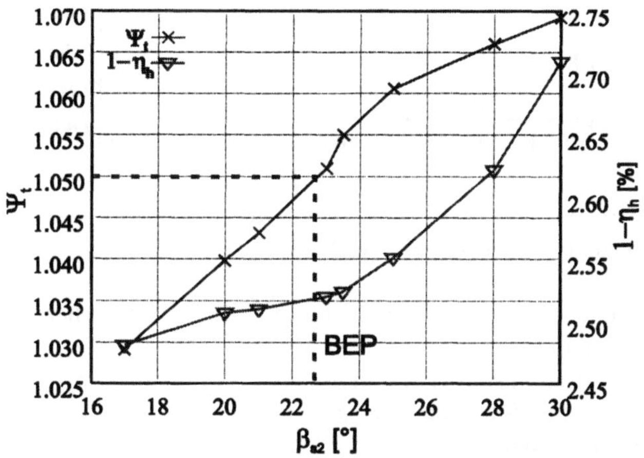

Fig. 13. Optimum design of the outlet blade angle

4.3.3. Optimum Design of the Blade Length

After having evaluated the optimal values for the inlet and outlet blade angles, β_{s1} and β_{s2}, the blade length has been optimised by varying the wrapping angle $\Delta\varphi_w$ and keeping constant the leading and trailing edge blade angles, s. Fig. 14. Since the wrapping angle defines the blade length the variation of $\Delta\varphi_w$ is equivalent to the optimisation of the blade length.

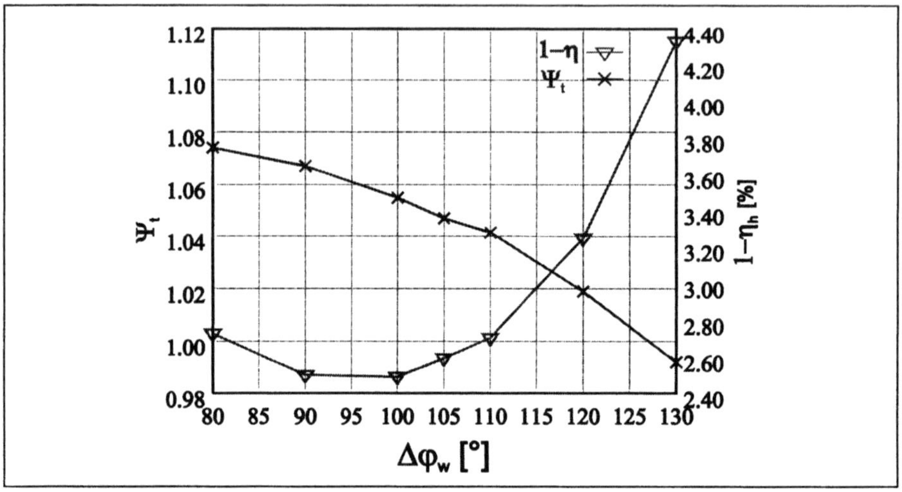

Fig. 14. Optimum design of the wrapping angle

Obviously, the design point Ψ_t is reached for $\Delta\varphi_w \cong 100°$ where the hydraulic losses ($1\text{-}\eta_h$) are minimal.

This final optimisation step yields an impeller design satisfying the design point criterion, optimal cavitation behaviour and energy conversion as well. The impeller efficiency is found to be:

$$\eta_{h,impeller} = 97.5\ \%$$

5. Conclusions

The comparison between the predicted and measured performance curves shows a fairly good agreement. Nevertheless the simulation of the tip clearance flow of axial fans yields unsatisfactory solutions since the efficiency decrease due to the relative tip clearance is overpredicted, especially for low loaded fans.

However, the developed real-time design system has proven to be an efficient tool for the optimum design of centrifugal pump impellers.

References

[1] Fletcher, C.A.J.: „Computational Techniques for Fluid Dynamics Vol. I, Vol. II", Springer Verlag Berlin Heidelberg, 1988.

[2] Wu, C.H.: „A general Theory of the 3D Flow in Subsonic and Supersonic Turbomachines of Axial Radial and Mixed Flow Type". NACA, TN 2604, 1952.

[3] Schilling, R.: „Numerical Calculation of the Q3D Incompressible Inviscid Flow in Turbomachines". 11th IAHR Symposium Amsterdam, 1982.

[4] Chorin, A. J.: „A numerical method for solving incompressible viscous flow problems." In: Journal of Computational Physics 2, S. 12, 1967.

[5] Riedel, N.: „Rotor-Stator Wechselwirkung in hydraulischen Maschinen." Ph.D. thesis, TU München, 1997.

[6] Ritzinger, S.: „Simulation realer Laufradströmungen." Ph.D. thesis, TU München,1997.

[7] Ferziger, J.H., Peric, M.: „Computational Methods in Fluid Dynamics." Berlin, Heidelberg: Springer Verlag, 1996.

[8] Müller, N.: „Entwicklung eines quasi-3D Navier-Stokes Codes zur Optimierung von Beschaufelungen hydraulischer Strömungsmaschinen." Interner Bericht des Lehrstuhls für Fluidmechanik, Abt. Hydr, 2003.

[9] Skoda, R.: „Numerische Simulation abgelöster und transitionaler Strömungen in Turbomaschinen." Ph.D. thesis, TU München, 2003.

[10] Maurer, W., Skoda, R., Frobenius, M., Schilling, R.: „Numerical Simulation of the 3D Flow Field Through a Slurry Pump", 4th International Conference on Multiphase Flow, ICMF-2001, New Orleans, May 2001.

[11] Thermann, H., Müller, M., Niehuis, R., Skoda, R., Schilling, R.: „Numerical Simulation of the flow in an annular compressor cascade with different turbulence and transition models", ETC 5, Prague, 2003.

[12] Skoda, R. Schilling, R. Thurso, J., Stoffel, B.: "Numerical simulation of unsteady and transitional flows pertaining to turbine cascades", Proceedings of the 5th Symposium on Engineering Turbulence Modelling and Experiments, Mallorca, 2002.

[13] Steinbrecher, Ch., Skoda, R., Schilling, R., Müller, N., Breitenbach, A., Mendler N.: „Numerical Simulation of a self-stabilizing rotor of a centrifugal pump", Proceedings of FEDSM'03, 4th ASME/JSME Joint Fluids Engineering Conference, Honolulu, Hawaii, USA, July 6-11, 2003.

[14] Temmerman, L.; Leschziner, M.A.: "Large Eddy Simulation of separated flow in a streamwise periodic channel construction", Int. Symp. on Turbulence and Shear Flow Phenomena, Stockholm, June 27-29, 2001.

[15] Mahesh, K.; Constantinescu, G.; Apte, S.; Iaccarino, G.; Moin, P.: "Large-Eddy Simulation of gas turbine combustors", CTR Annular Research Briefs, Stanford University, 2001.

[16] Spalart, P.R.: "Strategies for turbulence modelling and simulations", Proceedings of the 4th Symposium on Engineering Turbulence Modelling and Experiments, Corsica, 1999.

[17] Durbin, P.A.: "On the k-ε-stagnation point anomaly", Int. Journal of Heat and Fluid Flow 17, 1996.

[18] Pettersson Reif, B.A.; Durbin, P.A.; Ooi, A.: "Modeling rotational effects in eddy-viscosity closures", Int. Journal of Heat and Fluid Flow 20, 1999.

[19] Kato, M.; Launder, B.E.: "The modelling of turbulent flow around stationary and vibrating cylinders", 9th Symp. On Turbulent Shear Flows, Kyoto, 1993.

[20] Lien, F.S.; Chen, W.L.; Leschziner, M.A.: "Low-Reynolds number eddy-viscosity modelling based on non-linear stress-strain / vorticity relations", Proceedings of the 3rd Symposium on Engineering Turbulence Modelling and Experiments, Kreta, 1996.

[21] Craft, T.J.; Launder, B.E.; Suga, K.: "Prediction of turbulent transitional phenomena with a nonlinear eddy-viscosity model", Int. Journal of Heat and Fluid Flow 18, 1997.

[22] Durbin, P.A.: "Separated flow computations with the k-ε-v'^2 model", AIAA Journal 33, 1995.

[23] Sauer, J.: „Instationär kavitierende Strömungen – Ein neues Modell basierend auf Front Capturing (VOF) und Blasendynamik“, Ph.D. thesis, Universität Karlsruhe, 2000.

[24] Brodersen, S.: „Experimentelle Untersuchungen an hochbelasteten Axialventilatoren mit kleinem Nabenverhältnis", Ph.D. thesis, TU Braunschweig, 1986.

[25] Carey, C., Fraser, S.M., Shamsolahi, S., Mcewen, D.: „Studies of the flow of air in a model mixed-flow pump by laser doppler anemometry, Part 5: Velocity measurements in Mark 2 Pump", National Engineering Laboratory, East Kilbride, Glasgow (NEL Report 717), 1990.

[26] Bader, R.: „Simulation kompressibler und inkompressibler Strömungen in Turbomaschinen", Ph.D. thesis, TU München, VDI Fortschrittsberichte, Reihe 7, Nr. 396, 2000.

[27] Frobenius, M., Schilling, R , Bachert, R., Stoffel, B.: „Entwicklung und experimentelle Validierung eines Codes zur numerischen Berechnung kavitierender Strömungen in Kreiselpumpen“, Abschlussbericht zum AiF-Forschungsvorhaben 13325, 2003.

[28] Friedrichs, J.: „Auswirkungen instationärer Kavitationsformen auf Förderhöhenabfall und Kennlinineninstabilität von Kreiselpumpen.“, Ph.D. thesis, TU Braunschweig, 2003.

[29] Frobenius, M., Schilling, R., Friedrichs, J.; Kosyna, G.: "Numerical and Experimental Investigations of the Cavitating Flow in a Centrifugal Pump Impeller", ASME Fluids Engineering Division Summer Meeting, Montreal, 2002.

[30] Lepach, Th., Schilling, R.:„Weiterentwicklung des Real-time Design Systems zur interaktiven Optimierung von Laufrädern beliebiger Schnellläufigkeit." Interner Bericht des Lehrstuhls für Fluidmechanik, Abt. Hydr. Maschinen, TU München, 2003.

[31] Krämer, S.: „Entwurf optimaler Beschaufelungen hydraulischer Strömungsmaschinen mit Hilfe Künstlich Neuronaler Netze." Interner Bericht des Lehrstuhls für Fluidmechanik, Abt. Hydr. Maschinen, TU München, 2003.

Modelling Separation from Curved Surfaces with Anisotropy-Resolving Turbulence Closures
...and related observations on Large Eddy Simulation

Michael LESCHZINER
Department of Aeronautics, Imperial College London

Abstract Notwithstanding recent progress in Large Eddy Simulation, statistical modelling continues to be the principal approach to representing the effects of turbulence in CFD for engineering flows. This is likely to remain the case, at least for several decades, especially in relation to flows that are strongly affected by viscous near-wall processes. Flows featuring separation from curved surfaces fall into this category, and these are among the most difficult to model accurately because of their high sensitivity to the details of the turbulence field. This paper reviews some recent experience and current work on predicting separation from continuous surfaces with non-linear eddy-viscosity models and second-moment closure. Reasons for successes as well as failures are discussed by reference to several flows. The paper concludes with a brief discussion of the challenges facing Large Eddy Simulation in relation to separated high-Reynolds-number near-wall flows.

1. Introduction

Turbulent separation from plane or curved surfaces is one of the most difficult flow phenomena to predict with an acceptable level of accuracy, be it with statistical turbulence models or by way of simulation. In fact, this is currently regarded as one of the key limitations of CFD in practice, as the process is encountered in numerous engineering applications and has profound consequences to the operational characteristics of components such as highly-loaded compressor blades, high-lift wings, car bodies, ship hulls, tubes and pipes in cross flow and highly-curved ducts. Separation results in a dramatic increase in drag and losses, causes a precipitous drop in wing lift and blade loading, provokes increased blockage and deterioration in pressure recovery and induces vibrations.

From a predictive point of view, key challenges arise from the large temporal fluctuations in the position of the separation line, the high sensitively of the reattachment position or the recirculation length – and thus gross flow features - to the separation location, the dependence of the separation process on the evolution of the boundary layer as it decelerates and skews while being subjected to adverse pressure gradient, and the large-scale – often coherent – motions in the detached shear layer. The fact that the instantaneous separation occurs, over some proportion of time, well upstream of the mean location results in a turbulence structure just upstream of that location that does not scale with wall coordinates, because it is strongly affected by the large-scale structures emanating upstream. This is, of course, contrary to the assumption implicit in any steady-state statistical treatment, which links the state of the flow (i.e. the 'boundary layer') upstream of the mean separation line to the wall. Therein lies the origin of some major predictive defects displayed by turbulence models in separated flows. With this particular difficulty left aside, the separation process depends on the differentiated response of the turbulent stresses in the boundary layer as well as in the separated flow to shear irrotational and curvature-related straining, and turbulence anisotropy is likely to play an important role in the nature of this response. Similarly complex interactions are effective in the separated shear layer, in the intense streamwise vortices in 3d flow, in the reattachment process and in the post-reattachment recovery zone.

Much effort has gone, in recent years, into the investigation of turbulence closures for separated flows. Because of the complex flow conditions present, anisotropy-resolving closures have been advocated with considerable conviction. Specifically, a range of non-linear eddy-viscosity, explicit algebraic Reynolds-stress and full second-moment models have been applied to separated laboratory flows in an effort to identify optimal modelling methodologies. Most of this work has been done in a 2d framework, because of the relatively modest computational task involved, the more abundant availability of experimental data and the promise of greater insight into fundamental issues. However, the validity and usefulness of such studies are limited by issues of 3d contamination of the experimental flow and a substantial uncertainty on whether conclusions on the predictive characteristics derived for 2d conditions bear any relevance to 3d flows. This is an ongoing process of inquiry, and its outcome will dictate the pace at which elaborate closures are adopted for complex practical applications.

A general problem modellers face is the rarity of verifiably accurate and sufficiently detailed experimental data needed for reliable validation studies. Some experiments, designed with validation in mind and conceived to give prominence to some specific generic features, are available, but increasingly, highly-resolved simulations are used to provide data bases that are, arguably, more accurate and well-controlled and certainly more detailed than experiments. On the other hand, such simulations are limited, in terms of their geometric complexity and Reynolds number, by computer-resource constraints. Hence, there is a need to continue to exploit experimental as well as simulation data in an effort to arrive at sufficiently reliable conclusions.

In what follows, the primary focus is on a review of some recent turbulence modelling and validation studies, with emphasis being placed on separation from continuous surfaces, using both experimental and simulation results as a basis of assessing predictions. Against this background, and as a secondary topic, some of the challenges faced by Large Eddy Simulation in predicting separation from curved surfaces at high Reynolds numbers are discussed.

2. Review of Current Status in Modelling

Putting aside a few novel developments – for example, models that account for structural features [1], are based on the *turbulent potential* [2], adopt a multiple-time-scale representation [3] or are founded on two-point correlations [4] - the most complex closure forms used in practice are still based on conventional transport equations for the second moments (Reynolds stresses and scalar fluxes). Principal improvements in recent years have come from new model forms for the redistributive pressure-strain and the dissipative processes [5, 6, 7], as well as through detailed attention to modelling the effects of viscosity and wall blockage on the turbulence in the near-wall layer [8, 9, 10, 11]. Among the most elaborate models currently used for complex 3D engineering flows are forms that contain cubic approximations for the pressure-strain process, account for anisotropic dissipation and adhere to the two-component-turbulence limit at the wall [11, 12]. Another notable recent development in the area of second-moment closure is the *elliptic relaxation* technique [13], which includes, in addition to the stress-transport equations, a corresponding set of elliptic PDEs which 'relax' the pressure-strain terms to their correct wall-limiting levels.

Closure at a level higher than second order, or even one that includes transport equations for the dissipation-rate tensor or some structure-related tensor, does not, at present, appear to be a practically tenable proposition for general use, if only on the grounds of complexity and computational cost, and this position is unlikely to change in the near future. Indeed, the main thrust of research in recent years has been directed towards simplification and the adjustment of lower-order models so as to encourage the uptake of improved closures in industrial CFD. Thus, at the linear eddy-viscosity level, some improvements have been recorded through the replacement of the dissipation rate ε by other length-scale surrogates, such as the specific dissipation ω [14] and the turbulent time scale τ [15], and by the introduction of limiters that constrain the excessive generation of turbulence energy and hence stresses at high strain rates [16]. At an intermediate level of complexity, a whole category of *non-linear eddy-viscosity models* (NLEVMs) has emerged from the use of non-linear constitutive stress-strain/vorticity equations [17, 18, 19] or the inversion of implicit algebraic forms of the Reynolds-stress-transport equations to yield *explicit algebraic Reynolds-stress models* (EARSMs) that relate the Reynolds stresses, non-linearly, to the strain and vorticity components [20, 21, 22, 23].

Non-linear models involve various products of strain and vorticity tensors in the stress-strain relationship. A representative form that includes terms up to third order is:

$$\mathbf{a} = -2C_\mu \mathbf{s} + q_1(\mathbf{s}^2 - \frac{1}{3}\{\mathbf{s}^2\}\mathbf{I}) + q_2(\mathbf{ws}-\mathbf{sw})$$
$$+ q_3(\mathbf{w}^2 - \frac{1}{3}\{\mathbf{w}^2\}\mathbf{I}) - \gamma_1\{\mathbf{s}^2\}\mathbf{s} - \gamma_2\{\mathbf{w}^2\}\mathbf{s} \tag{2.1}$$
$$- \gamma_3(\mathbf{w}^2\mathbf{s} + \mathbf{sw}^2 - \{\mathbf{w}^2\}\mathbf{s} - \frac{2}{3}\{\mathbf{wsw}\}\mathbf{I}) - \gamma_4(\mathbf{ws}^2 - \mathbf{s}^2\mathbf{w})$$

where $\mathbf{a} = \overline{u_i u_j}/k - 2/3\delta_{ij}$; k is the turbulence energy; $\mathbf{s} \equiv s_{ij}$; $\mathbf{w} \equiv w_{ij}$; $\mathbf{I} \equiv \delta_{ij}$; $\{\mathbf{s}^2\} = s_{ij}s_{ij}$ and $\{\mathbf{w}^2\} = -w_{ij}w_{ij}$ are the second strain and vorticity invariants, respectively; and s_{ij} and w_{ij} are the strain and vorticity tensors, respectively, non-dimensionalised with the turbulent time scale τ (e.g. k/ε). For two-dimensional flow, relation (2.1) reduces to a (tensorially) quadratic form, as $\gamma_3 = \gamma_4 = 0$. While the terms associated with γ_1 and γ_2 arise from cubic bases (and will be referred to as "quasi-cubic"), they are, in fact, tensorially linear, because $\{\mathbf{s}^2\}$ and $\{\mathbf{w}^2\}$ are scalars. Hence, these terms, which can be shown to represent the effects of curvature-related strain on the stresses, may be viewed as additive corrections to the constant C_μ. All non-linear eddy-viscosity models are tensorially quadratic in two-dimensional flow. Differences in order (3rd or 4th) only arise in three-dimensional conditions, and these originate from differences in the derivation of relations of the form (2.1) and particular truncation choices.

Models that relate τ to k and ε, make use of the related turbulence-transport equations:

$$\frac{Dk}{Dt} = \nabla \cdot [(\nu + \frac{\nu_t}{\sigma_k})\nabla k] + (P - \tilde{\varepsilon} - D) \tag{2.2}$$

$$\frac{D\tilde{\varepsilon}}{Dt} = \nabla \cdot [(\nu + \frac{\nu_t}{\sigma_\varepsilon})\nabla \tilde{\varepsilon}] + (C_{\varepsilon 1} f_1 P - C_{\varepsilon 2} f_2 \tilde{\varepsilon})\frac{\tilde{\varepsilon}}{k} + S_\varepsilon \tag{2.3}$$

where $\nu_t = C_\mu \dfrac{k^2}{\tilde{\varepsilon}}$ and P is the production rate of k. In the above, $\tilde{\varepsilon}$ is the homogeneous dissipation rate which is related to the actual dissipation ε through $\tilde{\varepsilon} = \varepsilon - D$, where $D = 2\nu(\dfrac{\partial k^{1/2}}{\partial x_i})^2$. Additional source terms, deemed collected in S_ε, may be included to control the growth of the turbulent length scale and to procure the correct behaviour of the near-wall viscous sublayer. A frequent (and influential) additional term is the length-scale correction:

$$S_{\varepsilon,l} = \max[0.83(\gamma - 1)\gamma^2 \frac{\tilde{\varepsilon}^2}{k}, 0], \quad \gamma = \frac{k^{3/2}}{c_l \tilde{\varepsilon} y_n}, \quad c_l = 2.5 \tag{2.4}$$

Models that represent the time (and hence the length) scale via the specific dissipation $\omega = \varepsilon/k$ make use of the following transport equations [14]:

$$\frac{Dk}{Dt} = \nabla \cdot [(\nu + \frac{\nu_t}{\sigma_k})\nabla k] + (P - \beta^* \omega k) \tag{2.5}$$

$$\frac{D\omega}{Dt} = \nabla \cdot [(\nu + \frac{\nu_t}{\sigma_\omega})\nabla \omega] + \frac{\alpha}{\nu_t} P - \beta\omega^2 + S_\omega \tag{2.6}$$

where $\nu_t = \alpha^* \dfrac{k}{\omega}$ and α^*, α, β^* and β are constants or functions of the turbulent Reynolds number or an equivalent universal wall distance. There are considerable (and influential) differences between variants of Eq. (2.6) in different ω-based models. These include the use of different constants, damping functions and fragments in S_ω, the last rooted in the retention or omission of certain terms following the transformation of the ω-equation from that governing ε.

A major challenge posed by non-linear eddy-viscosity models (NLEVMs) is how to determine the large number of coefficients they involve. This calibration process, undertaken by reference to a range of relatively simple key flows, is a source of a considerable variability in model performance. In the case of explicit algebraic Reynolds-stress models (EARSMs), the coefficients arise naturally from the parent stress-transport equations and do not need to be determined separately. As the transport of stress anisotropy has to be removed in order to arrive at the algebraic form, any EARSM is, however, inevitably less general than the associated full stress-transport model.

Two recent models of the NLEVM/EARSM variety merit being highlighted here, for they are, arguably, the most elaborate representatives of this category. The model of Wallin and Johansson (identified in comparisons to follow by WJ) [23] was derived along a strict formal route from an implicit algebraic form of the Reynolds-stress equations. Apart from being a low-*Re* formulation, it differs from an earlier related model of Gatski and Speziale [20] in respect of its formal treatment of the production-to-dissipation ratio P_k/ε. Wallin and Johansson show that the non-linear system of equations in the stresses can be solved in the form of a linear system, complemented by a non-linear scalar equation for P_k/ε, whereas other models are based on the assumption that this ratio is given explicitly, allowing an easier inversion of the implicit algebraic stress equations. The default Wallin-Johansson form uses the k-ω set of equations for the velocity and length scales. However, a variant involving the use of a modified form of the ε-equation that is a part of the k-ε model of Launder and Sharma [25] has also been implemented [28], after minor adjustments to the damping functions designed to ensure the correct decay of grid turbulence and a satisfaction of the local-equilibrium constraint.

The model by Abe *et al.* (AJL) [24] is a quadratic low-*Re* formulation that differs in two important respects from others. First, it augments the basic quadratic constitutive form of (2.1) by two additive fragments intended to account, respectively, for high rates of normal straining and strong near-wall anisotropy. Second, it uses a form of the ω-equation that is much closer to the (transformed) ε-

equation than Wilcox's original proposal [14]. Specifically, it includes products of k and ω gradients and coefficients for the production and destruction terms, which are directly equivalent to $C_{\varepsilon 1}$ and $C_{\varepsilon 2}$ normally used in the ε-equation. One important feature of the model is that it represents correctly the trend towards two-component turbulence as the wall approached, as is observed in DNS. This asymptotic decay cannot be captured solely by the use of terms combining the strain and vorticity, and there is a need to introduce a tensorially correct term that takes into account the wall orientation. In the present model, the wall-direction indicator is of the form:

$$d_i = \frac{N_i}{\sqrt{N_k N_k}} \qquad N_i = \frac{\partial l_d}{\partial x_i} \tag{2.7}$$

where l_d is the wall-normal distance. This is then used in a wall-anisotropy correction to the basic quadratic variant of Eq. (2.1), which is of the form:

$$a_{ij}^{w} = C_d f_w(R_t)\left(d_i d_j - \frac{\delta_{ij}}{3} d_k d_k \right) + f\left(\tau_d S_{ij}, \tau_d \Omega_{ij}, \text{invariants}\right) \tag{2.8}$$

with f_w being a viscosity-related damping function. In Eq. (2.8), a composite time scale is used, which combines the macro-scale k/ε with the Kolmogorov scale $\sqrt{\nu/\varepsilon}$. The damping function f_w then provides a smooth transition between the two scales across the near-wall layer.

3. Model Performance

While recent efforts have undeniably led to real improvements in predictive capabilities for an ever-broadening range of flows and to a much better understanding of the sources of a number of recurring defects, the performance of even the most elaborate models has often been observed to be puzzlingly variable, with distinct improvements typically confined to a limited range of flows. Often, these improvements arise from the fact that higher-order models tend to return lower levels of shear stress and hence higher sensitivity to adverse pressure gradient and more vigorous separation. In part, this lack of performance consistency undoubtedly reflects the fundamental limitations of Reynolds-averaging and the fact that the calibration of model constants is undertaken in a decoupled (non-interactive) manner by reference to a few simple flows. However, there are many other 'peripheral' sources for inconsistent predictive behaviour, including numerical errors associated with insufficient resolution, errors in the prescription of boundary conditions and sensitivity to model-implementation practices, especially in models operating in conjunction with wall-laws accounting for processes in the semi-viscous wall layer.

The most reliable and insightful studies of model performance, or "validations", have been those undertaken within the framework of collaborative projects or as part of a wider workshop activity, such as ERCOFTAC's series of Workshops on Refined Turbulence Modelling. Such studies allow critical comparisons to be made of solutions contributed by several computor groups, and they frequently involve an 'iterative loop' in which solutions are re-examined and recomputed. However, this route is all too rarely offered.

Over the past few years, the writer has been involved in a number of studies, some within collaborative projects and others undertaken in isolation, which have included separation from curved surfaces as one key feature. Some of the flows investigated are shown in Fig. 3.1. A few of the flows in Fig. 3.1 are compressible, with separation being induced by strong shock waves. While these are obviously of substantial interest in the context of high-speed aerodynamics, they are not as challenging as low-speed flows in which the separation is more ill-defined, being induced by a much gentler adverse pressure gradient than that associated with a shock wave.

A conclusion that has emerged from many studies, exemplified by Figs. 3.2-3.9, is that second-moment closure and, to a lesser extent, (some) NLEVMs and EASTMs give a considerably better representation of separated flows than do linear eddy-viscosity models – unless the latter include various *ad-hoc* corrections and limiters and/or 'modified coefficients' of one sort or another. However, the favourable performance of anisotropy-resolving models is by no means all-encompassing, and the picture is quite fuzzy at the edges.

In highly simplified terms, second-moment and non-linear eddy-viscosity models derive their superior performance in separated flows from the fact that they tend to predict lower levels of turbulent transport. This arises from several interactions which these closures represent more realistically than do linear EVMs. First, at a high strain rate, they link the stresses more weakly to the strain than is implied by $\overline{u_i u_j} = C_\mu (k^2/\varepsilon) S_{ij}$, with C_μ =0.09. This relationship gives rise to a production of turbulence energy $P \propto s_{ij} s_{ij}$, as opposed to reality in which the relationship is essentially linear. Second, they establish a stronger sensitivity of the stresses to curvature, which is, in almost all separated flow, of a sign that attenuates turbulence. Third, they return, in accord with reality, a weak response of the stresses and the turbulence energy to the normal straining associated with deceleration and acceleration. Linear EVMs can be 'fixed' to mimic these interactions, to some degree, through the addition of correction terms, changes in the numerical model constants, the introduction of limiters and a functional dependence of C_μ on the strain or production-to-dissipation ratio, and alternative choices of the length-scale-governing equation. However, the generality of these fixes is inevitably limited.

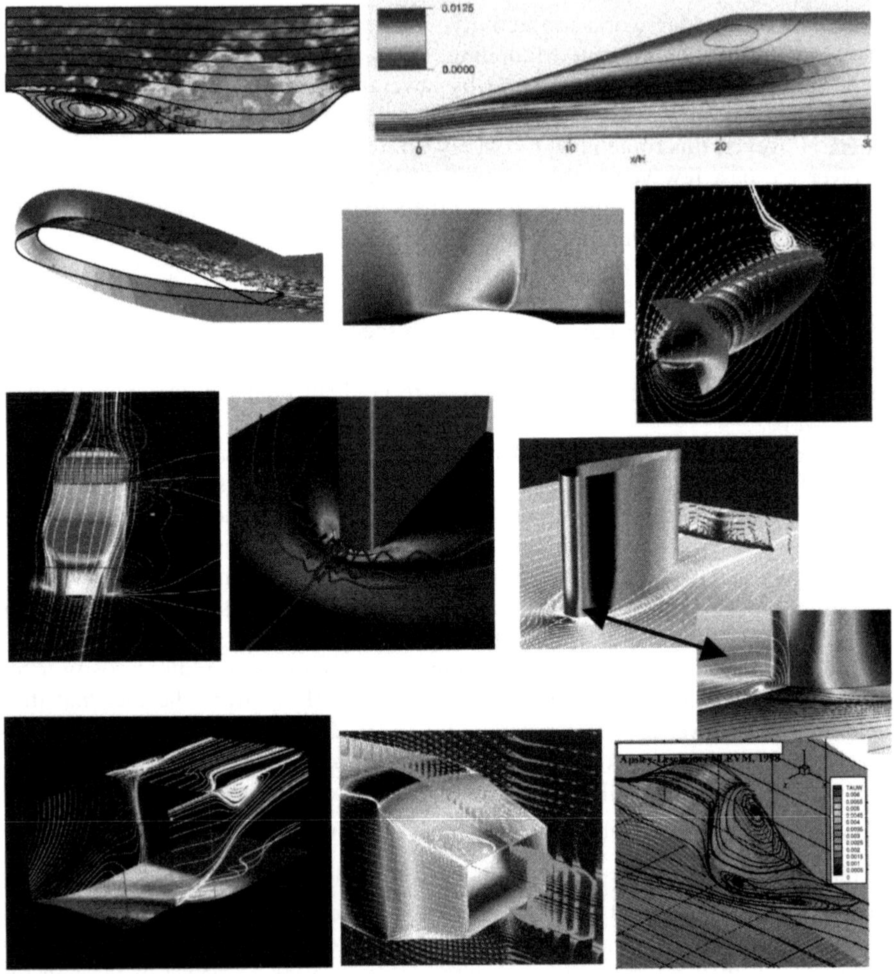

Fig. 3.1. Some separated flows investigated with anisotropy-resolving closures

It turns out that the above fundamental superiority of anisotropy-resolving models and their tendency to predict a lower level of turbulence activity are not helpful in all circumstances, in so far as they do not always translate to better predictive performance. This applies, for example, to the first flow (top, l.h.s. plot) in Fig. 3.1: a periodic segment of a channel with constrictions that provoke massive separation. This is an exceptionally 'rewarding' test case, because the baseline data have been generated by two highly resolved large-eddy simulations undertaken separately [26, 27] and yielding virtually identical time-mean solutions. It is also a highly taxing test case, because of the streamwise periodicity involved and the fact that a shift by an increment δ in the position of the separation point translates to a change of about 7δ in the recirculation length – a ratio that emerges from both simulations and model solutions. This case has recently been examined [28] using a wide range of NLEVMs, EARSMs and also Reynolds-stress-transport models (RSTMs). The flow has been computed both as a periodic configuration and as a sequence of 3 hill segments [29], with prescribed inlet conditions taken from the LES solution. The latter was done to examine the rate of approach of the flow to the periodic state. Figs. 3.7-3.9 give an impression of the breadth of the investigation, involving a searching examination of the turbulence structure. Of particular fundamental interest is Fig. 3.9, because it compares the ability of the investigated NLEVMs and EARSMs to predict the behaviour of the anisotropy near walls, especially the wall-asymptotic approach to the two-component state. This is done in Fig. 3.9 by reference to Lumley's "realisability map" which relates the variation of the second and third anisotropy invariants $A_2 = \{\mathbf{a}^2\}$ and $A_3 = \{\mathbf{a}^3\}$, respectively. The conclusion emerging from this study is that all models displayed some important defects, with none giving a satisfactory description in all respects. However, of the model of Abe *et al.* [24] performed best, especially in relation to the prediction of the anisotropy and of the wall-asymptotic approach of turbulence to the two-component state.

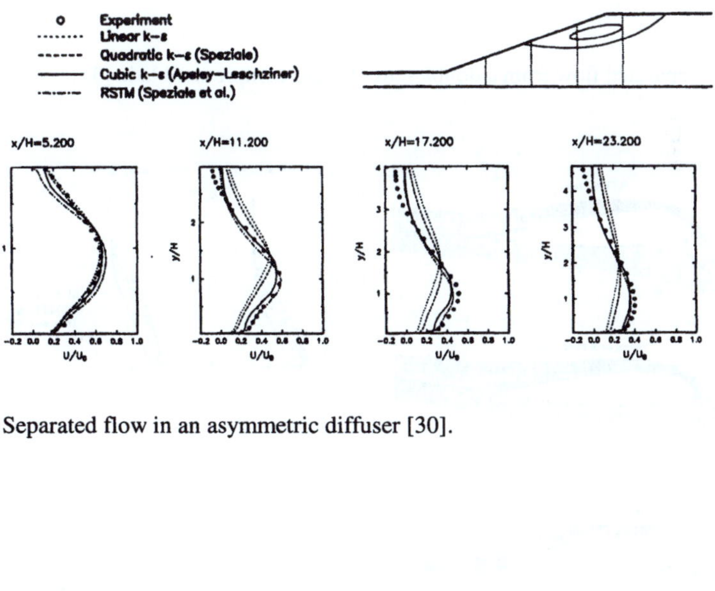

Fig. 3.2. Separated flow in an asymmetric diffuser [30].

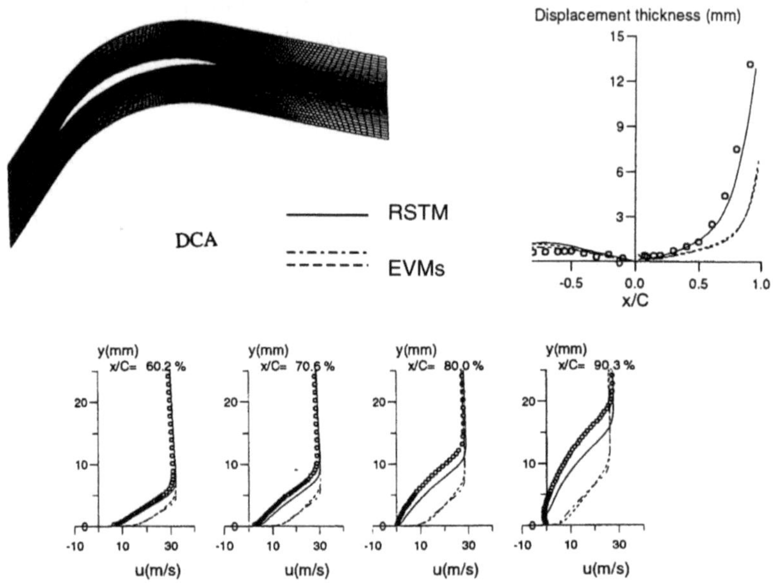

Fig. 3.3. Separated flow from a double-circular-arc compressor blade [31].

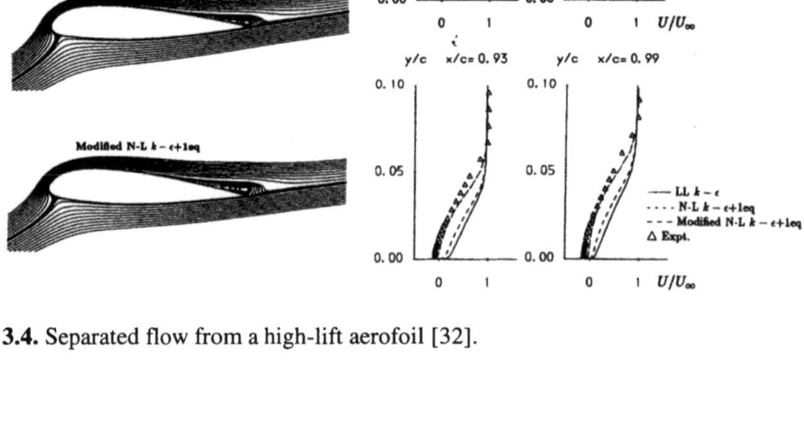

Fig. 3.4. Separated flow from a high-lift aerofoil [32].

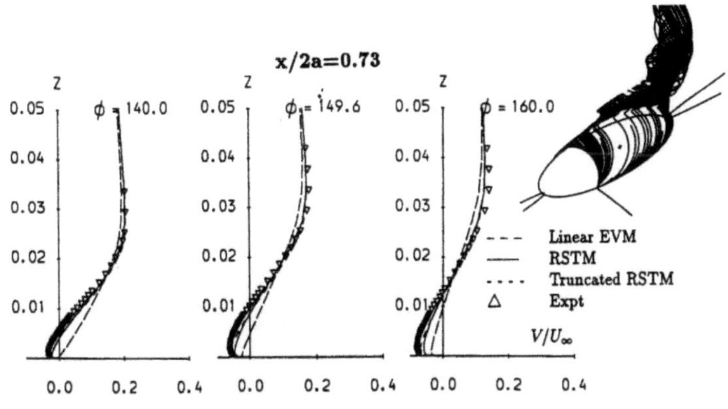

Fig. 3.5. Separated flow from a prolate spheroid at high incidence; profiles are for azimuthal velocities within the leeward vortex [33].

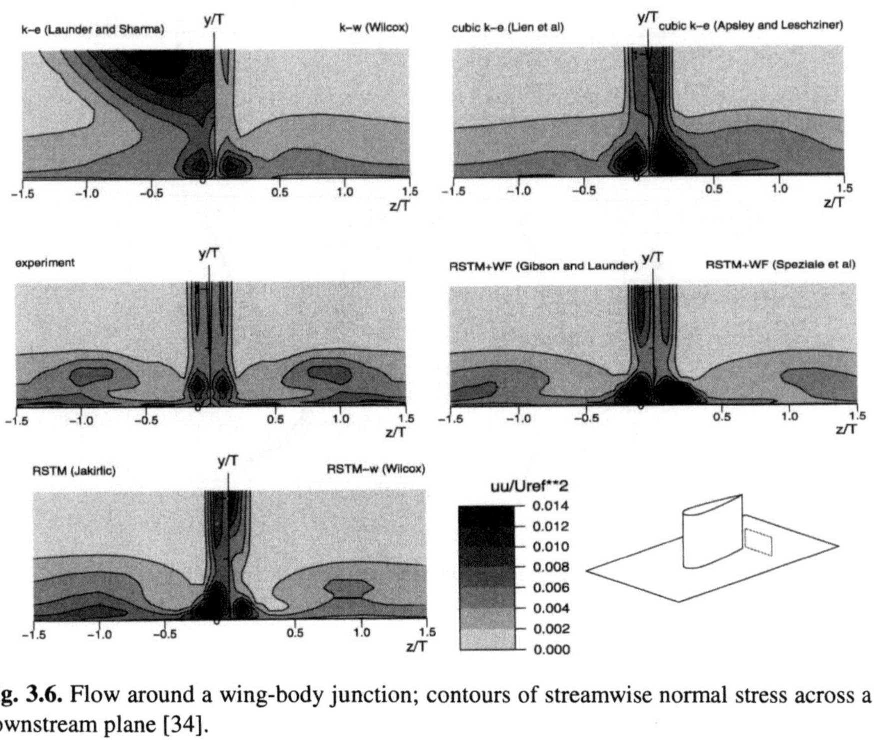

Fig. 3.6. Flow around a wing-body junction; contours of streamwise normal stress across a downstream plane [34].

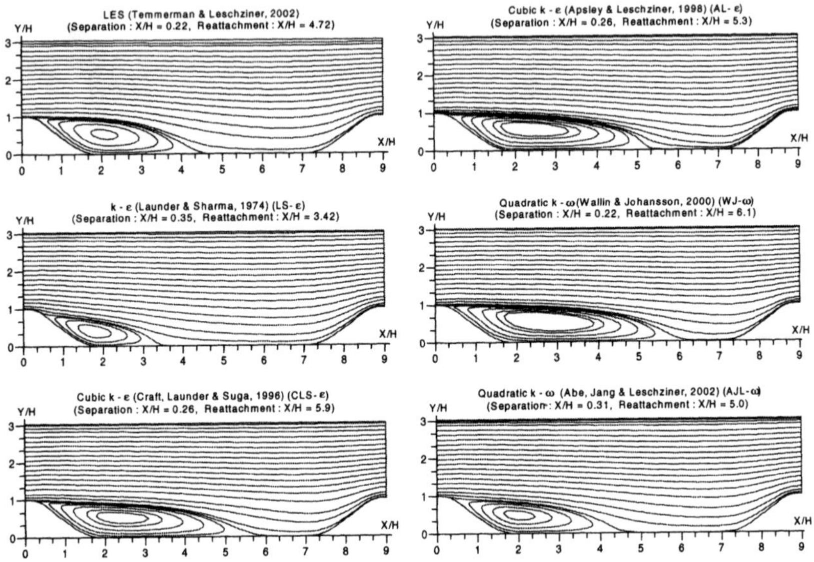

Fig. 3.7. Separated flow in a periodic duct segment with hill-shaped constrictions; streamwise contours [28].

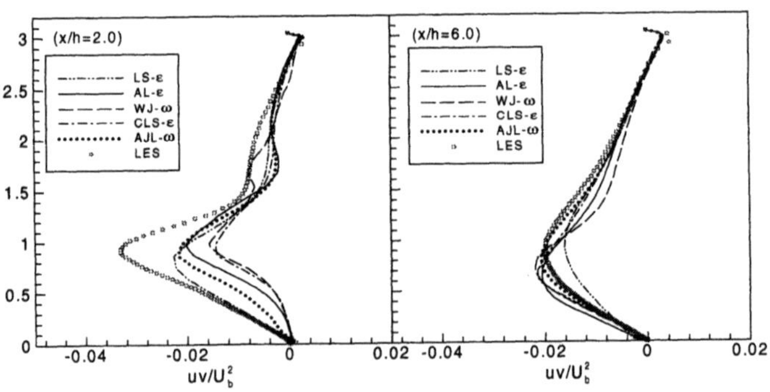

Fig. 3.8. Separated flow in a periodic duct segment with hill-shaped constrictions, shear-stress profiles [28].

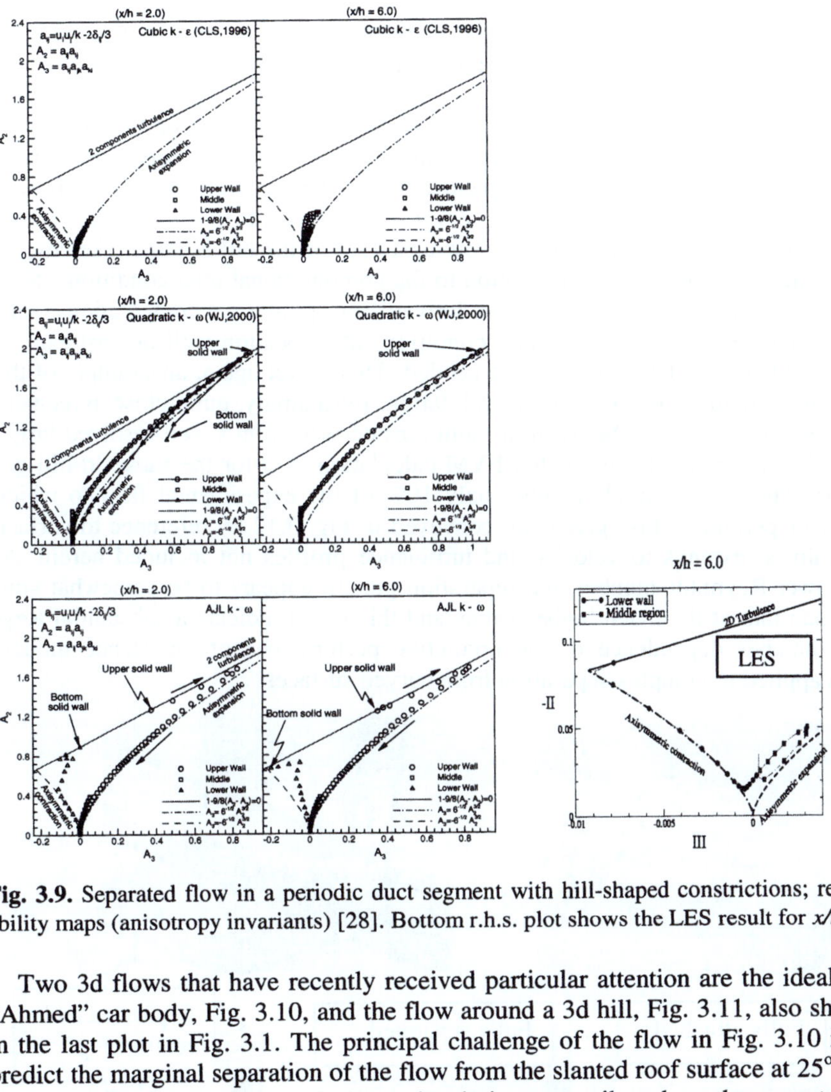

Fig. 3.9. Separated flow in a periodic duct segment with hill-shaped constrictions; realisability maps (anisotropy invariants) [28]. Bottom r.h.s. plot shows the LES result for x/h=6.

Two 3d flows that have recently received particular attention are the idealised "Ahmed" car body, Fig. 3.10, and the flow around a 3d hill, Fig. 3.11, also shown in the last plot in Fig. 3.1. The principal challenge of the flow in Fig. 3.10 is to predict the marginal separation of the flow from the slanted roof surface at 25° and the reattachment on that roof. Several solutions contributed to the recent 10th ERCOFTAC/IAHR Workshop of Refined Turbulence Modelling in Poitiers, France [35], illustrate that non-linear EVMs wrongly predict a fully separated flow with a large recirculating wake, contrary to experiments. On the other hand, the linear k-ε model predicts, again wrongly, a fully attached flow. However, a very recent second-moment-closure computation [36] has yielded a solution in which flow separation from the slanted roof surface is followed by reattachment on the same surface, in broad agreement with the experimental observation. A judgment on whether this result is trustworthy needs to await similar computations by others. An argument that has been made in respect of this flow, however, is that the separation from the roof corner may be highly intermittent, perhaps even periodic, in which case any Reynolds-averaged model would not be expected to perform well. Hence, caution is justified at this stage in relation to the present second-moment-closure result.

As regards the 3d-hill flow, Fig. 3.11 shows that neither second-moment closure nor non-linear eddy-viscosity modelling yields a credible representation of the highly complex, multi-vortex structure that is derived from flow-visualisation experiments [37]. However, one problem this flow presents to the modeller is that the experimental description of the flow upstream of the hill is rather sketchy, so that some uncertainty exists in relation to the computational inlet conditions. Specifically, these conditions need to be derived from pre-curser computations, subject to matching with experimental conditions at a location well downstream of that at which the inlet conditions are needed. This is, yet again, an example of the frequent difficulties encountered in validation. Intriguingly, quite close agreement is observed in Fig. 3.12 between the solutions derived from a well-resolved large-eddy simulation and from a cubic EVM calculation [45] for the same 3d-hill geometry, but at a Reynolds number only 10% of the experimental flow to which Fig. 3.11 pertains. This agreement, conveyed in Fig. 3.12 by reference to surface streamlines, extends to velocity and turbulence profiles not included herein. At this lower Reynolds number, the separation pattern appears to be somewhat simpler than that at the much higher value, and this might indicate a substantial Reynolds-number dependence of the predictive performance of turbulence models when applied to complex separation from curved surfaces.

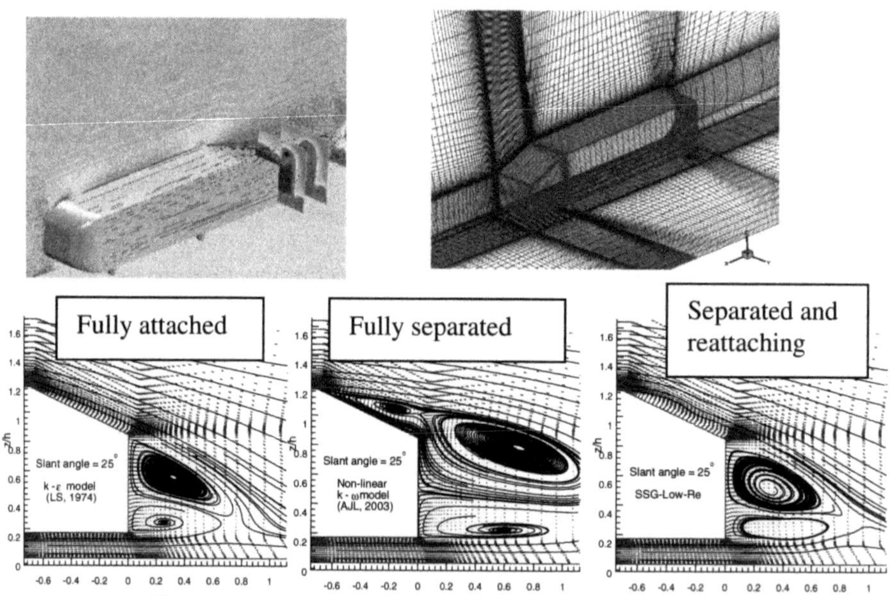

Fig. 3.10. Predicted flow at the rear of the Ahmed body at 25 degrees (rhs) computed with linear and non-linear EVMs and full Reynolds-stress-transport model [36]

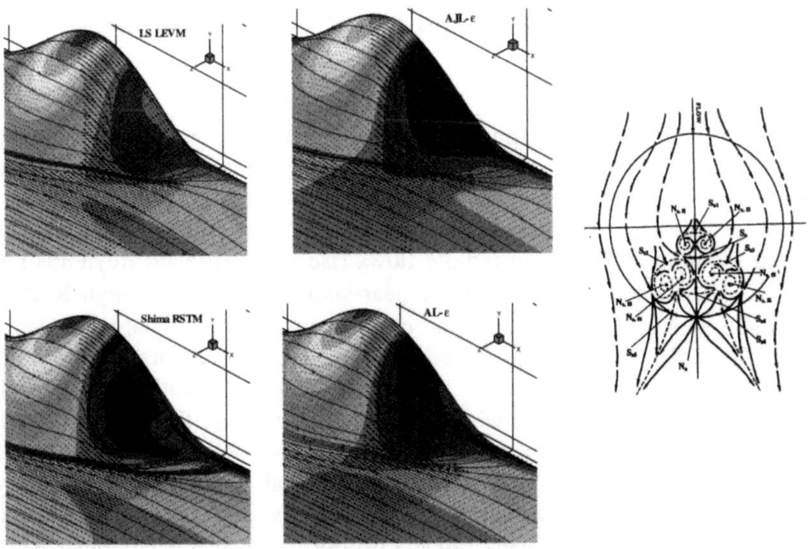

Fig. 3.11. Separated flow around a 3d hill; predicted surface-flow topology with four models (two non-linear EVMs and one second-moment closure), relative to experimental observations [29].

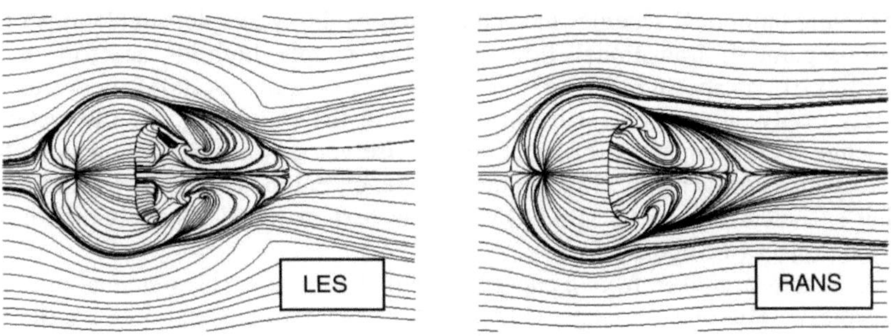

Fig. 3.12. Separated flow around a 3d hill; predicted surface-flow topology at Reynolds number of 13,000 (10% of that in Fig. 3.11) with LES and cubic EVM [45].

4. Some Comments on Large Eddy Simulation for Separated Flow

LES is often claimed to be an approach that will eventually replace RANS modelling. While LES is well suited to free flows, it faces major challenges in high-Reynolds-number shear flows that are strongly affected by viscous near-wall processes. Separation from curved surfaces is a case in point. While the mesh-resolution requirements for wall-remote flows rise modestly with Reynolds number as $N \propto O(Re^{0.4})$, the requirements for near-wall shear layers at high Reynolds numbers rise as $N \propto O(Re^2)$, because the dynamically important 'large' scales become small as the wall is approached and also because accuracy-related constraints demand relatively low cell-aspect ratios (typically, $\Delta x^+/\Delta y^+/\Delta z^+=O(100/1/20)$ for z-wise homogeneous channel flow, with x,y being the streamwise and wall-normal directions, respectively). Near-wall flows are also much more sensitive to the quality of the subgrid-scale model, especially the asymptotic decay of the subgrid-scale viscosity as the wall is approached. This decay should be $O\{(y^+)^3\}$, but many models do not return this behaviour [27], the consequence being a significant impact on the log-law representation.

High resolution, especially near the wall, is particularly important in boundary layers approaching separation along curved surfaces. At very high Reynolds numbers, the resource implications for such flows are frightening, quite apart from the numerical problems associated with the imperative of using non-diffusive, energy-conserving numerical schemes in such conditions. For example, simulations of the mildly-separated, transitional flow around a simple high-lift aerofoil at $Re_{chord}=2.2\text{x}10^6$, as shown in Fig. 4.1, is estimated to require O(100M) nodes to adequately resolve a spanwise slab of only 15% of the chord. Fig. 4.1 illustrates the challenge. This compares profiles of streamwise velocity on the suction side of the "Aerospatiale A" aerofoil [38], obtained by seven computor groups (including the writer's – identified as "UMIST/QMW") as part of the European Commission-funded project LESFOIL [39].

The results are taken from a paper [40], which reports on the outcome of part of the project. The only simulations approaching the experimental data are those of ONERA, but this was obtained with 7M nodes covering most of the domain by a 2d grid, confining 3d resolution to a small portion close to the aerofoil surface and the wake, and extending over a spanwise slab (almost a sliver) of only 1.2% of chord – very likely to be insufficient to resolve properly the three-dimensional turbulent structures in the separated region. Clearly, the use a 2d mesh to cover most of the domain outside the immediate near-wall region (through a single spanwise plane) is appropriate only to a practically inviscid flow approaching a 2d obstacle. Were the mesh 3d throughout the domain, the number of nodes would have risen to about 20M. It is interesting to compare the LES results in Fig. 4.1 with the RANS solutions for the same geometry in Fig. 3.4, the latter performed with a mesh of 25000 nodes at a cost of about 1 CPU hour on a workstation, relative to a resource equivalent to about 20000-30000 CPU hours on a Cray T3E for the LES computations.

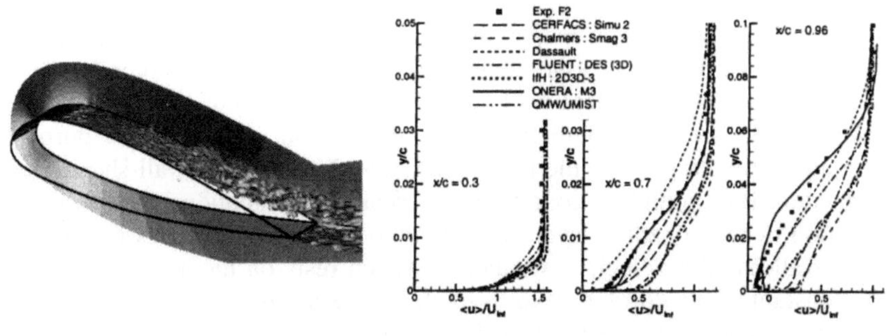

Fig. 4.1. LES of high-lift aerofoil at Re=2.2×10^6 (rhs plot shows velocity profiles on suction side [40]

A further illustration is provided in Fig. 4.2. This flow, already considered in Figs. 3.7-3.9, is much less challenging than the aerofoil flow: it is fully turbulent, is much easier to mesh adequately and, above all, has a much lower Reynolds number of 21500 (based on the channel height). Yet, it is also difficult to simulate because the separation process is sensitive to resolution and the treatment of the near-wall layer. Thus, Fig. 4.2 includes three sets of velocity profiles at one streamwise station derived from simulations [27] with grids containing between 660K and 4.7M nodes.

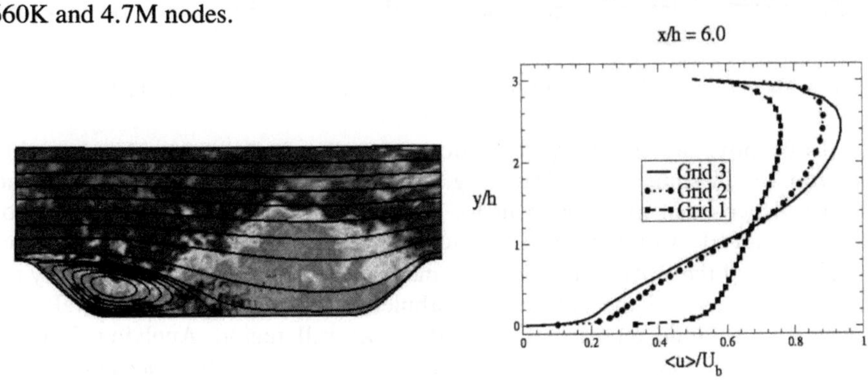

Fig. 4.2. LES of separated flow in a periodically constricted channel; r.h.s. plot shows velocity profiles 6 hill-heights downstream, computed with three grids [27].

All three simulations were undertaken with the same subgrid-scale model and no-slip conditions applied at the lower wall. Only at the highest resolution are the mean flow and the turbulence fields essentially insensitive to resolution. Reference is also made to a study [41], which reports simulations for the separated flow in the plane asymmetric diffuser shown in Fig. 3.2. This serves as a further example of the difficulties encountered when simulating separation from continuous surfaces.

As noted above, the prohibitive resolution requirements for near-wall flows at high Reynolds numbers is a major obstacle to achieving credible simulations for complex engineering flows. One conventional route taken to overcoming this impasse is to use semi-empirical wall-laws in the semi-viscous near-wall layer and couple these to the LES solution in the outer region. The principal purpose of these wall laws is to provide the LES solution with the correct wall-shear stress as a momentum-flux boundary condition. Alternative formulations are based on the assumption of two- or three-layer logarithmic or power-law profiles for the instantaneous wall-parallel velocity. Another approach rests on the assumption that the ratio of instantaneous-to-time-mean wall-shear stresses is linearly related to the ratio of instantaneous-to-time-mean wall-parallel velocities at the wall-nearest LES computational node which is supposed to lie outside the viscous sublayer.

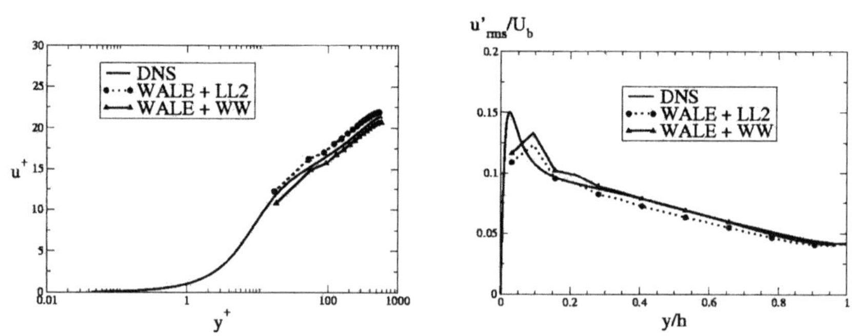

Fig. 4.3. Plane channel flow – results for two wall functions for LES at *Re*=11000 [26, 27]

All the above are crude approximations of reality, but can provide acceptable compromises in some cases. Thus, Fig. 4.3 shows solutions of the velocity and one turbulence-intensity component for a channel flow at a Reynolds number of 11000, obtained with two wall-functions, both covering the near wall layer $0 \le \Delta y^+ < 40$, and the WALE subgrid-scale model [42]. Both treatments correctly return the log law and the anisotropic turbulence state away from the wall, but clearly give a crude representation of the near-wall region. Applying different wall-laws to more complex flows can have a considerably more profound effect on the solution. This is demonstrated in Fig. 4.4 which shows velocity profiles obtained with five near-wall treatments applied in conjunction with a fairly coarse grid of 660K nodes covering the periodic segment shown in Fig. 4.2.

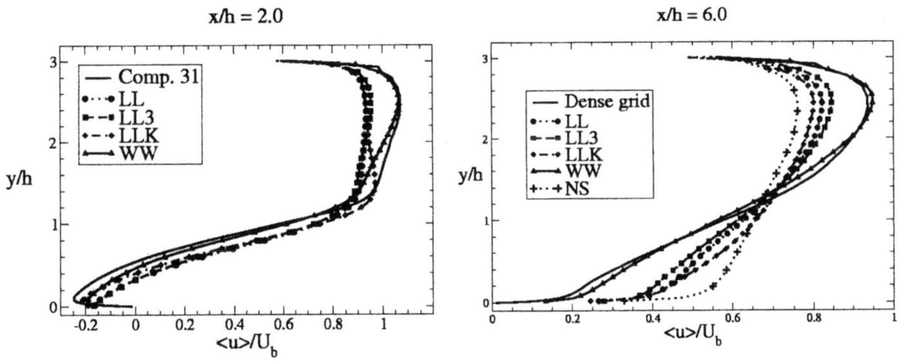

Fig. 4.4. Periodically-constricted channel (see Fig. 4.2) – sensitivity of flow to wall-function treatment in LES with a grid of 660k nodes [26, 27]

An alternative proposal to that above - also beset with difficulties and uncertainties - is to use a conventional low-Reynolds-number turbulence model (operating in unsteady mode) in the near-wall layer. Although these models require very dense wall-normal meshes, down to y^+=O(1), substantial savings in resource arise from the fact that low-*Re* turbulence models are held to permit cell-aspect ratio O(100-500) - that is, far in excess of those required for a wall-resolved simulations. The best-known method, termed *Detached Eddy Simulation* (DES), was proposed Spalart *et al.* [43], but there are several other techniques (see [44] for a review). None is a panacea. For example, when DES is applied to high Reynolds-number near-wall layers, it is observed that the high turbulent viscosity generated by the model in the RANS layer persists at a high level, in the form of subgrid-scale viscosity, well into the LES outer domain, resulting in a strong damping of the resolved motion in the LES layer and in a misrepresentation of the flow in the region bordering the RANS layer. RANS-LES hybrid strategies are subject to much current research, but their discussion would go well beyond the framework of this paper.

The objective of all LES/RANS hybrid formulations is to diminish the unsteady activity near the wall, delegate the representation of a large proportion of the turbulence activity to the RANS model and allow much higher grid aspect ratios. The DES method, in particular, uses a one-equation model near the wall, with the length-scale proportional to the distance from the wall, and that same formulation as a subgrid-scale model in the LES region, except that the length scale is replaced by a grid-distance norm. The switch between the two regions is dictated by the intersection of the two length-scales. One of several disadvantages of the method is that it does not allow any explicit control over the location of the interface. Given grid constraints in the LES domain outside the near-wall region, this location can be very close to the wall at relatively low Reynolds numbers. Another disadvantage is that it tends to over-estimate the turbulence activity, especially in the buffer region, as a consequence of the 'double-counting' arising from the RANS turbulence stresses being added to the non-negligible resolved stresses.

In a method proposed recently by Temmerman et al [46], the LES/RANS interface may be specified and the coupling conditions at the interface are dynamically adjusted the on the basis of an examination of the turbulence activity predicted by the LES, as indicated in Fig. 4.5. At the interface, the requirement is imposed that the sum of the two viscosity contributions arising, respectively, from the turbulence model and the resolved motion should be continuous:

$$\nu_{RANS}^{mod} + \nu_{RANS}^{res} = \nu_{LES}^{mod} + \nu_{LES}^{res} \tag{4.1}$$

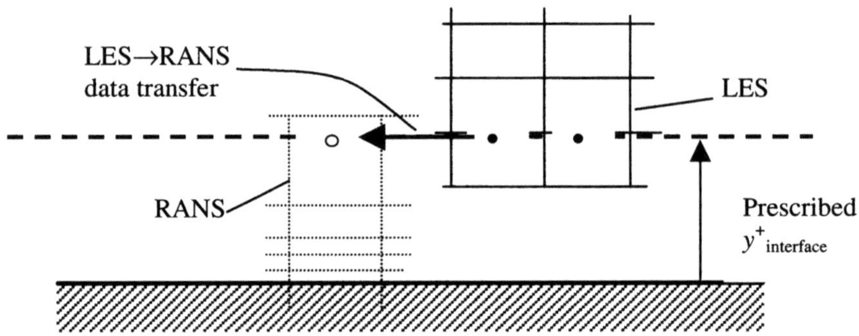

Fig. 4.5: Two-layer near-wall region with LES-RANS overlap

Continuity in the resolved (suitably averaged) contributions across the interface then imply the need for equality of the modelled contributions in Eq. (4.1). If, for the sake of simplicity, a one-equation model is used in the RANS layer,

$$\nu_t = C_\mu k^{0.5} l_\mu \tag{4.2}$$

continuity can be secured by modifying the interface value of C_μ as follows:

$$C_{\mu,int} = \frac{\nu_{LES}^{mod}}{l_\mu k_{RANS}^{0.5}} \tag{4.3}$$

This value is determined from the subgrid-scale model at LES nodal positions shown as solid dots in Fig. 4.5 and used by the RANS scheme at the node indicated by an open circle. It now needs to be 'relaxed' in the RANS region towards the value 0.09 normally used in the turbulence model. The choice of the relaxation function is necessarily arbitrary. One function that has been investigated is:

$$C_\mu = 0.09 + \left(C_{\mu,int} - 0.09\right)\frac{\left(1-\exp(-y/\Delta)\right)}{\left(1-\exp(-y_{int}/\Delta_{int})\right)} \tag{4.4}$$

Temmerman et al [47] show, on the basis of a-priori studies for channel flow, that the absence of the above dynamic approach leads to a serious over-estimation of the total turbulence activity, because the resolved motion is insufficiently attenuated by the modelled component.

The above methodology has been applied to channel as well as separated flows, using the subgrid-grid-scale model of Yoshizawa and Horiuti [48] in the LES region and the Wolfshtein model [49] in the RANS layer. In Fig. 4.6 results are included for a channel flow at $Re_\tau = 2000$ ($Re_{h/2} = 42200$) with the interface placed at y^+_{int} =135 and 318. A 64x64x32 grid covering $2\pi h \times 2h \times \pi h$ with cell dimensions $(\Delta x^+, \Delta y^+, \Delta z^+) = (196, 0.8 - 222, 196)$ has been used.

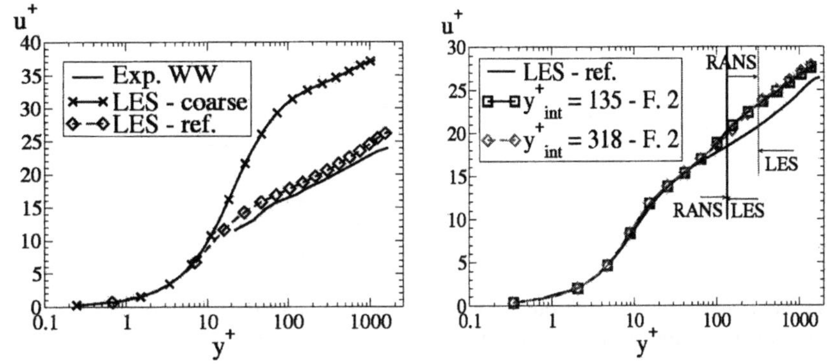

Fig. 4.6: Pure LES (l.h.s) and hybrid LES-RANS solutions for channel flow at Re_h=42200.

Two pure LES solutions are contained in the left-hand-side plot of Fig. 4.6: a highly-resolved solution (LES-ref.) with a 512x128x128 grid and an under-resolved solution (LES-coarse) with the same 64x64x32 mesh used for the hybrid LES-RANS solutions contained in the right-hand-side plot. As seen, the LES-RANS hybrid leads to solutions which are far superior to the one obtained with the3 under-resolving grid.

The above method has now also been applied to separated flow, and one set of results for the case, also shown in Fig. 4.2, is given in Fig. 4.7. The reference solution ("Ref") was obtained with almost 5 million nodes, while the remaining three solutions were obtained with about 0.6 million nodes. The DES solution is given alongside the solution arising from hybrid RANS-LES method introduced above. Both are seen to give recirculation zones which are too long, but in the latter the interfaces were chosen arbitrarily and are much further away from the wall than in the DES solution. It must be acknowledged here that the Reynolds number of this flow is too low (at 21560) for any hybrid RANS-LES method to manifest its true characteristics. Testing for much higher Reynolds numbers is in progress as this paper is being written.

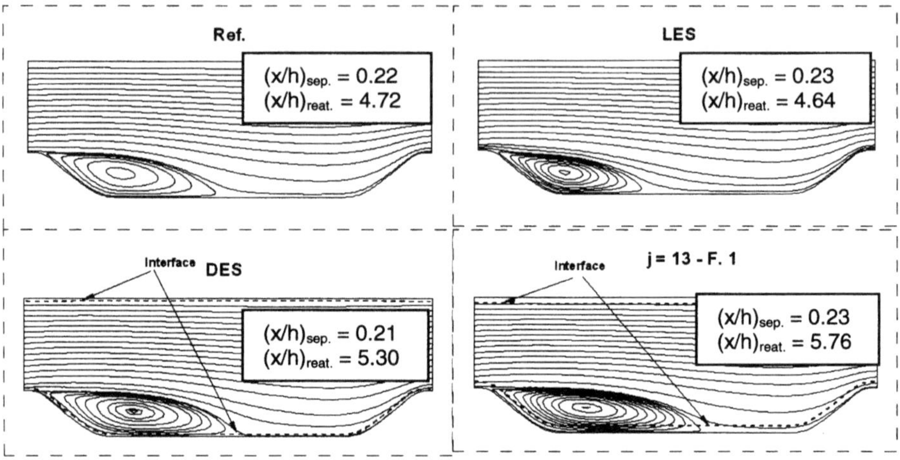

Fig. 4.7: Separated flow in periodic channel with hill-shaped constrictions: averaged streamlines for the reference simulation, LES, DES and RANS-LES cases

5. Concluding Remarks

Has turbulence modelling reached the end of the road? The answer is a qualified *No.* Research continues on a number of fronts: second-moment closure (differential as well as algebraic), structure-based and multi-scale modelling and two-point closure, to name some. This is aided by the steadily increasing diagnostic power of simulation, DNS as well as LES, which permits searching a-priory testing of new modelling proposals to be undertaken. However, in the absence of a '*big idea*', progress is likely to be measured in small increments, and it may well be that the subject is approaching an asymptotic state bounded by the need for compromise between accuracy, simplicity and economy.

Can current models predict separation from curved surfaces with sufficient accuracy? The answer is again a qualified *No.* There are a number of cases where anisotropy-resolving closures are demonstrably superior to simpler models; they are certainly fundamentally sounder. However, one important limitation appears to be the inability of models to capture the consequences of the major spatial and temporal fluctuations in the separation location, specifically the extremely high stress levels that exist around the mean separation location, due to large-scale detached structures generated when the instantaneous separation location moves upstream of the mean location.

Is Large Eddy Simulation a panacea? The answer is an emphatic *No.* LES undoubtedly offers clear predictive advantages in bluff-body aerodynamics in which large-scale unsteadiness is a key feature. It performs well when separation is provoked at sharp edges and when the principal features of interest are associated with shear layers remote from wall and are not affected by viscous wall processes. LES is much more problematic in near-wall flows, especially when wall curvature induces marginal separation. LES continues to be very costly, typically two orders of magnitude more costly than RANS, places considerably more stringent constraints than RANS on numerical accuracy and grid quality and requires the spectral content of boundary conditions to be specified. While LES and LES/RANS hybrid strategies will increasingly be used in practical applications, it is unlikely that they will replace RANS, especially not in circumstances in which the flow properties of primary interest are governed or strongly affected by near-wall shear.

References

[1] Kassinos SC, Langer CA, Haire SL, Reynolds WC (2000) Structure-based modelling for wall-bounded flows. J Heat and Fluid Flow 21:599-605

[2] Perot JB (1999) Turbulence Modeling Using Body Force Potentials. Int J Physics of Fluids 11

[3] Schiestel R.(1987) Multiple time scale modelling of turbulent flows in one point closures. J Phys. Fluids. 30:722

[4] Cambon C, Scott JF (1999) Linear and nonlinear models of anisotropic turbulence. Annual Rev Fluid Mech 31:1-53

[5] Shih TH, Lumley JL (1985) Modelling of pressure correlation terms in Reynolds-stress and scalar-flux equations. Report FDA-85-3, Sibley School of Mech. and Aerospace Eng, Cornell University

[6] Speziale CG, Sarkar S, Gatski TB (1991) Modelling the pressure-strain correlation of turbulence: an invariant dynamical systems approach. J Fluid Mech. 227:245.

[7] Fu S., Launder B E, Tselepidakis DP (1987) Accommodating the effects of high strain rates in modelling the pressure-strain correlation. Report TFD/87/5 Mechanical Engineering Dept., UMIST, Manchester

[8] Launder BE, Shima N (1989) Second-moment closure for the near-wall sublayer, J. AIAA, 27:1319-1325

[9] So RMC, Lai YG, Zhang HS, Hwang BC (1991) Second-order near-wall turbulence closures: Review J AIAA 29:1819-1835

[10] Jakirlic S, Hanjalic K (1995) A second-moment closure for non-equilibrium and separating high- and low-Re-number flows. Proc. 10th Symp. on Turbulent Shear Flows, Pennsylvania State University 23.25

[11] Craft T.J, Launder BE (1996) A Reynolds stress closure designed for complex geometries. Int J Heat Fluid Flow 17:245

[12] Batten P, Craft TJ, Leschziner MA, Loyau H (1999) Reynolds-stress-transport modelling for compressible aerodynamic flows. J AIAA, 37:785-796

[13] Durbin P (1995) Separated flow computations with the k-ε-v^2 model. J AIAA 33: 659-664

[14] Wilcox DC (1994) Simulation transition with a two-equation turbulence model. J AIAA 32:247-255

[15] Speziale CG, Abid R., Anderson EC (1992) Critical evaluation of two-equation models for near-wall turbulence. J AIAA 30:324-331

[16] Menter FR. (1994) Two equation eddy viscosity turbulence models for engineering applications. J AIAA 32:1598-1605

[17] Shih T-H, Zhu J, Lumley JL (1993) A realisable Reynolds stress algebraic equation model. NASA TM105993

[18] Craft TJ, Launder BE, Suga K. (1996) Development and application of a cubic eddy-viscosity model of turbulence. Int J Num Meth In Fluids 17:108-115

[19] Apsley DD, Leschziner MA. (1998) A new low-Reynolds-number nonlinear two-equation turbulence model for complex flows. Int J Heat and Fluid Flow 19:209-222

[20] Gatski T.B, Speziale CG (1993) On explicit algebraic stress models for complex turbulent flows. J Fluid Mech 254:59-78

[21] Taulbee DB., Sonnenmeier JR, Wall KM (1993) Application of a new non-linear stress-strain model to axisymmetric turbulent swirling flows. In Engineering Turbulence Modelling and Experiments 2 Rodi W, Martelli F (eds) Elsevier 103-112

[22] Rung T, Fu S, Thiele (1999) On the realisability of non-linear stress-strain relation for Reynolds-stress closures. Flow, Turbulence and Combustion, 60:333-359

[23] Wallin S, Johansson AV (2001) Modelling of streamline curvature effects on turbulence in explicit algebraic Reynolds stress turbulence models. Proc 2dn Int Symp on Turbulence and Shear Flow Phenomena, Stockholm 223-228

[24] Abe K, Jang YJ, Leschziner MA (2003) An investigation of wall-anisotropy expressions and length-scale equations for non-linear eddy-viscosity models. Int J Heat and Fluid Flow, 24:181-198

[25] Launder BE, Sharma BI (1974) Application of the energy-dissipation model of turbulence to the calculation of flow near a spinning disc. Letters in Heat and Mass Transfer, 1:131-138

[26] Temmerman L, Leschziner MA, Mellen C, Froehlich J (2003) Investigation of subgrid-scale models and wall-function approximations in Large Eddy Simulation of separated flow in a channel with streamwise periodic constrictions. Int J Heat Fluid Flow, 24: 157-180

[27] Temmerman L, Leschziner MA (2001) Large eddy simulation of separated flow in a streamwise periodic channel constriction. Proc 2nd Symp on Turbulence and shear flow phenomena, Stockholm 399-404

[28] Jang YJ, Leschziner MA, Abe K, Temmerman L (2002) Investigation of anisotropy-resolving turbulence models by reference to highly-resolved LES data for separated flow. Flow, Turbulence and Combustion, 69: 161-203

[29] Chen W, Jang YJ, Leschziner MA (2003) Modelling 2d and 3d separation from curved surfaces with anisotropy-resolving turbulence closures. Proc. 3rd Symposium on Turbulence and Shear Flow Phenomena, Sendai, Japan, 257-262.

[30] Apsley DD, Leschziner MA (2000) Advanced turbulence modelling of separated flow in a diffuser. Flow, Turbulence and Combustion 63:81-112

[31] Chen WL, Leschziner MA (1999) Modelling turbomachine-blade flows with non-linear eddy-viscosity models and second-moment closure. Proc. 3rd European Conference on Turbomachinery, IMechE Conference Transactions 1B:189-199

[32] Lien FS, Leschziner MA (1995) Modelling 2D separation from high-lift aerofoils with a non-linear eddy-viscosity model and second-moment closure. The Aeronautical Journal, 99:125-144

[33] Lien FS, Leschziner MA (1997) Computational modelling of separated flow around streamlined body at high incidence. The Aeronautical Journal 101:269-275

[34] Apsley DD, Leschziner MA (2001) Investigation of advanced turbulence models for the flow in a generic wing-body junction. Flow, Turbulence and Combustion, 67: 25-55

[35] Manceau R., Bonnet JP (2000) (eds.), Proc. of 10th ERCOFTAC/IAHR Workshop on Refined Turbulence Modelling, Poitiers, France

[36] Jang Y-J, Leschziner MA (2004) An investigation of higher-order closures in the computation of the flow around a generic car body, European Congress on Computational Methods in Applied Sciences and Engineering, ECCOMAS 2004, Jyväskylä, Finland (to be presented)

[37] Simpson R.L, Long CH, Byun G (2002) Study of vortical separation from an axisymmetric hill. Int J Heat and Fluid Flow 23:582-591

[38] Piccin O, Cassoudesalle D (1987) Etude dans la soufflerie F1 des profils AS239 et AS240. ONERA Technical Report PV 73/1685 AYG

[39] Davidson L, Cokljat D, Froelich J, Leschziner MA, Mellen C, Rodi W. (Eds.) LESFOIL: Large Eddy Simulation of Flow a High-Lift Airfoil,. Notes on Numerical Fluid Mechanics and Multidisciplinary Design 83 Springer

[40] Mellen P, Froehlich J, Rodi W (2002) Lessons from the European LESFOIL Project on LES of Flow around an Airfoil, AIAA 0111

[41] Kaltenbach HJ, Fatica M, Mittal R, Lund TS, Moin P (1999) Study of flow in a planar asymmetric diffuser using large-eddy simulation. J Fluid Mech, 390:151-185

[42] Ducros F, Nicoud F, Poinsot T (1998) Wall-adapting local eddy-viscosity models for simulations in complex geometries. Proc 6th ICFD Conference on numerical methods for fluid dynamic 293-299.

[43] Spalart PR., Jou WH, Strelets M, Allmaras SR (1997) Comments on the feasibility of LES for wings and on the hybrid RANS/LES approach. In Advances in DNS/LES, 1st AFOSR Int Conf On DNS/LES (Greden Press)

[44] Piomelli U, Balaras E (2002) Wall-layer models for Large Eddy Simulations, Annual Review of Fluid Mechanics 34:349-374

[45] Temmerman L, Wang C, Leschziner MA, A comparative study of separation from a three-dimensional hill using large-eddy simulation and second-moment-closure RANS modeling, European Congress on Computational Methods in Applied Sciences and Engineering, ECCOMAS 2004, Jyväskylä Finland (to be presented)

[46] Temmerman L, Leschziner MA, Hanjalic K (2003), A combined RANS-LES strategy with arbitrary interface location for near-wall flows, Proc 3rd Int Symp on Turbulence and Shear Flow Phenomena, 929-934

[47] Temmerman L, Leschziner MA, Hanjalic K (2002), A-priori studies of a near-wall RANS model within a hybrid LES/RANS scheme, Engineering Turbulence Modelling and Experiments 5, W. Rodi and N. Fueyo (Eds.), 317-326

[48] Yoshizawa A, Horiuti K (1985), A statistically-derived subgrid-scale kinetic energy model for the large-eddy simulation of turbulent flows, J Phys Soc Japan, 54: 2834-2839

[49] Wolfshtein M (1964), The velocity and temperature distributions in one-dimensional flow with turbulence augmentation and pressure gradient, Int J Heat Mass Transfer, 12: 301-318

Development of Fluid Mechanics Methods in the 20th Century and the Application to Laminar and Turbulent Flow Investigations

Prof. Dr. Franz DURST*, Dr. Dubravka MELLING
Institute of Fluid Mechanics, University of Erlangen-Nuremberg

Prof. Simeon OKA
Ministry of Science, Technology and Development

*Corresponding author: Prof. F. Durst, Cauerstr. 4, D-91058 Erlangen, Germany, Tel.: +49 (0) 9131 85 295 00, Fax: +49 (0) 9131 85 295 03, e-mail: durst@lstm.uni-erlangen.de

Abstract The development of fluid mechanics is briefly reviewed and the importance of fluid flows to heat and mass transport in nature as well as in science and engineering is outlined. The early theoretical developments are explained and it is indicated that the basic equations were already available at the end of the 18th century. Methods to solve these equations for engineering flows were not, however, developed until the second half of the 20th century. This was an important period for fluid flow research during which all the experimental fluid mechanics methods, particularly the optical methods, available today were also developed. The same is true for all the numerical methods that are used very successfully nowadays to solve scientific and engineering fluid flow problems. The Institute of Fluid Mechanics at the University of Erlangen-Nürnberg has contributed extensively to these developments.
Fig. 1 provides a chronological summary of important scientists in fluid mechanics. It gives an impression when certain contributions to the development of the theory of fluid mechanics were made. But the development of experimental or numerical techniques are not indicated in this presentation.
Key Words: Fluid mechanics development, experimental methods, computational methods

Nomenclature

c_{μ}	turbulence constant
c_D	drag coefficient
$c_{\varepsilon 1}, c_{\varepsilon 2}$	turbulence constant
$d_p\ [m]$	particle diameter
$E\ (\varpi)[s]$	energy spectral density function
$g_j \left[\frac{m}{s^2}\right]$	gravitational acceleration
$k \left[\frac{m^2}{s^2}\right]$	turbulence kinetic energy
$k_w, k_{air} \left[\frac{W}{mK}\right]$	thermal conductivity (wall, air)
m	refractive index
$m_p\ [kg]$	particle mass
Pr	Prandtl number
Re	Reynolds number
S_{Φ}	source term of Φ
$t[s]$	time
$u_p, u_f \left[\frac{m}{s}\right]$	particle, fluid velocity
$U_{ij},\ u \left[\frac{m}{s}\right]$	velocity
$x_i\ [m]$, $x, y, z\ [m]$	Cartesian coordinates
$\underline{x}\ [m]$	object location
$X^+,\ Y^+$	wall coordinates
$\underline{X}\ [m]$	image location
$\beta\ [^{\circ}]$	angle (fig. 11)
δ_{ij}	Kronecker delta
Γ_{Φ}	diffusivity of Φ
ε	dissipation rate of kinetic energy
$\lambda\ [m]$	wave length

$\rho\left[\frac{kg}{m^3}\right]$	density
$\nu_D\,[Hz]$	Doppler frequency
$\varphi[°]$	angle (fig. 11)
$\theta\,[°]$	angle (fig. 12)
Φ	transport variable
$\Delta\Phi\,[°]$	phase shift
$\mu\left[\frac{kg}{ms}\right]$	viscosity
$\Psi[°]$	angle (fig. 12)
$\Omega\,[Hz]$	frequency
$\omega\left[\frac{Rad}{s}\right]$	angular velocity
$\Omega[ster]$	solid angle
$\tau_{ij}\left[\frac{N}{m^2}\right]$	shear stress
$\sigma_k,\sigma_\varepsilon$	turbulence constants

1. Introduction

Fluid flows are present everywhere in nature and are widely experienced by those people who observe nature with open eyes. They see that many processes in our natural environment are vitally dependent on the convective transport of heat and mass. Hence, without fluid motion, life in the form we know it on earth could not exist.

The technical importance of fluid flows is also easily observable in many engineering fields where heat and mass transfer processes are strongly controlled by fluid motions. The rate of conversion of chemically bonded energy into heat by non-premixed combustion is mainly controlled by the flow transport of the chemically reacting species and heat engines, for example, could not fulfil their function without the corresponding flow. Hence, fluid flows are essential in many fields where engineering equipment is used.

This publication summarises the development of fluid flow investigations that can be carried out using modern experimental and numerical techniques as they were developed in the last fifty years. It is pointed out that these techniques can be used successfully for the solution of flow problems in science and engineering. An outlook is given on the golden age of fluid mechanics that lies ahead of us, as it has been pointed out in recent summaries of fluid mechanics developments covering the last fifty years.

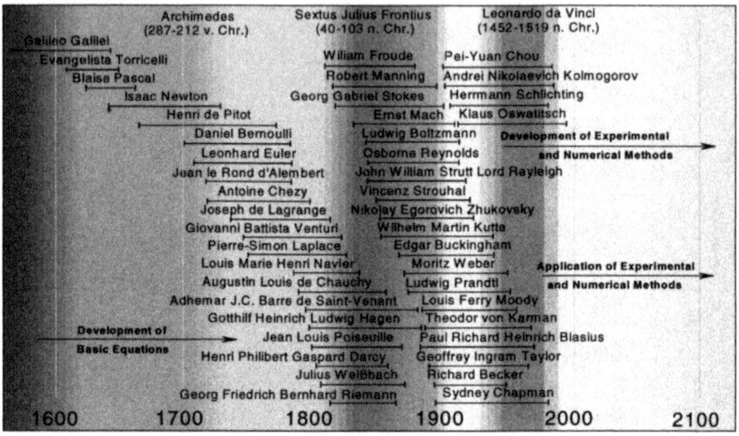

Fig. 1. Historical record of scientists contributing to the development of fluid mechanics

The above introductory remarks that stress the importance of fluid flows might explain why fluid mechanics has a history that goes back as long as records of human culture exist. Although there were contributions by Archimedes and Sixtus Julius Francius as well as first flow visualisations by Leonardo da Vinci, fluid mechanics as a science only began under G. Galilei in the 17th century. This is indicated in Fig. 1 which shows the long list of researchers who have made contributions to the subject over the last four centuries. Among the fluid mechanics scientists of the 17th to the 19th century one finds the names of famous mathematicians and physicists indicating that fluid mechanics as a basic engineering subject is fairly new in a historical context. Nowadays, any serious engineering education in fields such as aeronautical engineering, mechanical engineering and chemical engineering is heavily based on a sound fluid mechanics education. This will also continue in the future but emphasis in the education will change. Instead of teaching the integral form of the energy equation (Bernoulli equation), the integral form of the continuity and momentum equation, education will concentrate on the differential form of the Navier-Stokes equations. Advanced teaching will embrace experimental methods in fluid mechanics and also methods of numerical fluid mechanics to solve the differential form of the basic fluid mechanics equations.

Major contributions were made by Newton, Euler, Lagrange, Navier and Stokes to the derivation of the basic equations of fluid mechanics. From the dates in Fig. 1 one can see that by the end of the 18th century the equations of fluid mechanics were basically known in the general form given below for ρ = constant:

Continuity equation:

$$\frac{\partial \rho}{\partial t}+\frac{\partial(\rho U_i)}{\partial x_i}=0 \tag{1}$$

Momentum equation (j =1, 2, 3):

$$\rho\left(\frac{\partial U_j}{\partial t}+U_i\frac{\partial U_j}{\partial x_i}\right)=-\frac{\partial P}{\partial x_j}-\frac{\partial \tau_{ij}}{\partial x_i}+\rho g_j$$

$$\tau_{ij}=-\mu\left(\frac{\partial U_j}{\partial x_i}+\frac{\partial U_i}{\partial x_j}\right)+\frac{2}{3}\delta_{ij}\mu\frac{\partial U_k}{\partial x_k} \tag{2}$$

For ρ = constant, the three velocity components and the pressure can be named as the unknowns of the above set of four differential equations. Hence four unknowns and four partial differential equations exist and so all fluid flow problems seem to be soluble if the appropriate initial and boundary conditions exist.

Although these equations were known by the end of the 18th century, methods for their solution did not exist. The present paper stresses that all methods to solve the above set of equations for engineering flow problems were developed in the second half of the 20th century following tremendous advances in numerical methods and computer power. As Fig. 2 shows, high-performance computer developments have provided an increase in computational speed by a factor 10 every 5 years. In the last three decades, a factor of 10 every 8 years was also achieved by advanced numerical techniques. All these developments together now permit numerical solutions of the above set of equations for engineering problems.

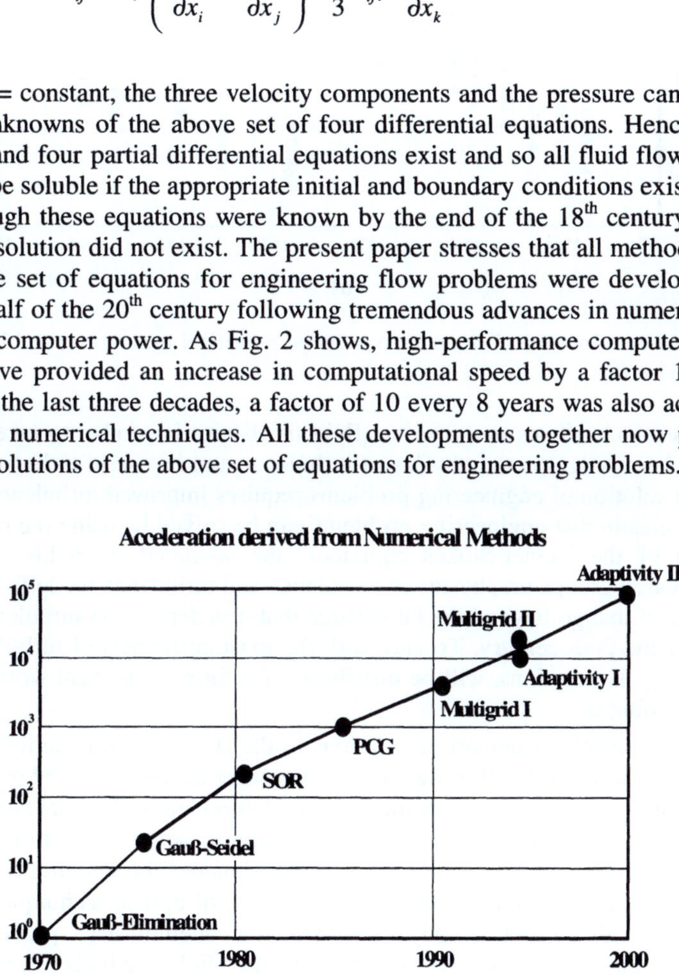

(a)

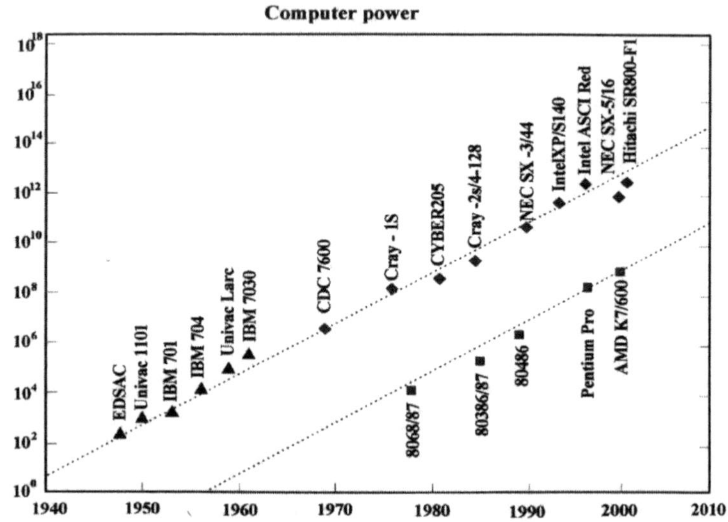

(b)

Fig. 2. Speed-up of numerical computations due to numerical methods (a) and increase in computer power (b)

Major research efforts in numerical fluid mechanics have also to take into account new knowledge on the physics of turbulence and its use in turbulence modelling. The solution of engineering problems requires improved turbulence models in order to ensure that engineering problems can be solved by using the time averaged form of the Navier-Stokes equations, the so-called Reynolds equations. Hence, research groups on physics of turbulence and turbulence modelling have to be established that go beyond the knowledge that was derived on turbulence in the first half of the 20th century. To stay with the existing models of turbulence, i.e. using isotropy assumptions, will be insufficient for future numerical work to solve fluid flow problems.

The developments in numerical solutions to the above set of equations of fluid mechanics in the second half of the 20th century were accompanied by tremendous developments in the field of experimental fluid mechanics. The development of fast electronic components, lasers, integrated optics, sensors, micro-techniques, *etc.*, has resulted in a wide range of measuring techniques for studying fluid flows. The present paper concentrates on the development of optical techniques such as photography and cinematography, laser-Doppler anemometry, phase-Doppler anemometry, particle image velocimetry and other field methods. The developments are summarised and emphasis is given to those developments carried out in the research groups initially at the University of Karlsruhe and later at the University of Erlangen-Nürnberg.

2. Experimental methods

2.1. Flow visualization: photography and cinematography

It is difficult to define clearly where and when fluid mechanics as a science started. However, the first contribution that had an impact on the subject as it is treated today is the flow visualisation carried out by Leonardo da Vinci (1452-1519). Since this early work, flow visualisation has advanced with major developments being based on modern illumination techniques, as well as developments in photography and cinematography. This is indicated in Fig. 3, which compares photographic records of flows with the early sketches of observed vortex motions by Leonardo da Vinci.

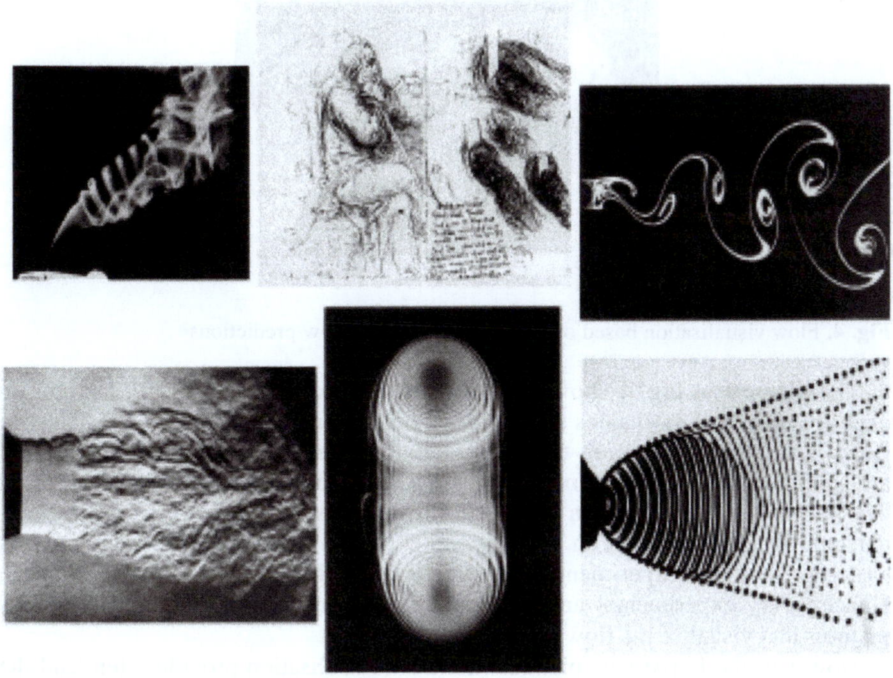

Fig. 3. Flow visualisation is the start of good fluid mechanics research

A summary of flow visualisation results is given in the Album of Fluid Motion assembled by van Dyke [1]. This assembly of pictures makes clear that photographic records of fluid motion can result in path lines, streak lines or streamlines of the flow. All of them are difficult to analyse in terms of local velocity information or to obtain results in the terminology of the basic equations of fluid mechanics. Nevertheless, they provide a good physical insight into the flow so that it has become common practice also to visualize the results of flow predictions, *e. g.* see Fig. 4.

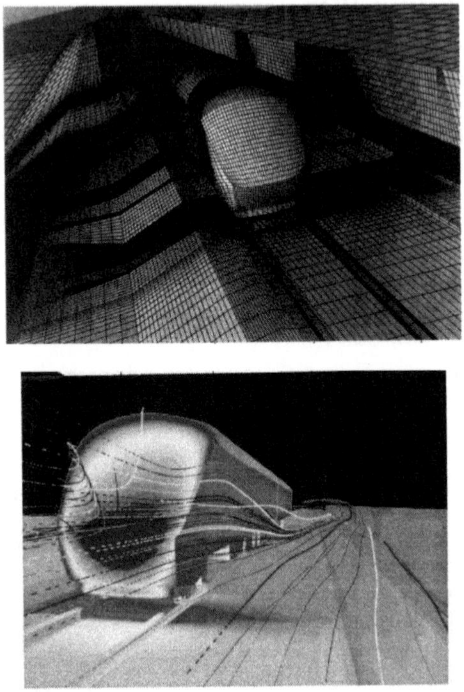

Fig. 4. Flow visualisation based on results of numerical flow predictions

The pictures in Fig. 4 show that flow visualisation is a tool that is helping in experimental fluid mechanics to get a good insight into what happens physically in flows. One should be aware that the results obtained numerically on fluid flows are similar to those of experimental investigations and therefore flow visualisation can also be a powerful tool in numerical flow studies. Usually it is the basis to understand what happens in an investigated flow and with this understanding it is easier to interpret experimental or numerical results of certain flow problems. Hence, every experimental and numerical investigation should aim first at clear pictures that visualise the flow behaviour.

However, the impressive pictures that flow visualisation provide often lead the flow researcher to forget that it is very difficult to interpret correctly the resultant flow motions. This was outlined by Hama [2] and demonstrated by Eckelmann [3], from which the example in Fig. 5 is taken.

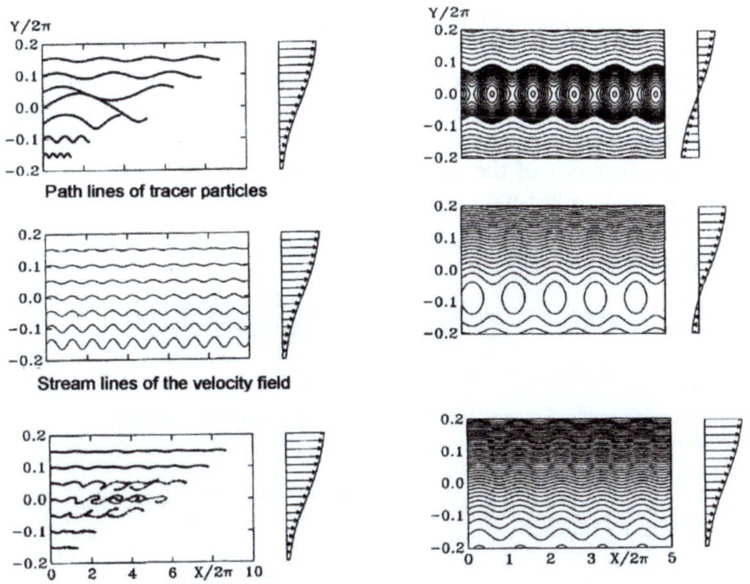

Fig. 5. Demonstration of complex flow information in path, streak and stream lines

Although Leonardo da Vinci showed the power of flow visualisation as a basis for physical insight into flows, and modern methods of flow illumination and recording of fluid motion have become available, flow visualisation is not applied to its best in modern fluid mechanics. The turbulent flows that are being studied these days yield complex flow visualisation pictures and quantitative information is difficult to obtain by flow visualisation.

2.2. Laser-Doppler anemometry

During the second half of the 20th century, rapid developments in electronics and optics components provided the basis for the advancement of methods to measure local and time resolved flow velocities. Fast operating electronic feedback amplifiers permitted the development of constant temperature hot wire anemometry, e. g. see Bruun [4]. With the help of this technique, first detailed velocity information became available about turbulent flows, providing an insight into the complexity of turbulence. Detailed turbulent flow studies provided the basis for advanced analytical treatments of flows, e. g. see Lumley [5]. However, the application of hot wire anemometry was, and still is, limited to flows with low levels of turbulence.

The method also is intrusive, yielding flow disturbances that are unacceptable in recirculating flows. These shortcomings of hot-wire anemometry triggered new developments. Laser-Doppler anemometry emerged from these development efforts, providing new means to study fluid flows.

Laser-Doppler Anemometry is an interferometric velocity measure technique, i.e. it uses the coherence properties of laser light. Hence, the development of the laser provided the basis for a rapid development of this measuring technique and for its extensive usage in fluid mechanics. A clear physical understanding of the measuring technique was developed in the years 1970-1980. Good instrumentation developments followed in the years 1980 to 1990. The last 10-15 years brought out extensive applications of the measuring technique in fluid flows where measuring techniques such as hot-wire anemometry could not be applied.

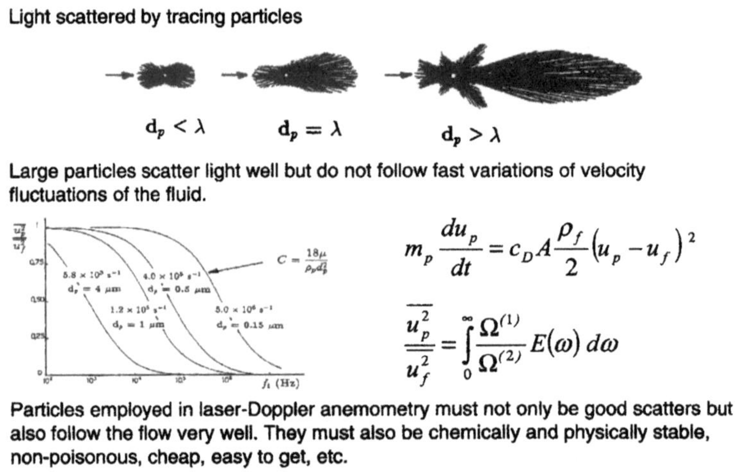

Fig. 6. Scattering particles with high scattering efficiency and good flow velocity response

Laser-Doppler anemometry is a well documented measuring technique, *e. g.* see Durst, Melling and Whitelaw [6]. Light scattering particles are needed that follow the flowing fluid so that the particle velocity is close to that of fluid, *e. g.* see van de Hulst [7], Kerker [8] and Hjelmfelt and Mockros [9]. These basic requirements for LDA scattering particles are summarised in Fig. 6. From the many scattering mechanisms that can be used to deduce the local particle velocity from laser frequency shift information, *e. g.* see Durst [10], the dual scattering beam laser-Doppler effect was shown to be the most efficient and most robust for use in LDA velocity measurements. Hence optical systems of the type shown in Fig. 7 are these days employed to measure local flow velocities by laser-Doppler anemometry.

The laser-Doppler system shown in figure 7 contains a so-called Bragg cell module that permits light frequencies in the two beams to be shifted so that a difference frequency between both beams is produced. In this way, the imaginary fringe pattern in the crossing region of the two beams is moving and particle velocities are measured relative to this moving fringe pattern. Hence, not only the magnitude of the velocity perpendicular to the axes of the optical system can be measured but also the sign of the measured velocity component can be obtained.

Measurements are possible in regions with reversed flows. Multi-component anemometers, as indicated in Fig. 8, are available.

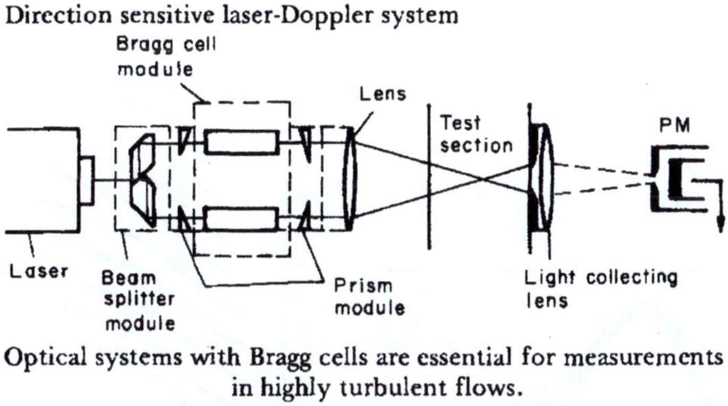

Fig. 7. Dual beam laser-Doppler optical system

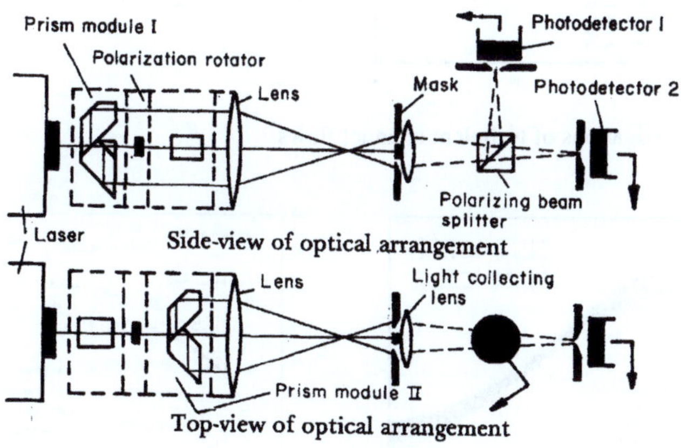

Fig. 8. Two-component laser-Doppler optical system

Laser-Doppler anemometers can be applied to carry out flow measurements that are not feasible with any other fluid flow measuring technique. Figure 9 shows a test facility providing a fully developed channel flow. With the help of a laser-Doppler anemometer, detailed flow measurements were carried out, yielding the velocity information sketched in Fig. 9. Analysing these data revealed a Reynolds number dependence of the wall value of the turbulence intensity, as shown in Fig. 10

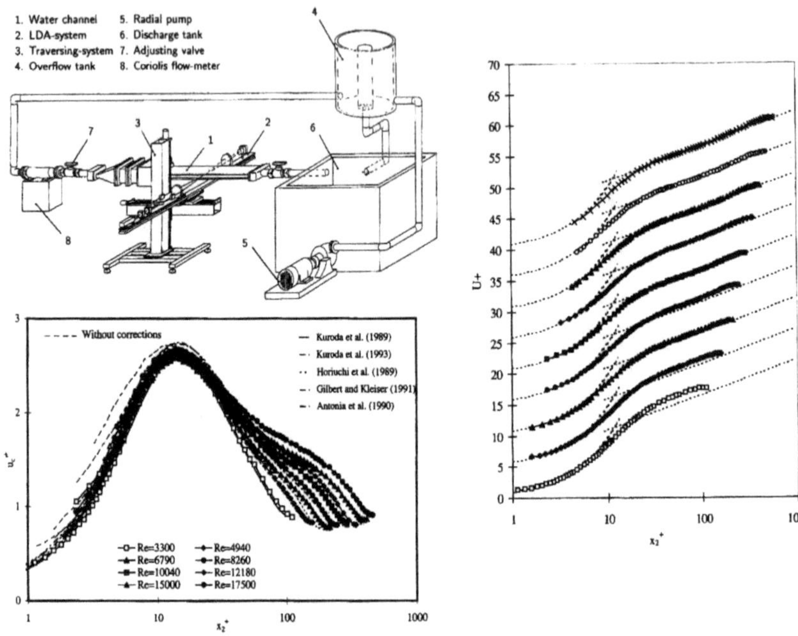

Fig. 9. LDA investigations of turbulent channel flows

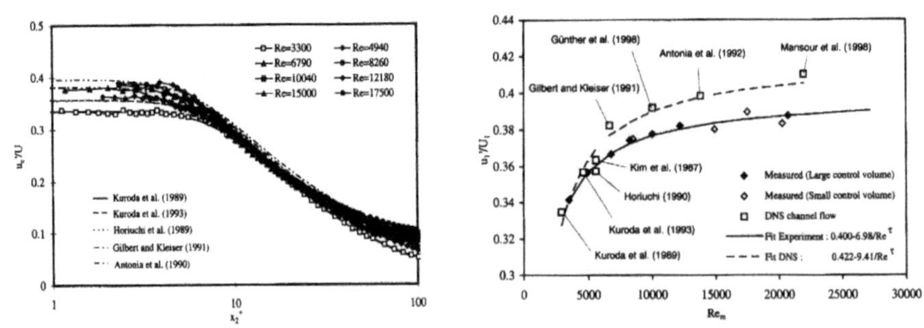

Fig. 10. Wall values of turbulence levels in fully developed channel flows

It is interesting to note that at present the Reynolds number dependence of the wall value of the turbulence intensity is not understood. Both experimental and numerical studies show this dependence, but there are differences between the experimental and numerical results which cannot readily be explained. They may be due to the finite sizes of the computational grids and the measuring control volume of the LDA systems. The evaluation of the final data requires finite size volume corrections; the experimental data in Fig. 10 were obtained with optical systems with measuring control volumes of different sizes.

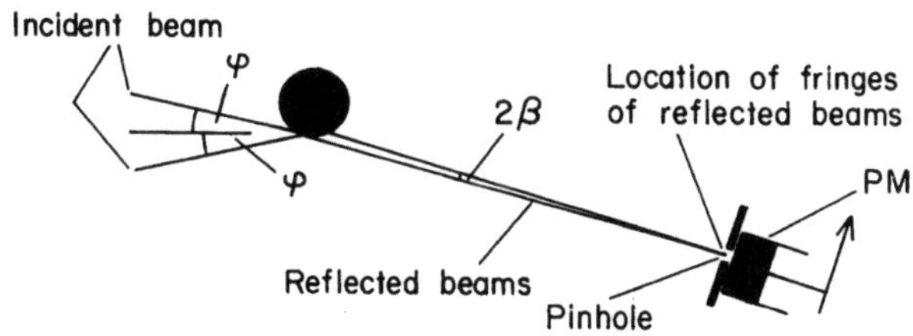

Fig. 11. Phase-Doppler measurements in two-phase flows

2.3. Phase-Doppler anemometry

When light from two inclined laser beams is refracted or reflected from a particle, interference fringes result in space, *e. g.* see Durst and Zaré [11]. For a given particle location, the fringes exist only in the direction in which the beams are reflected or diffracted. As the particle moves, the fringe system changes its location and, for this reason, a mask in front of the photomultiplier is crossed by fringes that cause varying light intensity at the pinhole. The fringes change their shape as they move through space, and are linear in the backward direction and highly non-linear in the forward direction.

Irrespective of the shape of the fringes, the derivation by Durst and Zaré [11] revealed that the resultant frequency due to the fringes crossing a mask in front of the photodetector is given by an equation that is independent of the photomultiplier location:

$$\nu_D = \frac{2}{\lambda}\left[U_\perp \cos\beta \pm U_\parallel \sin\beta\right]\sin\varphi$$

This equation indicates that the frequency of the resultant signal is sensitive to the velocity components perpendicular and parallel to the axis of the two incident beams respectively. For most practical cases, however, the angle β is fairly small and, hence, the term $U_\parallel \sin\beta$ is small in comparison to the term $U_\perp \cos\beta$ so that the evaluation equation reads

$$\nu_D \cong \frac{2U_\perp \sin\varphi}{\lambda}$$

The angle β is a function of the angle between the two incident beams and the ratio of detector distance to particle diameter. For large values of L/R, where L is the detector distance and R is the particle radius, the angle β tends to zero; β also decreases with decreasing angle φ. The same arguments apply to light beams refracted by transparent particles. In this case, linear interference fringes result in the forward direction and the Doppler frequency is not dependent on the velocity component parallel to the axis of the two incident light beams. The equation for the Doppler frequency reads in this case

$$\nu_D \cong \frac{2U_\perp(\sin\varphi - \sin\beta)}{\lambda} \tag{3}$$

In addition to measuring the particle velocity, phase Doppler systems also permit the measurement of particle size by measuring the phase difference between signals from two detectors, as indicated in Fig. 12. The resultant equation used to deduce the particle diameter from the measured phase difference depends on the location of the photodetector with respect to the transmission optics, *i. e.* whether reflected or refracted light is used for signal detection. Hence information of the kind shown in Fig. 13 is obtainable with phase-Doppler systems at every location in a flow field. Therefore, particulate two-phase flows can be studied by phase-Doppler anemometry.

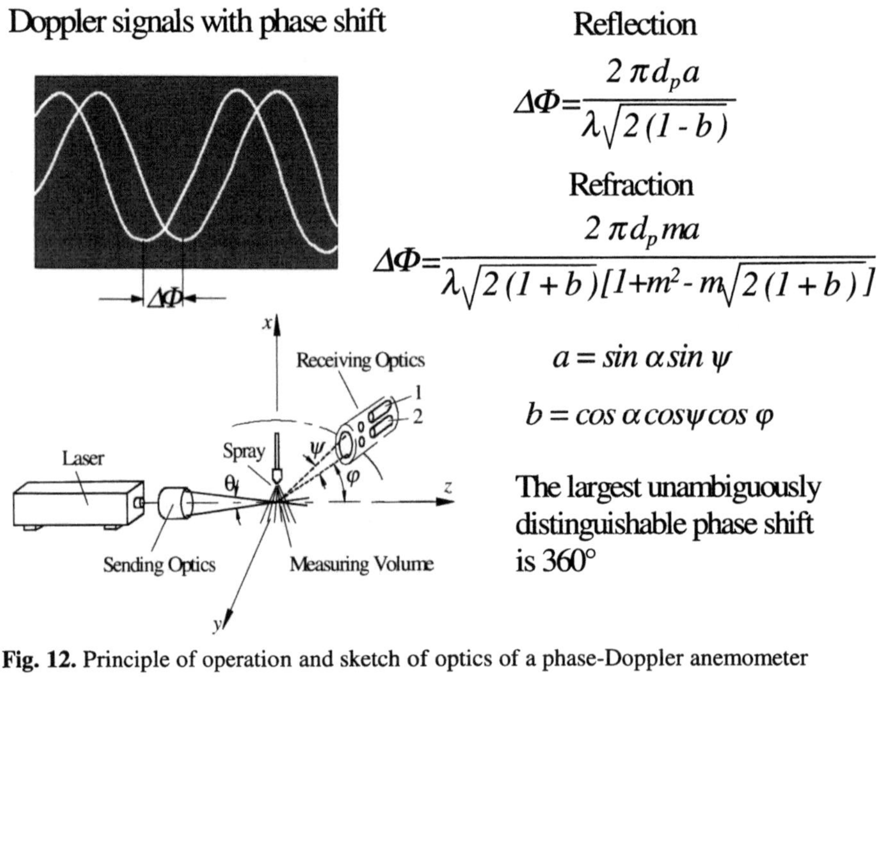

Fig. 12. Principle of operation and sketch of optics of a phase-Doppler anemometer

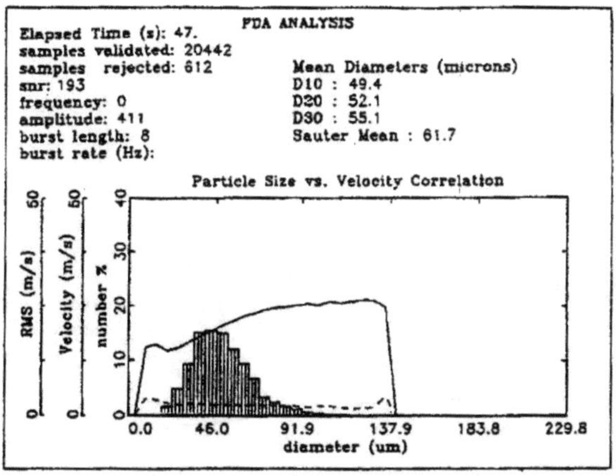

Information achieved:

- **Drop size number distribution**
- **Size / velocity correlation**
- **Mean / RMS velocity correlation**
- **Moments of the drop size spectrum**

Fig. 13. Typical result of local time and velocity measurements in sprays

2.4. Spray measurement results

Phase-Doppler Anemometry is a measuring technique that can be applied in a wide range of two-phase flows where the particulate phase in a gas or liquid flow shows a different refractive index and where the particles are spherical. This indicates that phase-Doppler anemometry provides the basis for extending the detailed investigations of single-phase flows to particulate two-phase flows. Of course, only those two-phase flows with spherical particles can be studied, but studies of this kind will provide a basic understanding of the fluid particle interactions that occur in particulate two-phase flows. The developments of phase-Doppler anemometry have therefore provide the basis for detailed studies of two-phase flows.

The phase-Doppler method has also been extended to yield information on the refractive index of the light refracting particles in a flow. Extended phase-Doppler anemometers have been suggested for this purpose using four detectors located in different directions. This is indicated in Fig. 14, which also provides the equation used for particle refractive index measurements utilising the ratio of two phase difference measurements. Results of application of an extended phase-Doppler system are shown in Fig. 15.

Phase shift measured redundantly with EPDA

$$\Delta\Phi_1 = f(m, d_p, \theta, \varphi_1\, \psi_1)$$

$$\frac{\Delta\Phi_1}{\Delta\Phi_2} = \frac{\sin\psi_1}{\sin\psi_2}\sqrt{\frac{\left(1+\cos\frac{\theta}{2}\cos\psi_2\cos\varphi_2\right)\left[1+m^2-m\sqrt{2\left(1+\cos\frac{\theta}{2}\cos\psi_2\cos\varphi_2\right)}\right]}{\left(1+\cos\frac{\theta}{2}\cos\psi_1\cos\varphi_1\right)\left[1+m^2-m\sqrt{2\left(1+\cos\frac{\theta}{2}\cos\psi_1\cos\varphi_1\right)}\right]}}$$

$$\Delta\Phi_2 = f(m, d_p, \theta, \varphi_2\, \psi_2)$$

Calculation of refractive index

$$m = -\frac{1}{2}\frac{\sqrt{f_2} - A\sqrt{f_1}}{A-1} + \sqrt{\left(\frac{1}{2}\frac{\sqrt{f_2} - A\sqrt{f_1}}{A-1}\right)^2 - 1}$$

$$f_1 = 2\left(1+\cos\frac{\theta}{2}\cos\psi_1\cos\varphi_1\right)$$

$$f_2 = 2\left(1+\cos\frac{\theta}{2}\cos\psi_2\cos\varphi_2\right)$$

$$A = \left(\frac{\Delta\Phi_1\sin\psi_2}{\Delta\Phi_2\sin\psi_1}\right)^2\frac{f_1}{f_2}$$

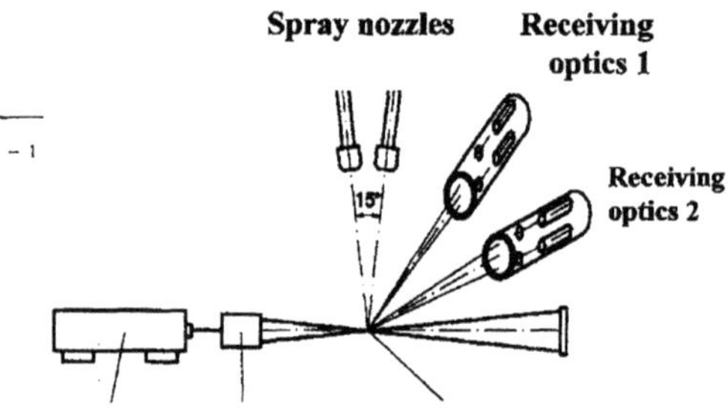

Fig. 14. Optical system for extended phase-Doppler measurements

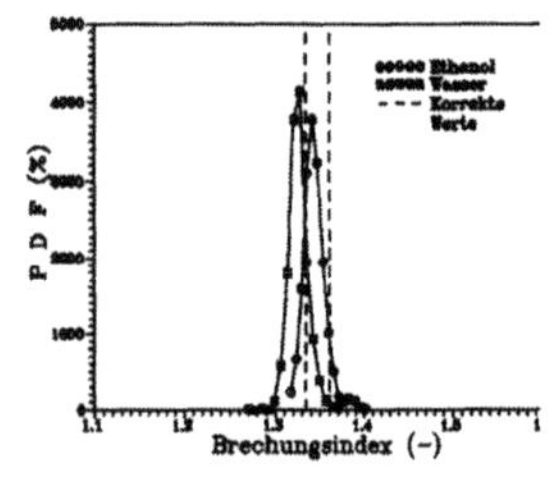

Refractive index distributions measured in monodisperse drop streams.
Water and ethanol can be distinguished.

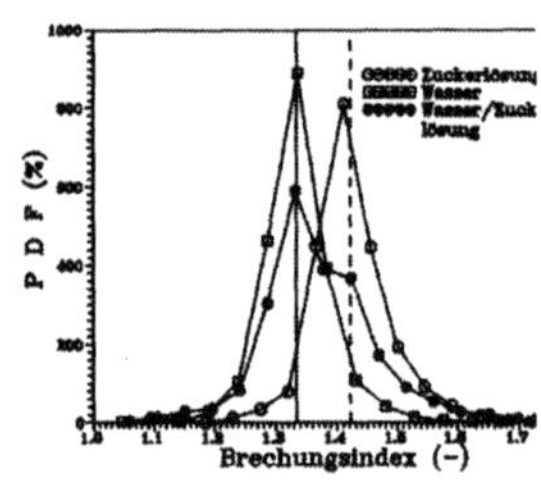

Water $m_w = 1.334$
Ethanol $m_e = 1.361$
Sugar water 50/50 $m_z = 1.421$
Refractive indices in mixed polydisperse sprays.

Fig. 15. Refractive index measurements using extended phase-Doppler anemometer

Without any doubt, phase-Doppler anemometers work best with spherical particles. The application of PDA systems could lead to interesting results on laminar and turbulent particulate two-phase flows. Fluid mechanics research should concentrate on using the phase-Doppler anemometers that are available rather than placing emphasis on the extension of existing phase-Doppler anemometers to yield information on particle material, measurements of non-spherical particles, *etc.*

There have been numerous contributions to the development of phase-Doppler anemometry and a good summary of the work was given by Hirleman [12], with useful contributions by Naqwi *et al.* [13] and Gouesbet *et al.* [14].

2.5. Particle image velocity and other field methods

When flows are described in terms of the local time varying velocity field, *i. e.* as $U_j(x_i, t)$, there are basically two ways to analyse experimental and/or numerical data.

- *Local analysis:* This approach fixes the measuring location x_i = constant and analyses the velocity data as a time series. This is the usual way in which hot wire and laser Doppler anemometers are operated.
- *Spatial analysis:* This approach fixes the measuring time t = constant and analyses the velocity data as a distribution of velocities in space. This is the usual way in which particle image velocimetry works.

Without any doubts, particle image velocimetry has become an important measuring technique in fluid mechanics. However, its success in this field will heavily depend on useful results that come out by applying the technique to answer basic questions in fluid mechanics or to provide solutions to engineering problems. The development work has been going on for years and it is important that the application of developed instruments has to go beyond demonstrations of the measuring technique. Multi-point information on flow fields can be easily provided by particle image velocimetry and appropriate signal process techniques should be provided to yield this kind of information.

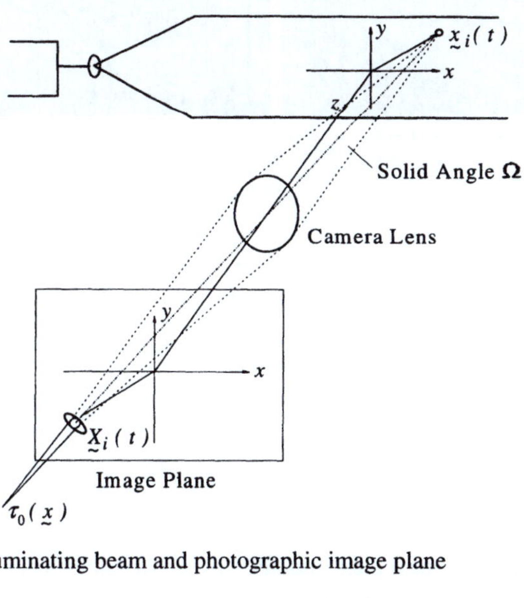

Fig. 16. Pulsed illuminating beam and photographic image plane

Particle image velocity (PIV) uses an optical set-up of the kind shown in Fig. 16, as described by Adrian [16]. Field information results are sketched in Fig. 17. In the illuminated plane of thickness Δs, all those pairs of components of the velocity fields are recorded that were represented by a scattering particle and fulfilled the following requirements:

- the particle was large enough to scatter sufficient light to yield a good record of the particle,
- the particle was small enough, considering its size, density and shape, to follow the flow,
- the velocity component of the particle perpendicular to the light sheet was small so that the following relationship holds:

$$U_{\perp} \leq \frac{\Delta s}{\Delta t}$$

where $U_{\perp}$ = velocity component perpendicular to light sheet, Δs = thickness of illuminated sheet and Δt = time between particle images. Hence conditional information on the velocity field is obtained.

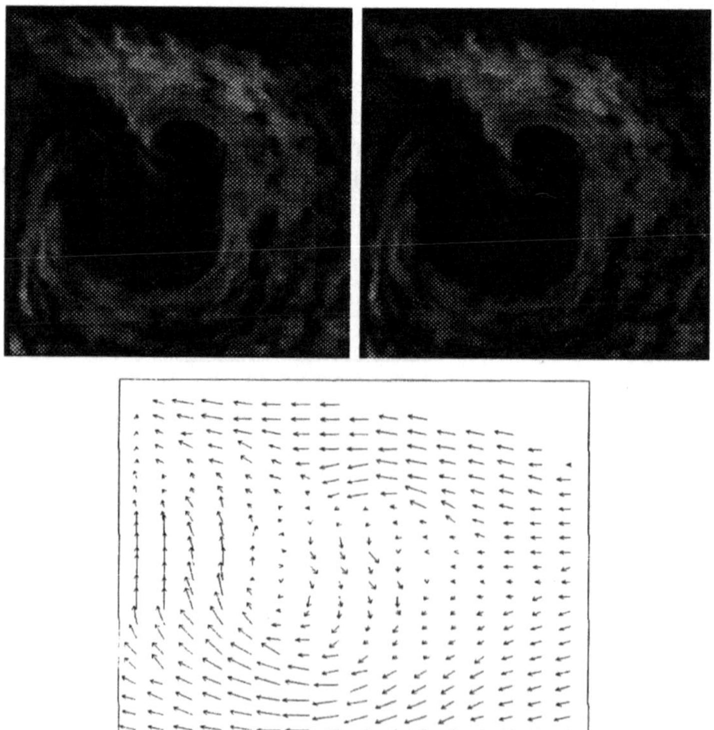

Fig. 17. Flow against a vertical plate visualised by smoke and velocity distribution measured by PIV, results by Uemura *et al.* [15]

Another method that should be mentioned as a field method is Doppler global velocimetry (DGV), *e. g.* see Meyers and Lee [17]. With this technique the absorption properties of iodine are used to obtain direct Doppler shift information for the scattered light optical systems of the kind shown in Fig. 18.

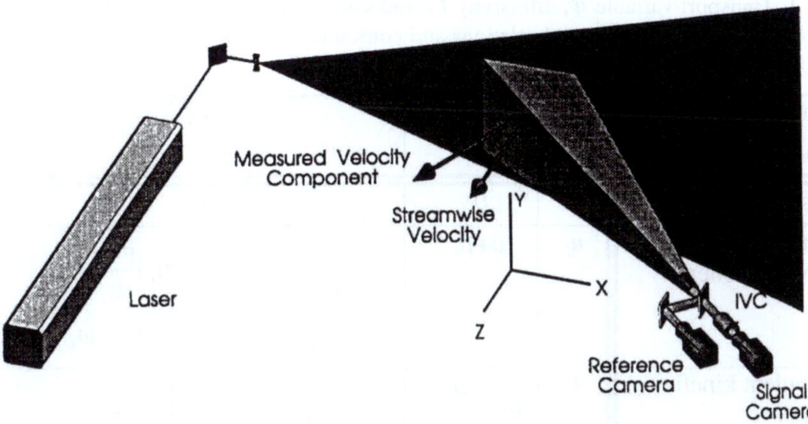

Fig. 18. Sketch of optical arrangement of Doppler global velocimeter, Meyers and Lee [17]

3. Computational Methods

3.1. Computational methodology

The governing equations of fluid mechanics given in the introduction are a set of partial differential equations. For solving turbulent flows in industrial applications these equations are often combined with a statistical turbulence model, *e. g.* the standard k-ε model. The complete set of equations can be brought into a common transport equation:

$$\rho\frac{\partial\Phi}{\partial t}+\frac{\partial}{\partial x_i}\left(\rho u_i\Phi-\Gamma_\Phi\frac{\partial\Phi}{\partial x_i}\right)=S_\Phi, \tag{4}$$

where Φ denotes the transport variable, xi the i-th Cartesian coordinate, ui the i-th Cartesian velocity component in the rotating reference frame, ρ the density of the fluid, ΓΦ the diffusivity of Φ, and SΦ the source term of Φ. Table 1 gives the values of Φ, ΓΦ and SΦ for all transport equations applied in the model. This equation is applicable to both a stationary and a rotating frame of reference. The velocity vector in a stationary reference frame U and the velocity vector in a rotating reference frame $\underline{u}$ are related by $\underline{U}=\underline{u}+\underline{\omega}\times\underline{x}$ where $\underline{\omega}=(\omega_1, \omega_2, \omega_3)$ represents the angular velocity vector and $\underline{x}=(\omega_1, \omega_2, \omega_3)$ the position relative to the rotation axis.

The numerical solution method is based on a fully conservative finite volume discretization on non-orthogonal boundary fitted grids with a non-staggered arrangement of the variables. Second order discretization is used for all terms (central differences, linear interpolation) together with flux blending and a deferred correction approach for the convective fluxes.

Table 1. Transport variable Φ, diffusivity Γ_Φ and source term S_Φ for all transport equations applied in the model. Additional relations and constants not mentioned in the text: $\mu_t = c_\mu \rho(k^2/\varepsilon)$, $c_\mu = 0.09$, $\sigma_k = 1.0$, $\sigma_\varepsilon = 1.3$, $c_{\varepsilon 1} = 1.44$, $c_{\varepsilon 2} = 1.92$

Conservation variable	Φ	Γ_Φ	S_Φ
Mass	1	0	0
Momentum (j-th component)	u_j	$\mu+\mu_t$	$-\frac{\partial p}{\partial x_j}+\frac{\partial}{\partial x_j}\left((\mu+\mu_t)\frac{\partial u_j}{\partial x_i}\right)$ $+2\varepsilon_{mnj}u_m\omega_n-\varepsilon_{mnj}\omega_m(\varepsilon_{pqn}\omega_p x_q)$
Turbulent kinetic energy	k	$\mu+\frac{\mu_t}{\sigma_k}$	$\mu_t\left(\frac{\partial u_i}{\partial x_k}+\frac{\partial u_k}{\partial x_i}\right)\frac{\partial u_i}{\partial x_k}-\rho\varepsilon$
Dissipation rate of k	ε	$\mu+\frac{\mu_t}{\sigma_\varepsilon}$	$\frac{\varepsilon}{k}c_{\varepsilon 1}\mu_t\left(\frac{\partial u_i}{\partial x_k}+\frac{\partial u_k}{\partial x_i}\right)\frac{\partial u_i}{\partial x_k}-c_{\varepsilon 2}\rho\frac{\varepsilon^2}{k}$

Based on the continuity equation, a pressure correction equation is derived according to the SIMPLE algorithm. The linearized equations for the velocity components, the pressure correction and other scalar variables are assembled and solved sequentially, where the ILU approach of Stone is employed as a linear system solver. Outer iterations are performed to take into account the non-linearities, the coupling of the variables, and the effects of grid non-orthogonality, which are treated explicitly in all equations. In the case of unsteady computations, the discretization of the time derivatives is done by a three time level, second order fully implicit scheme.

Block structured grids, where the blocks are globally unstructured but the grids are locally structured, are used as they can be viewed as a compromise between the high geometric flexibility of fully unstructured grids and the high numerical efficiency achieved on globally structured grids. The coupling of the blocks along the block interfaces, *i. e.* the transfer of information among neighboring blocks, is realized by adding auxiliary control volumes along the block interfaces containing the corresponding boundary values of the neighboring block. To ensure the coupling of the subdomains, the boundary data in the auxiliary control volumes of neighboring blocks is updated after each inner iteration of the iterative linear system solver. More details can be found in [18, 19].

The numerical simulation of practically relevant flows often involves the handling of complex geometries and complex physical and chemical phenomena, requiring the use of very fine grids and small time steps in order to achieve the necessary numerical accuracy. In recent years, intensive research has been undertaken to improve the performance of flow computations in order to extend their applicability to a cost effective solution of practically relevant flow problems. These improvements concerned both acceleration by the use of more efficient solution algorithms such as multi-grid methods and acceleration by the use of more efficient computer hardware such as high performance parallel and vector parallel computers. Care has to be taken in order to benefit from both the efficient numerical techniques and parallel computing. In the present method, high numerical efficiency is obtained by a global nonlinear multi-grid method with a pressure correction smoother also ensuring only slight deterioration of the convergence rate with increasing processor numbers. The results presented here illustrate that the high performance of the underlying sequential multi-grid algorithm can be largely retained in the parallel implementation.

The three dimensional flow around a circular cylinder in a square channel at Re = 20 was considered; this represents one of the test cases for benchmark computations gathered at LSTM-Erlangen. The problem was computed with the multi-grid method with nested iteration for different grid sizes, where the coarsest grid with 768 CVs and up to five grid levels were used. With increasing number of fine grid CVs, the number of processors was also increased (not linearly). In Table 2, the computing times and numbers of fine grid iterations are given for the computations, and for comparison the corresponding values for the single grid computation are also indicated.

Table 2. Computing times and numbers of fine grid iterations (in parentheses) for single grid (SG) and full multi-grid (MG + NI) methods with corresponding acceleration factors (with respect to computing time) for different numbers of processors and control volumes for the flow around a circular cylinder

Method	6144 CV P = 1	49152 CV P = 8	393216 CV P = 32	3145728 CV P = 128
SG	1679 (58)	6041 (207)	40604 (675)	–
MG + NI	624 (25)	783 (27)	1738 (26)	2630 (25)
Acceleration factor	2.7	7.7	23.4	> 70

Several conclusions can be drawn from the results. The multi-grid method is significantly superior to the corresponding single grid computation. Whereas the single grid method shows a typical linear increase in iteration numbers with grid refinement, the iteration numbers for the multi-grid method change only slightly. When the number of processors is increased with an increasing number of control volumes, the multi-grid method gives a very good scale-up in the computing time. Because of the increase in iteration numbers caused by the decoupling due to the grid partitioning this is not the case for the single grid computations. If one compares, for instance, the results for P = 8 and P = 32, which correspond to grid sizes per processor differing by a factor of two, one can see that a good parallel efficiency is obtained. We have scale-ups of 97% for SG and 90% for MG + NI. The lower value when using MG is due to the increase in work on coarser grids, where the portion of communication relative to the arithmetic operations is larger.

3.2. Examples for the solution of engineering problems

The CFD techniques developed at LSTM Erlangen have been used for the solution of a large variety of practical applications, *e. g.*:

- side wind effects on high speed trains, [20,21]
- turbulent flow in stirred vessels, [22-24]
- fluid-structure interaction for civil engineering applications, [25,26]
- flow and heat transfer in the melt of a Czochralski crucible, [27-30]
- turbulent flow around bluff bodies, [31-36]
- heat transfer of hot wires in near wall configuration, [37,38]

These examples include internal and external laminar as well as turbulent flows. For the turbulent case all three relevant techniques are applied, namely Reynolds averaged Navier-Stokes equations (RANS), direct numerical simulation (DNS), and last but not least large eddy simulation (LES). In the present paper one example for a laminar flow and one example for turbulent case including heat transfer are considered in detail.

Flow and heat transfer of hot wire in near wall configuration

Hot wire near wall correction plays an important role in flow measurement. Since several decades, it is not clear in the literature whether no correction, a small positive correction or a negative correction is needed in the case of walls consisting of poorly conducting materials. The physical mechanism concerned also remains not fully understood.

At LSTM Erlangen, a detailed numerical investigation of the two dimensional laminar flow and heat transfer around a single circular cylinder located close to walls of different materials was performed to study the hot wire near wall correction. The conjugated heat conduction in the solid wall was taken into account to bridge the discrepancy between the previous theoretical models and the practical situation. Simulations were carried out for several realistic wall materials (aluminium, with thermal conductivity ratio to air $k^*=k_w/k_{air}$=9186; glass, k^*=29.6; plexiglas, k^*=7.2) and some theoretical materials with k^*=1, 0.1, and 0.01, respectively. The computed results show good agreement with experimental data in the literature. Accurate correction curves for hot wire anemometers were obtained with respect to different wall materials. Based on the present study, the physical mechanism of the wall effect has been revealed and the existing confusions can be clarified.

The predicted velocity corrections expressed by the correction factor $C_U=U_0/U_{appa}$ are presented in Fig. 19 for walls consisting of mirror glass and plexiglas (poorly conducting materials). Here U_0 is the actual and U_{appa} the apparent velocity value, respectively. As demonstrated, the velocity correction curves can be divided into three regions: (a), no correction (C_U = 1); (b), negative (C_U > 1) and (c), positive (C_U < 1) corrections with respect to different wire to wall distances. This phenomenon could also be clearly recognized in some experimental data (see Fig. 19) in the literature but was ignored up to now.

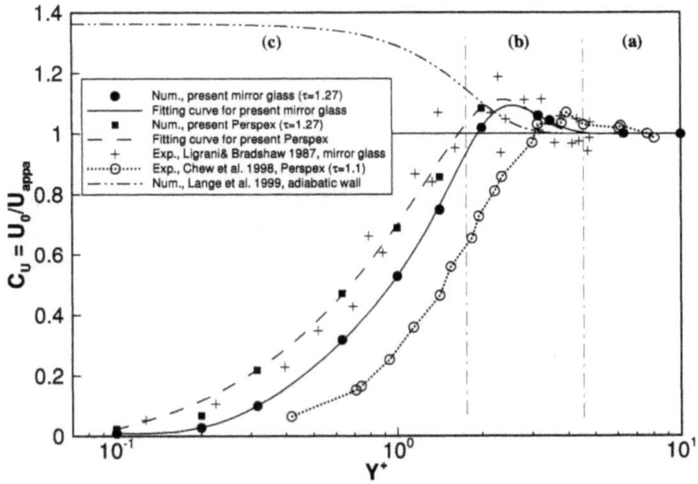

Fig. 19. Figure 19. Comparison of numerical and experimental values of the velocity correction factor C_U in cases of walls with low conductivities

(a) $C_U = 1.0$ (b) $C_U > 1.0$ (c) $C_U =< 1.0$

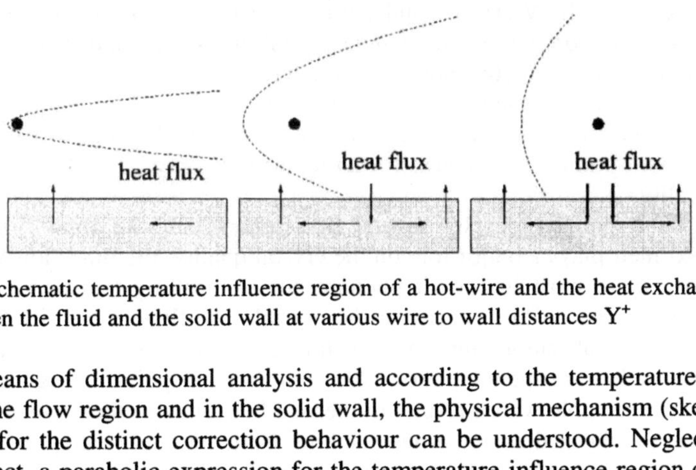

Fig. 20. Schematic temperature influence region of a hot-wire and the heat exchange process between the fluid and the solid wall at various wire to wall distances Y^+

By means of dimensional analysis and according to the temperature isolines both in the flow region and in the solid wall, the physical mechanism (sketched in Fig. 20) for the distinct correction behaviour can be understood. Neglecting the shear effect, a parabolic expression for the temperature influence region of a wire in the near wall measurement is obtained:

$$\frac{X}{D} = \Pr\left(\frac{Y}{D}\right)(Y^+)^2 - \frac{1}{4\Pr \mathrm{Re}_D} \tag{5}$$

When a hot wire comes close to a wall below a certain value of Y^+, heat transfer from the fluid into the wall material occurs at the interface between the temperature influencing region of a hot wire and the solid wall. The heat flux from the fluid is then conducted both in the upstream and downstream directions in the solid wall and fed back into the flow when the fluid temperature is lower than that in the solid wall. In case (a), this „temperature influencing region-wall“ interaction occurs far away downstream from the wire location and the heat „feed-up“ through the wall conduction has no evident effect on the hot wire heat loss due to large values of Y^+. As a result, no correction is required ($C_U \approx 1.0$). In case (b), the interaction between the temperature influencing region and the wall occurs at a smaller distance downstream of the wire. The feed-back heat flux from the up-stream conduction in the wall warms up the oncoming fluid which will flow over the wire. As a result, heat loss from the wire is reduced, *i. e.* $C_U > 1$. In case (c), the parabolic influencing region collides with the wall in the proximity of the wire location. The heat loss from the wire is significantly enhanced due to the much higher conductivity of the wall material compared with the fluid. The enhancing effect becomes dominant over the restraining effect arising from the heat „feed-up“. Thus, positive corrections are observed ($C_U < 1$). In conclusion, based on this detailed numerical study, a long existing confusion on hot wire near wall corrections was finally resolved (see also [37] and [38]).

3.3. Buoyancy driven flow in a Czochralski crucible

Czochralski (CZ) crystal growth is one of the most popular methods of producing large single crystals for electronic and photonic devices. Most industries producing silicon crystals, III-V crystals and crystals of oxide materials use the CZ process. Due to the importance for the single crystal production and its technological impact on the semi-conductor industry, there have been extensive research studies in the past couple of decades for thorough fundamental understanding of mechanisms governing the Czochralski process. The objective has been to improve as well as further optimize the process for larger crystals (*e. g.*, 0.3 - 0.4 m diameter) of high quality.

It has been pointed out by several researchers that the fluid flow in the Czochralski melt plays a major role on the crystal quality; the fluid flow in the CZ melt is the single most important phenomenon controlling transport of species such as oxygen and, in turn, purity of the crystal. The flow in the silicon melt is governed by several interacting forces such as buoyancy, Coriolis and centrifugal forces. The melt is maintained above the solidification point through controlled heating of the crucible which results in thermal buoyancy in the melt. In order to homogenize the thermal asymmetries resulting from external heating, the crucible is rotated. The uniformity at the crystal/melt interface is also improved by rotating the crystal while pulling to generate von Kàrmàn-Cochran centrifugal flow with its constant boundary layer thickness. The most important and interesting aspect of the flow in a CZ melt is the breakdown of the axisymmetric nature of the field leading to a host of interesting alternative flow patterns including non-periodic time dependent flows.

The fluid flow and heat transfer in an industrial Czochralski melt was analyzed by solving the time dependent three dimensional Navier-Stokes equations on curvilinear boundary fitted grids in a rotating frame of reference. The numerical methodology implemented in FASTEST-3D was extensively validated based on two different test cases [27, 29]. In order to represent the ellipsoidal crucible, a grid with about 1.73 million control volumes was generated in six blocks. Using the natural advantage of block structuring, computations were carried out on a parallel vector machine (NEC SX-4) with an optimal load balancing efficiency of 100% using four processors with a total performance of about 3.0 GFlops. The simulation shown here was performed for a laboratory scale crucible with a diameter D_c=0.339 m and height H_c=0.1 m, while the crystal has a diameter D_s=0.1 m. The crucible rotates with an angular velocity of Ω_c = 5 rpm, whereas the crystal rotates in the opposite direction with Ω_c = -20 rpm. A measured temperature distribution was prescribed at the crucible walls, while at the crystal/melt interface the solidification temperature was assumed. At the free surface of the melt, surface tension was not taken into account and heat loss took place by radiation. The most important non-dimensional parameters for this flow are the Rayleigh number Ra = $5.3 \cdot 10^6$, the Reynolds number Re = $7.9 \cdot 10^3$ and the Prandtl number Pr = 0.0106, which is very low for liquid silicon.

Simulations of the flow field were performed with and without a buoyancy extended k-ε turbulence model. It was found that this turbulence model suppresses the fluid mechanical instabilities caused by the interaction of buoyancy, centrifugal and Coriolis forces leading to an axisymmetric flow and thermal field, while the simulations without any turbulence model (semi-DNS: The synonym „semi-DNS" implies a fine computational grid, which, however, is insufficient for the resolution of all scales down to the Kolmogorov length) were found to predict the three dimensional and time dependent features of the melt flow well. As an example, Figs. 21 and 22 show the instantaneous flow structure and thermal field in a horizontal and a vertical cut through the centre of the crucible.

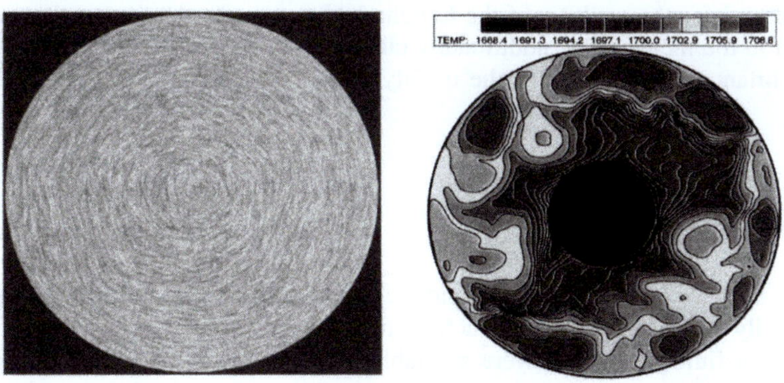

Fig. 21. Three dimensional, time dependent turbulent flow visualized by LIC (left) and temperature distribution (right) in an industrial Czochralski crucible, horizontal plane directly below the free surface

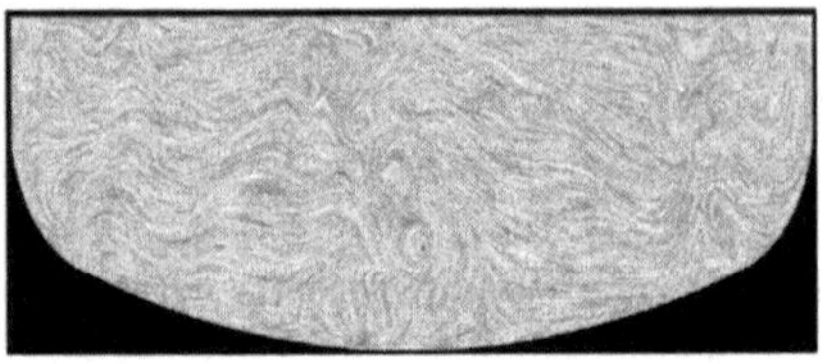

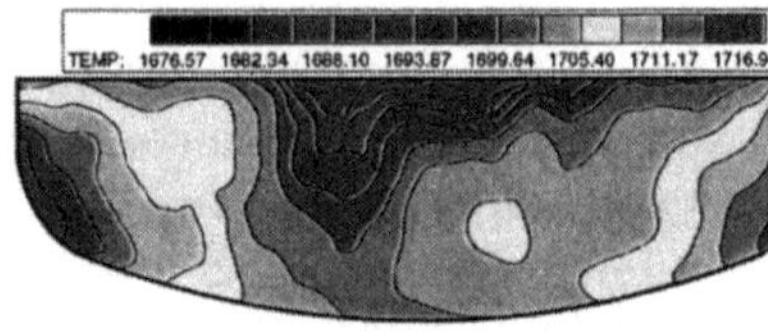

Fig. 22. Three dimensional, time dependent turbulent flow visualized by LIC (left) and temperature distribution (right) in an industrial Czochralski crucible, vertical plane through the centre of the crucible

As expected from experimental observations, the predicted flow and temperature field is highly non-axisymmetric consisting of multiple vortices and buoyant plumes which rise from the bottom of the crucible up to the free surface visible in Fig. 21. Looking at animations of the simulations leads to the conclusion that these plumes rotate with an angular velocity similar to the crucible rotation rate Ω_c. Strong temperature fluctuations are also observed in the vicinity of the crystal which may influence the growth rate and finally the quality of the single crystal. Unfortunately, it is very difficult to measure the temperature distribution in the entire liquid melt. Therefore, comparisons between predictions and measurements are restricted to selected locations in the crucible. For that reason, the time history of the temperature at several positions in the melt was actually recorded in order to compare with measurements carried out by the Institute of Materials Science, University of Erlangen-Nürnberg. Based on mean values, standard deviations and spectra of the measured and predicted signals, a detailed validation of the computational approach can be performed. Preliminary results point out a reasonable agreement between measurements and predictions. Since the computations provide three dimensional, time resolved data of the entire melt flow, they are a very powerful tool for the analysis of the Czochralski system. In the future, the oxygen transport in the melt will be additionally taken into account in order to investigate this important phenomenon for the quality of the growing crystal (see also [27-30]).

4. Summary

In this review we have shown that fluid mechanics research has a long history and was initially embedded in fields such as mathematics and physics. Hence, basic equations of fluid mechanics were available at the end of the 17th and the beginning of the 18th century, but the equations could not be solved for engineering types of flows. The necessary methods were developed in the second half of the 20th century to yield experimental and numerical techniques which can now be used to investigate laminar and turbulent flows. There is basically no flow these days below a Reynolds number of 50 000 which cannot be studied in detail by either experimental or numerical techniques. Higher Reynolds number flows, however, cannot be investigated in detail because the available experimental techniques cannot resolve the velocity gradients and numerical methods require computer powers that are not yet available to solve directly the Navier-Stokes equations. Turbulence at high Reynolds number is the only remaining basic problem to be solved in fluid mechanics. Low Reynolds number flows can be investigated with the methods that have been developed in the second half of the 20th century.

Without any doubt, developments of experimental and numerical methods will continue in the years ahead. There will be a change in fluid mechanics research, however, as can already be observed today, towards the application of methods to study fluid flows. The golden age of fluid mechanics has started. Fluid flows can be studied using well developed, currently available experimental and numerical techniques.

References

[1] Van Dyke, M., 1982, "An Album of Fluid Motion", Parabolic Press, Stanford, California

[2] Hama, R.F., 1962, "Streaklines in a perturbed shear flow", Physics of Fluids 5, pp. 644-650

[3] Eckelmann, H., 1967, "Einführung in die Strömungsmesstechnik", Teubner Studienbücher Mechanik, B.G. Teubner, Stuttgart

[4] Bruun, H.H., 1995, "Hot-Wire Anemometry: Principles and Signal Analysis", Oxford University Press Inc. New York

[5] Lumley, J.L., 1970, "Stochastic Tools in Turbulence", Academic Press, New York

[6] Durst, F., Melling, A., Whitelaw, J.H., 1981, "Principles and Practice of Laser-Doppler Anemometry", Second Edition, Academic Press, New York

[7] van de Hulst, H.C., 1957, "Light Scattering by Small Particles", Wiley Press, New York

[8] Kerker, M., 1969, "The Scattering of Light and oher Electromagnetic Radiation", Academic Press, New York

[9] Hjelmfelt, A.T., Mockros, L.F., 1966, "Motion of discrete particles in a turbulent fluid", Appl. Sci. Res. 16, p. 149

[10] Durst, F., 1973, "Scattering phenomena and their application in optical anemometry", Zeitschrift für Angewandte Mathematik und Physik 24, p. 619

[11] Durst, F., Zaré, M., 1975, "Laser-Doppler measurements in two-phase flows", Proceedings of the LDA-Symposium Copenhagen, pp. 403-429

[12] Hirleman, E.D., 1996, "History of development of the phase-doppler particle-sizing velocimeter", Part. Part. Syst. Charact. 13, pp. 59-67

[13] Naqwi, A., Durst, F., Liu, X., 1991, "Two optical methods for simultaneous measurement of particle size, velocity and refractive index", Applied Optics, 30, pp. 4949-4959

[14] Gouesbet, G., 1994, "Generalized Lorenz-Mie theory and applications", Part. Part. Syst. Charact. 11, pp. 22-34

[15] Uemura, T., Yoshimoto, M., Tatumil, M., Kaga, A., 1997, "Simultaneous multiple pixel processing algorithms for PTV and PIV", Developments in Laser Techniques and Fluid Mechanics, Selected Papers from the 8th Int. Symp. on Applications of Laser Anemometry to Fluid Mechanics, Ladoan, Portugal

[16] Adrian, R.J., 1988, "Statistical properties of particle image velocimetry measurements in turbulent flow", Laser Anemometry in Fluid Mechanics – III, Selected Papers from the Third Int. Symp. on Applications of Laser Anemometry to Fluid Mechanics, Ladoan, Portugal

[17] Meyers, J.F., Lee, J.W., 1993, "Three component Doppler global velocimeter measurements of the flow above a delta wing", Procedings of the 6th Int. Symp. Lisbon, Portugal, 20-23 July 1992, Springer-Verlag, Berlin, pp. 345-363

[18] Durst, F., Schäfer, M., Wechsler, K., 1996, „Efficient simulation of incompressible viscous flows on parallel computers", In: Flow Simulation with High-Performance Computers II, Notes on Numerical Fluid Mechanics, vol. 52, Vieweg, Braunschweig, pp. 87-101

[19] Durst, F., Schäfer, M., 1996, „A parallel block-structured multigrid method for the prediction of incompressible flows", Int. J. Num. Methods Fluids, 22, pp. 549-565

[20] Khier, W., Breuer, M., Durst, F., 2000, „Flow structure around trains under side wind conditions: a numerical study", Int. J. Computers and Fluids, 29, pp. 179-195

[21] Khier, W., Breuer, M., Durst, F., 2001, „Numerical computation of 3-D turbulent flow around high speed trains under side wind conditions", TRANSAERO - a European initiative on transient aerodynamics for railway system optimisation, B. Schulte-Werning, R. Gregoire, A. Malfatti and G. Matschke (eds.), Proc. of the Brite/Euram Project Symp. `Transient Aerodynamics for Railway System Optimisation', May 4--5, 1999, Paris, France, Notes on Numerical Fluid Mechanics, Springer Verlag, vol. 79

[22] Wechsler, K., Breuer, M., Durst, F., 1999, „Steady and unsteady computations of turbulent flows induced by a 4/45° pitched blade impeller", J. Fluids Engineering, 121 (2), pp. 318-329

[23] Bartels, C., Breuer, M., Wechsler, K., Durst, F., 2001, „CFD-applications on parallel-vector computers: Computations of stirred vessel flows", Int. J. Computers and Fluids, 31, pp. 69-97

[24] Bartels, C., Breuer, M., Durst, F., 2000, „Comparison between direct numerical simulation and k-ε model prediction of the flow in a vessel stirred by a Rushton turbine", 10th European Conference on Mixing, Delft University of Technology, The Netherlands, July 2-5, 2000

[25] Glück, M., Breuer, M., Durst, F., Halfmann, A., Rank, E., 2001, „Computation of fluid-structure interaction on lightweight structures", Fourth Int. Colloq. on Bluff Body Aerodynamics & Applications, Bochum, Germany, Sept. 11-14, 2000, Int. J. of Wind Engineering and Industrial Aerodynamics, 89, pp. 1351-1368

[26] Glück, M., Breuer, M., Durst, F., Halfmann, A., Rank, E., 2001, „Computation of wind-induced vibrations of flexible shells and membranous structures", subm. to J. Fluids and Structures

[27] Basu, B., Enger, S., Breuer, M., Durst, F., 2000, „Three-dimensional simulation of flow and thermal field in a Czochralski melt using a block-structured finite-volume method", J. Crystal Growth, vol. 219/1-2, pp. 123-143

[28] Basu, B., Enger, S., Breuer, M., Durst, F., 2001, „Effect of crystal rotation on the three-dimensional mixed convection in the oxide melt for Czochralski growth", 3rd Int. Workshop on Modeling in Crystal Growth, Stony Brook, New York, USA, Oct. 18-20, (2000), J. Crystal Growth, 230/1-2, pp. 148-154

[29] Enger, S., Basu, B., Breuer, M., Durst, F., 2000, „Numerical study of three-dimensional mixed convection due to buoyancy and centrifugal force in an oxide melt for Czochralski growth", J. Crystal Growth, 219/1-2, pp. 144-164

[30] Enger, S., Gräbner, O., Müller, G., Breuer, M., Durst, F., 2001, „Comparison of measurements and numerical simulation of melt convecti convection in Czochralski crystal growth of silicon“, 3rd Int. Workshop on Modeling in Crystal Growth, Stony Brook, New York, USA, Oct. 18-20, 2000, J. Crystal Growth, 230/1-2, pp. 135-142

[31] Breuer, M., 1998, „Large eddy simulation of the sub-critical flow past a circular cylinder: numerical and modeling aspects“, Int. J. Num. Methods Fluids, 28, pp. 1281-1302

[32] Breuer, M., 2000, „A challenging test case for large eddy simulation: high Reynolds number circular cylinder flow“, Int. J. Heat Fluid Flow, 21, pp. 648-654

[33] Breuer, M., Bernsdorf, J., Zeiser, T., Durst, F., 2000, „Accurate computations of the laminar flow past a square cylinder based on two different methods: Lattice-Boltzmann and Finite-Volume“, Int. J. Heat Fluid Flow, 21, pp. 186-196

[34] Breuer, M., 2001, „Direct Numerical Simulation and Large-Eddy Simulation of Turbulent Flows with High-Performance Computers“ (in German), Habilitation Thesis, University of Erlangen--Nürnberg, Germany

[35] Breuer, M., Jovicic, N., 2001, „Separated flow around a flat plate at high incidence: an LES investigation“, Second Int. Symposium on Turbulence and Shear Flow Phenomena, Stockholm, Sweden, June 27-29, 2001, vol. III, pp. 393-398, J. of Turbulence, in press

[36] Breuer, M., Jovicic, N., 2001, „An LES investigation of the separated flow past an airfoil at high angle of attack“, Proc. of the 4th Workshop on Direct and Large Eddy Simulation, Enschede, The Netherlands, July 18-20, 2001, ERCOFTAC Series, vol. 8, Direct and Large-Eddy Simulation IV, eds. B.J. Geurts, R. Friedrich, O. Mètais, Kluwer Academic Publishers, Dordrecht

[37] Durst, F., Shi, J.M., Breuer, M., 2001, „Numerical prediction of hot-wire corrections near walls“, J. Fluids Engineering, in press

[38] Shi, J.M., Breuer, M., Durst, F., 2002, „Wall effect on heat transfer from a micro cylinder in near-wall shear flow“, Int. J. Heat and Mass Transfer, 45/6, pp. 1309-1320

Theodore von Kármán: A Global Life

Tibor FRANK
Eötvös Loránd University, Budapest, Hungary

One of the most successful Hungarians abroad was Theodore von Kármán (1881-1963). Though he started his American career in the late 1920s, von Kármán was one of those whose excellence was nationally recognized by his contributions to the war effort. Indeed, many of the outstanding émigré professionals from Hungary attained their greatest recognition in the U. S. during, and as a consequence of, World War II and the subsequent Cold War period.

Von Kármán came from a distinguished Jewish-Hungarian professional background [1]. His father Mór Kármán (1843-1915) was one of Hungary's most renowned educational experts, a pedagogical reformer who established Hungary's high school system following the German models. Kármán Sr., who studied philosophy and classical philology at the University of Vienna and received his Ph.D. in Budapest in 1866, was sent by Baron József Eötvös to Leipzig, Germany, in 1869 to become a professor of pedagogy [2] and to study the modern methods of training high school teachers. Upon returning from Germany, Eötvös's successor Tivadar Pauler helped him found the University Institute for Teacher Training, as well as a practicing high school, which he directed himself. All four of his sons went to this high school in Budapest.

Hungary's foremost expert on education, Mór Kármán was elevated to the Hungarian nobility in 1907 [3], and became a full professor at Budapest University in 1909. He belonged to the assimilated Jewish upper-middle class of Hungary, and married into a well-connected family through which he was distantly related to the titled Jewish aristocracy of Hungary [4]. Mór Kármán felt himself close to Hungarian culture, which he also served by studying pieces of Hungarian literature that were part of the national literary canon [5].

The son of this great Hungarian educational innovator, commented historian William O. McCagg, Jr., von Kármán "grew up in the leading circles of the Budapest late nineteenth-century literati, living in comfortable flats and villas, schooled as a child by private tutors, meeting everyone who counted even in Vienna, having entrée into aristocratic circles" [6] because his father held a high position. Well before his formal education started, this sophisticated atmosphere exerted a specific influence on his intellectual development. At the age of nine young Kármán was sent to the school of his father. One of the characteristics of this education was for the students to develop informal social contact with some of Hungary's best teachers as well as an emphasis on learning from experience. As von Kármán remembered towards the end of his career,

At no time did we memorize rules from a book. Instead, we sought to develop them ourselves. I think this is a good system of education, for in my opinion how one learns the elements of reasoning in primary school will determine his later capacity for intellectual pursuit. In my case, the Minta [the "model" high school of his father] gave me a thorough grounding in inductive reasoning that is, deriving general rules from specific examples--an approach that remained with me throughout my life [7].

As a student of the high school directed by his father, young von Kármán participated in the competition for the Eötvös Loránd Prize, and won it in 1897, a distinction that added to the special care devoted to his initial grounding in science [8].

He graduated from the Technical University of Budapest in 1902. He spent four years in Hungary upon graduation, partly teaching as an assistant professor at the Technical University, partly as a research engineer at Ganz and Co. His father, however, wanted him to continue his education abroad. Von Kármán remembers his father in 1906 as a man who

had grown tired and bitter. I remember his saying one day that he had devoted his life to Hungary, but was unappreciated, and even tormented by petty university politics. . . He didn't want me to make the same mistake he had made. If I were going to make something of myself as a scientist and an independent thinker, he insisted that I would have a greater chance outside Hungary [9].

For his graduate university studies Theodore von Kármán went to Göttingen in 1906 and remained strongly connected with Germany for a quarter of a century. Göttingen attracted him in many ways. He could work there with illustrious scientists such as Ludwig Prandtl, a widely recognized expert of fluid mechanics, Felix Klein who applied mathematics to engineering, and David Hilbert, one of the great figures of pure mathematics. It was during his stay in Göttingen that he made the decision to choose aeronautics as his principal field of study [10]. His dissertation was written on the buckling strength of straight columns and in it he developed the Kármán-Engesser double-modulus theory of the behavior of columns under load. He received his Ph.D. from Göttingen in 1908 and stayed there as a Privat-Dozent until 1912.

It was in Göttingen that he became seriously interested in fluid mechanics, his most important contribution being the quantitative analysis based on the theory of vortices. Ever since his Göttingen studies the pattern of alternating eddies which form behind a circular cylinder in fluid flow is referred to as the Kármán vortex street (trail) [11]. The von Kármán theory of vortex streets was published in three papers in 1911-12, while in Göttingen. This classic theory of the unsymmetrical vortex arrangement in the wake of a cylinder provided the sound mathematical foundation and formulae on which aircraft designers have depended ever since. [12].

In 1912 von Kármán accepted the invitation of the Technical University of Aachen where he became Professor of Aeronautics and Mechanics and the Director of the Institute of Aerodynamics until 1929. In fact he helped build up the new Institute where he continued his Göttingen research in fluid mechanics. Gradually, a friendly rivalry developed with the Aeronautical Institute at Göttingen [13]. His Aachen subjects included the statistical theory of turbulence, wing theory, the stability of laminar flow, pressure distribution on airship hulls, the theory of lift, the approximate solutions of problems in boundary layer flow and skin friction (the Kármán integral relation) [14]. He developed the Kármán-Trefftz method of computing potential flow about given wing sections. Von Kármán was instrumental in showing the interrelationship between pure and applied mathematics and to utilize these concepts in the solution of problems then found almost hopeless [15].

In his years in Germany

he already gave a sympathetic hearing to the first estimates that were being made as to the possibility of propelling a device away from the Earth. A rocket enthusiast at a technical meeting in Berlin had suggested that chemical combinations produced sufficient energy to propel a rocket away from the Earth; he was severely taken ton task by a well-known German physicist for talking nonsense. Von Kármán defended the young man, for during the discussion he had made a quick, rough estimate and concluded that it was theoretically possible [16].

"I am not a Fantast," he stated. *"It is just a simple fact that one pound of kerosene has more energy than is necessary to take ... one pound out of the gravitational field. It is only a question of technology and progress and time* [17]."

Von Kármán's immense knowledge of his several fields, his many languages, his great sense of humor attracted large numbers of eminent students to Aachen which during his years as Director of the Institute became a highly acclaimed German and international center of aeronautics, astronautics, aerodynamics, and applied sciences.

During his long years at the University of Aachen, von Kármán also served as a consultant of several large airplane companies in and out of Germany, such as the Junkers Airplane Works (1912-1928), the Luftschiffbau Zeppelin (1924-28), Handley-Page Ltd., England (1926-30), the Guggenheim Aeronautical Laboratories, California Institute of Technology (1926-30), and the Kawanishi Airplane Company, Japan (1927-29). He designed the Kobe Wind Tunnel in Japan in 1927 and helped build the 10-foot Wind Tunnel of the Guggenheim Lab at CalTech (1926-27) [18].

At the invitation of the Guggenheim Fund for the Promotion of Aeronautics and of the California Institute of Technology, von Kármán came to the United States first in 1926, and more regularly after 1928. He settled permanently in California in 1930, becoming Director of the Guggenheim Aeronautical Laboratory at Pasadena, CA, and of the Guggenheim Airship Institute, Akron, OH. At the age of eighty von Kármán remembered the United States in the 1920s as a country with "very little military aviation", and where, he had thought, he "would work in civil aeronautics. It is ironic that for the last 25 years I have devoted most of my time to problems connected with military aviation [19]."

Von Kármán left Germany before Hitler came to power. With his Jewish family background, he was not in a position to return there and maintained a sharply critical distance from the Nazi regime right from the beginning. When he was approached from Berlin in the Summer of 1933 "suggesting that I take up my activities over there in the fall," he commented from CalTech ironically to his longtime Göttingen mentor and friend, Professor Prandtl: *"I do not think I will do this: I find my situation here quite satisfactory. The German academic life has some advantages, for instance a definitely better beer than here, but I think you will agree with me that this is not sufficient reason for me to neglect the disadvantages* [20]." In 1937 he did travel *"through Germany,"* but could not meet Prandtl. He wrote to him in February 1938, *"A few days ago, I received a very impressive volume of German scientific papers including General Göring's photograph, and I have the intention to study the volume with due respect* [21]."

During his 19 years at Caltech, von Kármán published some fifty papers on the problems of high subsonic, transonic and supersonic flow, on buckling problems, and on applied mathematics for engineers. Von Kármán was instrumental in establishing what was to become the Jet Propulsion Laboratory sponsored jointly by the U. S. Army and the Air Force. Several of his papers dealt with the statistical theory of isotropic turbulence. In 1944, General H. H. ("Hap") Arnold asked him to organize and chair a U. S. Air Force Scientific Advisory Group to study "the use of science in warfare by the European nations and to interpret the significance of the new developments in rockets, guided missiles, and jet propulsion for the future of the Air Force [22]." Von Kármán chaired the group actively until 1954 and was the founder of a similar body for NATO after 1951. Headquartered in Paris, the NATO Advisory Group for Aeronautical Research and Development (AGARD) had substantial impact on NATO countries in Europe through various projects.

After World War II, von Kármán had an active role in reorganizing German science. As head of the U.S. Army Air Force Scientific Advisory Board, he led a task force on a mission to Germany as well as to several other European countries. "During the European tour the name and international reputation of Dr. von Kármán was the 'Open Sesame' to considerable valuable information far beyond that obtained in normal interrogation. This was especially true in Germany where the respected name of von Kármán drew forth an integrated, intelligent picture of the German technical effort [23]." German scientists quickly found out that he was a key to rebuilding science in their country. In June 1945 he was approached, "through channels," by Professor Dr. Werner Osenberg, earlier Head of the Planning Bureau of the Reich's Research Council *(Leiter des Planungsamtes des Reichsforschungsrates),* who sent his essay on "Vorschläge zur Organisation der deutschen Forschung im Frieden und für Friedenszwecke" to von Kármán and asked for consultations with him on "the problems of future direction of research" [24]. The new information was assessed "on-the-spot" and channeled back to General Arnold in the U. S. to be used in the Air Force planning [25].

Helpful as he proved to be in the practical field, he was very reluctant to rejoin post-War German academic life. The Göttingen-based Akademie der Wissenschaften tried to win him back among its members in 1947-48 [26]. Refuted ("*abgelehnt... mit einer gewissen Schärfe*") by von Kármán, President Smend of the Göttingen Academy asked fellow member and Nobel Laureate James Franck to use his influence and convince von Kármán that he should rejoin the Adademy: "*... wir meinen, dass in dem entsetzlichen Trümmerfeld, das die Vergangenheit sittlich und geistig hinterlassen hat, jede Möglichkeit des Wiederaufbaus eines noch so kleinen Stücks geistiger und sittlicher Gemeinschaft von uns so lange auf das sorgfältigste verfolgt werden sollte, als eine solche Verfolgung noch irgend eine Aussicht auf Erfolg bietet* [27]." Franck himself rejoined the Academy, together with such eminent scientists as Rudolf Ladenburg and Lisa Meitner, and added in a letter to von Kármán: "*... I [myself] rejoined only after great hesitation. In fact, I did so only because I felt that if one does not help the people who want to work for a future Germany free of nationalism and racism, etc. the chances for such a Germany to develop become practically zero* [28]."

From the mid-1950s, however, von Kármán accepted a series of high German decorations such as the Federal Grand Cross of Merit with Star of the Federal Republic of Germany in 1955, the Ludwig Prandtl Ring Award of the WGL in Göttingen in 1957, as well as the Karl Friedrich Gauss Medal in Braunschweig, Germany, in 1960, both "for scientific accomplishment in fluid mechanics." He was also showered with honorary doctorates at several universities, including German schools such as the Technische Hochschule in Aachen (1953), and the Technische Universität Berlin-Charlottenburg (1953) [29].

What was the most significant result of von Kármán's research? As von Kármán had come from the vigorous school initiated by Felix Klein in Germany at the beginning of the 20th century which sought to modernize engineering science by relating it to basic sciences and mathematics, von Kármán considered his main task that of bridging the gap between basic science and practical aeronautical engineering [30].

Discovering the configuration of vortices in the wake of cylindrical bodies immersed in a flow made him an expert not only of air currents, but also a word-class specialist in the diagnosis of aerodynamic factors affecting the stability of bridges.

Several of his scientific papers were the results of consulting contracts with many firms, to which von Kármán contributed numerous engineering reports. Some of his work on the aerodynamics of airships rose most probably from his work for Luftschiffbau Zeppelin, in 1924-28. His interest in rocketry also dates back to the 1920s in Germany, when he headed the Aachen Aeronautic Institute. He continued his rocket experiments in the 1930s, with his California Institute of Technology students in the area now known as the Rose Bowl. Von Kármán was considered to be the only scientist of stature with the foresight to believe in the development of rocketry when most leading scientific figures were skeptical of its potential [31]. General H. H. Arnold, then Chief of the Army Air Force, asked him to investigate the application of rockets to lift heavy bombers off short runways. This lead to the development of the JATO (Jet Assisted Take-Off) rockets, one of the first products of the Jet Propulsion Laboratory [32].

As a member of the special committee appointed by the U. S. Navy, he investigated the Akron and Macon Dirigible disasters in 1933-37 [33]. In 1942 he investigated the collapse of Tacoma Narrows Bridge in Washington State, which enabled him to put his long-familiar subject, the "Kármán Vortex Trail," to a practical test [34]. As consultant to the Ballistic Research Laboratory of the U. S. Army at Aberdeen Proving Ground in Maryland between 1938 and 1952, he promoted the early construction of supersonic wind tunnels in the U. S. Other wind tunnels he built earlier in Japan, China, and Italy, helped him study the flow of air across fighters and bombers and led to the development of giant jet airliners, such as the 707 and the DC8, which could cross the United States in five hours [35].

Well before World War II, the U.S. Army Air Corps commissioned the National Academy of Science in Washington, D.C., to investigate the feasibility of assisting the take-off of heavily loaded aircraft with some form of auxiliary power. A member of the special academy committee appointed in 1938, von Kármán started to investigate, at Caltech, the possibility of using rockets. He was 57 when he first started working in the field of jet propulsion and rocket flight, which he continued until his death at 82 [36].

Many of his other contributions will have to be discovered once the files of the companies he worked for are available and declassified. A serious biography of von Kármán, still missing at this point, will reveal the role he played in transforming the results of basic research into applied aerodynamics, of which he was an uncontested pioneer.

It was for his contributions "*to improve and speed the defenses of this Nation and of the Atlantic Community*" that the U. S. Senate expressed its gratitude to Dr von Kármán. When presenting Senate Resolution 133 honoring Dr von Kármán on his 80th birthday and expressing "*the admiration and gratitude for the great contributions which Theodore von Karman has made to our country, and indeed, the entire world,*" Senator Jackson recalled the significant share of Hungarian scientists who built up modern science in the U. S.

The vigor of science in the United States today is due in large part to the contributions of brilliant and dedicated men who came to our shores from Europe... It is an interesting bit of history that five of the greatest of these men should have been born, and spent their childhood, in the same district of one city, Budapest, Hungary. I am, of course, thinking of Dr. Leo Szilard, Dr. John von Neumann, Dr. Edward Teller, Dr. Eugene Wigner, and finally, Dr. Theodore von Karman [37].

Senate Resolution 133 made it clear that Congress acknowledged particularly the work done by von Kármán "*in the cause of strengthening the military defenses of our country and our free world partners* [38]." The Hungarian-born scientist was cited as "*one of the most influential and respected advisers at the highest level of our Defense Establishment* [39]." Senator Jackson emphasized the leading role von Kármán played "*in speeding and strengthening aeronautical and space research in the NATO community* [40]." The NATO countries indeed valued von Kármán's work which led to the mobilization of the Western scientific effort [41].

President Kennedy, as well, cited von Kármán's "*outstanding. . . counsel and assistance to the United States Air Force and to the NATO Advisory Group for Aeronautical Research and Development,*" when selecting von Kármán as the first recipient of the National Medal of Science in 1963 [42]. The White House also issued an announcement regarding his being awarded the National Medal of Science, stating that his "*distinguished counsel to the Armed Services" came second only to his "leadership in the science and engineering basic to aeronautics* [43]." Presenting the nation's highest scientific award, President Kennedy said, "*I know of no one else who more completely represents all of the areas with which this award is appropriately concerned--science, engineering and education* [44]."

Part of the appreciation von Kármán earned was through his willingness to comment on political questions. His word had weight in a time of superpower confrontations. However, he refrained from commenting on issues such as America's lag in space and whether NASA should launch an effort to beat the Russians in the "Space Race," though many newsmen were invited by confidential invitation by the Institute of the Aerospace Sciences to hear him talk on this particular subject [45]. Von Kármán was willing, however, to address issues such as the U.S. future in manned flights and missiles, the feasibility of an ideal retaliatory system and new approaches to missile defense [46].

Nonetheless, von Kármán did not think of himself as a political figure, nor did he consider it appropriate for fellow scientists to play a political role. At a press conference on the eve of his 80th birthday von Kármán made it clear that to him scientists were "*not necessarily learned in statesmanship and politics*," and therefore should refrain from pontificating on television and elsewhere on the state of the world. He jabbed gently at fellow scientists Edward Teller and Leo Szilárd, commenting that he had watched them on TV programs talking about the future of the nation: "*I don't think that a man such as Edward Teller can say what is the good of the nation--I don't think this is a scientific question,*" von Kármán commented [47].

Von Kármán was asked at the press conference how it is that so many of America's top scientists are former Hungarians. Smiling, he said, they are not originally from Hungary, but from Mars. "*We decided to infiltrate the U. S., and were sent to Hungary--where some queer people live anyway--for conditioning for human life. There are many of us here from Hungary who are not quite earth people, like Zsa Zsa Gabor* [48]." More germanely, von Kármán summed up the secret of the Hungarian success, which he attributed to the "*Hungarian fundamental educational method and the U. S. liberal climate,*" which did the rest, he added, according to the *Washington Post & Times Herald* [49].

At the subsequent banquet honoring von Kármán's 80th birthday attended by more than 700 leading figures in the world of aerospace science and technology, Air Force Undersecretary Joseph V. Charyk used the example of the octogenarian, equipped with all the experiences of two World Wars, to note that technology has become inextricably linked with today's politics, economics, military affairs and international relations." Undersecretary Charyk warned scientists that they should no longer escape to ivory towers and evade responsibility to humanity. "*Fear, timidity, compromise and shirking of responsibility are the antithesis of the things that built this great country,*" Charyk said. The Air Force Undersecretary reminded the distinguished gathering of 1945, when von Kármán first presented General "Hap" Arnold his report entitled "Toward New Horizons" which laid the foundations of research and development programming for several decades. Air Force research development funds gradually increased to the level urged by von Kármán in that report 15 years ago, Charyk pointed out [50].

Von Kármán actually (and perhaps also symbolically) died in Aachen, Germany of a heart attack at the age of eighty-one, in 1963, during a European tour [51].

References

[1] Biographical details referring to Theodore von Kármán are based on the Theodore von Kármán Papers at the California Institute of Technology Archives in Pasadena, CA. I made particularly good use of the following biographical sketches, #135.7-8: Hugh L. Dryden, "The Contributions of Theodore von Kármán: A Review," Astronautics and Aerospace Engineering, July, 1963. pp. 12-17; Frank J. Malina, "Theodore von Kármán, 1881-1963," Revue Française d'Astronautique, June 17, 1963; Hugh L. Dryden, "Theodore von Kármán, 1881-1963," American Philosophical Society Yearbook, 1963; "In Memoriam Theodore von Kármán, Technische Hochschule Aachen, May 28, 1963;" Hugh L. Dryden, "The Contributions of Theodore von Kármán to Applied Mechanics," Applied Mechanincs Reviews, Vol. 16., No. 8, August 1963, pp. 589-595. Consult also Theodore von Kármán with Lee Edson, The Wind and Beyond. Theodore von Kármán, Pioneer in Aviation and Path Finder in Space (Boston: Little, Brown & Co., 1967)

[2] B. József Eötvös, Minister of Religion and Education to Mór Kleinmann, Buda, July 20, 1869. No. 12039, Theodore von Kármán Papers, 142.10, California Institute of Technology Archives, Pasadena, CA.

[3] Mór Kármán had some responsibility for planning the education of one of the Habsburg Archdukes and he received his title partly for this reason. Cf. William O. McCagg Jr., Jewish Nobles and Geniuses in Modern Hungary, Boulder, CO., East European Monographs, 1972, repr. 1986. s, p. 209, note 46--it was this title that Theodore von Kármán used in a Germanized form.

[4] Theodore von Kármán, Untitled note on Mór Kármán, Theodore von Kármán Papers, 141.6, California Institute of Technology Archives, Pasadena, CA. Dr József Gerő, ed., A Királyi Könyvek (Royal Books) (Budapest, 1940), p. 100, William O. McCagg Jr, Jewish Nobles, op. cit., pp. 209. Péter Újváry, ed., Magyar Zsidó Lexikon (Budapest, 1929), pp. 453-454.

[5] Mór Kármán, "Az Ember Tragédiája. Elemző tanulmány," (Budapesti Szemle, No. 346, 1905)--It is interesting to note that the Tragedy of Man was also a source of inspiration for other émigrée scientists, such as Leo Szilárd.

[6] William O. McCagg, op. cit., p. 209.

[7] Theodore von Kármán with Lee Edson, The Wind and Beyond, pp. 21-22.

[8] William O. McCagg, op. cit., p. 210.

[9] Theodore von Kármán with Lee Edson, The Wind and Beyond, p. 33, quoted by William O. McCagg, op. cit., p. 211.

[10] Frank J. Malina, "Theodore Von Kármán (1881-1963)," MS for the Revue Française d'Astronautique, June 17, 1963. Theodore von Kármán Papers, 135.8, California Institute of Technology Archives, Pasadena, CA.

[11] Hugh L. Dryden, "Theodore Von Kármán (1881-1963)," MS for the American Philosophical Society Yearbook, September 4, 1963. Theodore von Kármán Papers, 135.8, California Institute of Technology Archives, Pasadena, CA.

[12] Hugh L. Dryden, "In Memoriam Theodore Von Kármán," Technische Hochschule Aachen, May 28, 1963. p. 6. Theodore von Kármán Papers, 135.8, California Institute of Technology Archives, Pasadena, CA.

[13] Hugh L. Dryden, "The contributions of Theodore von Kármán: A review," Astronautics and Aerospace Engineering, July 1963, p. 12. Theodore von Kármán Papers, 135.7, California Institute of Technology Archives, Pasadena, CA.
[14] Hugh L. Dryden, "Theodore von Kármán (1881-1963)," MS for the American Philosophical Society Yearbook, September 4, 1963. Theodore von Kármán Papers, 135.8, California Institute of Technology Archives, Pasadena, CA.
[15] "Dr. Theodore von Kármán," Theodore von Kármán Papers, 135.4, California Institute of Technology Archives, Pasadena, CA.
[16] Frank J. Malina, "Theodore von Kármán (1881-1963)," MS for the Revue Française d'Astronautique, June 17, 1963. Theodore von Kármán Papers, 135.8, California Institute of Technology Archives, Pasadena, CA.
[17] Hugh L. Dryden, "In Memoriam Theodore von Kármán," Technische Hochschule Aachen, May 28, 1963. p. 6. Theodore von Kármán Papers, 135.8, California Institute of Technology Archives, Pasadena, CA.
[18] Hugh L. Dryden, "In Memoriam ," op. cit., pp. 6, 8.
[19] Vern Haugland, "Von Karman," Associated Press report, Washington, May 12, [1961], Theodore von Kármán Papers, 137.4, California Institute of Technology Archives, Pasadena, CA.
[20] Theodor von Kármán to L. Prandtl, August 2, 1933. Theodore von Kármán Papers, 23.44, California Institute of Technology Archives, Pasadena, CA.
[21] Theodor von Kármán to L. Prandtl, February 2 2, 1938. Theodore von Kármán Papers, 23.44, California Institute of Technology Archives, Pasadena, CA.
[22] Hugh L. Dryden, "The Contributions of Theodore von Kármán to Applied Mechanics," Applied Mechanics Reviews, Vol. 16, No. 8, August 1963, pp. 14-15.
[23] Frank L. Wattendorf, "Theodore von Kármán, International Scientist," Zeitschrift für Flugwissenschaft, 4 (1956) Heft 5/6, p. 165. Theodore von Kármán Papers, 135.7, California Institute of Technology Archives, Pasadena, CA.
[24] Werner Osenberg to Theodore von Kármán, June 12, 1945. Theodore von Kármán Papers, 22.18, California Institute of Technology Archives, Pasadena, CA.
[25] Frank L. Wattendorf, op. cit., ibid.
[26] R. Smend to James Franck, December 23, 1947, James Franck to Theodore von Kármán, February 11, 1948, both in Theodore von Kármán Papers, 9.36, California Institute of Technology Archives, Pasadena, CA.
[27] R. Smend to James Franck, December 23, 1947. Theodore von Kármán Papers, 9.36, California Institute of Technology Archives, Pasadena, CA.
[28] James Franck to Theodore von Kármán, February 11, 1948. Theodore von Kármán Papers, 9.36, California Institute of Technology Archives, Pasadena, CA.
[29] Hugh L. Dryden, In Memoriam..." Theodore von Kármán Papers, 135.8, California Institute of Technology Archives, Pasadena, CA.
[30] Frank J. Malina, "Theodore von Kármán, 1861-1963," op. cit., pp. 3-4. Theodore von Kármán Papers, 135.8, California Institute of Technology Archives, Pasadena, CA.-- From the great figures of modern engineering L. Prandtl, G. I. Taylor, S. Timoshenko, N. E. Joukowski, G. Eiffel and G. A. Crocco belonged to Felix Klein's school in Germany.
[31] "Memo to News Editors," May 8, 1961. News from the Institute of Aerospace Sciences, New York, N. Y., Theodore von Kármán Papers, 137.4, California Institute of Technology Archives, Pasadena, CA.
[32] "Von Karman Busy at 80," The Sun, Baltimore, May 7, 1961.

[33] The Akron blew up in 1933, and Mason crashed in 1935. Marie D. Roddenbery to "June," April 14, 1961. Theodore von Kármán Papers, 135.9, California Institute of Technology Archives, Pasadena, CA.
[34] "A Tribute to Dr. von Karman," NATO Letter, July-August, 1963, p. 23. Cf. Judith Sz. Hódy, "Mindig tanítani akartam. Utolsó beszélgetés Kármán Tódorral," Irodalmi Újság, May 15, 1963, p. 3.
[35] Ralph Dighton, "Expert on Wind Called Father Of Supersonic Age," Columbus Dispatch, May 16, 1961.
[36] Frank J. Malina, op. cit., p.3.
[37] Congressional Record, Senate, May 3, 1961, pp. 6587-8.
[38] Congressional Record, Senate, May 3, 1961, pp. 6587-8.
[39] Congressional Record, Senate, May 3, 1961, pp. 6587-8.
[40] Congressional Record, Senate, May 3, 1961, p. 6587.
[41] "A Tribute to Dr. von Karman," NATO Letter, July-August, 1963, p. 23.
[42] John F. Kennedy to Theodore von Karman, The White House, Washington, February 8, 1963, Theodore von Kármán Papers, 135.4, California Institute of Technology Archives, Pasadena, CA.
[43] White House Announcement for Palm Beach, Florida, relative to the award of The National Medal of Science to Dr von Karman, December 31, 1962. Theodore von Kármán Papers, 135.4, California Institute of Technology Archives, Pasadena, CA.
[44] "President J.F. Kennedy Presents U. S. Science Medal to Dr. von Karman," The Aerojet Booster, Vol. II, No. 13., No. 1., March 1, 1963.
[45] "For Release: Memo to News Editors," May 8, 1961. News from the Institute of Aerospace Sciences, New York, N. Y., Theodore von Kármán Papers, 137.4, California Institute of Technology Archives, Pasadena, CA. John G. Norris, "Air-Breathing Orbital Planes Urged By Expert as Cheaper Than Rockets," Washington Post & Times Herald, May 10, 1961.
[46] "For Release: Memo to News Editors," May 8, 1961. News from the Institute of Aerospace Sciences, New York, N. Y., Theodore von Kármán Papers, 137.4, California Institute of Technology Archives, Pasadena, CA. John G. Norris, "Air-Breathing Orbital Planes Urged By Expert as Cheaper Than Rockets," Washington Post & Times Herald, May 10, 1961.
[47] "Von Karman," Associated Press, May 9, 1961. Theodore von Kármán Papers, 137.4, California Institute of Technology Archives, Pasadena, CA.
[48] Walter Wingo, "'Supersonic flight will do most...'" Washington News, May 10, 1961.
[49] John G. Norris, "Air-Breathing Orbital Planes Urged By Expert as Cheaper Than Rockets," Washington Post & Times Herald, May 10, 1961. – – Even in September 2003, Professor Dr.-Ing. Karl Otto Felsch of the University of Karlsruhe vividly remembered the polyglot von Kármán moderating international scientific conferences in four different languages consecutively.
[50] "For Release," May 12, 1961. News from the Institute of Aerospace Sciences, New York, N. Y., Theodore von Kármán Papers, 137.4, California Institute of Technology Archives, Pasadena, CA.
[51] Hugh L. Dryden, "In Memoriam ," op. cit., p. 6.

WORKSHOP SUMMARIES

Workshop on Complex Flows with Chemical Reactions

Leader: Prof. Dominique THÉVENIN,
University of Magdeburg "Otto von Guericke", Germany

The Workshop on *Complex Flows with Chemical Reactions* involved 5 presentations from France, the United Kingdom and Germany concerning experimental as well as theoretical and numerical aspects. The presentations have been followed by many interesting questions and lively discussions. Following conclusions can be given at the end of this Workshop:

- Practical flows with chemical reactions involve increasingly complex problems and multi-physics coupling. In recent years we observe in particular numerous studies pertaining to flows at micro-scales, leading to very high heat exchange rates, similar to what can be found for flows through porous media (Trimis *et al.* [1]). The question of flow stability and acoustic coupling (Paschereit *et al.* [2]) is also very important for a growing number of applications.
- The crucial importance of verification and validation for numerical codes has been emphasized. Getting pretty pictures and reasonable trends is by far not enough to assume that a specific CFD code is really doing a great job in a quantitative manner.
- This implies that, more and more, quantitative and simultaneous experimental measurements should be carried out as far as possible, since such results are the only way to really carry out meaningful comparisons and identify possible errors and mistakes associated to CFD results (in the numerics or, more often, in the employed models or simplification hypotheses).
- Therefore, an intensive collaboration between experimentalists and numericians, as far as possible inside the same research Lab, appears to be the only possible way to achieve considerable progress in modeling complex flows during the next few years. Such combined investigations are demonstrated for example in Kim *et al* [3].

References:

[1] Contribution by D. Trimis. Presentation: D. Trimis, and K. Wawrzinek, 2003, "Flame stabilization of highly diffusive gas mixtures in porous inert media," Proc. Conference on Modelling Fluid Flow (CMFF'03), September 2003, Budapest, Vol. 1, pp. 61-68.

[2] Contribution by C.O. Paschereit. Presentation: C.O. Paschereit, B. Schuermans, V. Belluci, P. Flohr, 2003, "Hydrogen autoignition in a non-premixed turbulent flow: experiments and modelling," Proc. Conference on Modelling Fluid Flow (CMFF'03), September 2003, Budapest, Vol. 1, pp. 71-78.

[3] Contribution by E. Mastorakos. Presentation: I.S. Kim, C. Markides, Y. Whright and E. Mastorakos, 2003, "Hydrogen autoignition in a non-premixed turbulent flow: experiments and modelling," Proc. Conference on Modelling Fluid Flow (CMFF'03), September 2003, Budapest, Vol. 1, pp. 79-86.

Numerical Modelling of Turbomachinery Aero- and Thermodynamics

Franco RISPOLI, Alessandro CORSINI
Dipartimento di Meccanica e Aeronautica
Università di Roma "La Sapienza"
Via Eudossiana, 18, I00184 Roma
Tel: (+39) 06 44585233, fax: (+39) 06 4881759,
e-mail: rispoli@dma.ing.uniroma1.it
corsini@dma.ing.uniroma1.it

Summary of the workshop organised at the Conference on Modelling Fluid Flow (CMFF'03), September 3 to 6, 2003, Budapest

Computational Fluid Dynamics (CFD) coupled to turbomachinery devoted thermo-fluid dynamic models, has greatly enhanced the understanding of flow features pertinent to turbomachinery. CFD has become a mature investigation tool, pushed by the efforts in improving numerical algorithms and computing platforms. However the effectiveness of CFD application is limited by some pacing items, and those linked to the turbulence modelling are considered critical. For this reason, the workshop aimed at the presentation of state-of-the-art strategies developed for the modelling of turbulence related phenomena peculiar of turbomachinery configurations. With respect to gas turbine framework, still remains a lack of generally applicable techniques with satisfactory accuracy in demanding flow conditions exploiting:

- on the turbomachinery bladings, boundary layer development in presence of transition [1] and severe non-equilibrium effects [2] (such as those related to large three-dimensional separation, or secondary-to-main flow interactions, &c);
- in the combustors, finite rate chemistry combustion phenomena in turbulent regime [3];
- in the gas turbines, heat transfer and film cooling [4].

The unsteady laminar-turbulent (l-t) transition is one of the most important and also most complex issues of the contemporary turbomachinery aerodynamics [1]. The unsteady flow character results not only from the presence of turbulent fluctuations of all the important flow quantities but also from the mutual rotor-stator blade row interactions. Despite the fact that even physics of these phenomena is not well understood, it is observed that the periodic flow unsteadiness affects strongly the location of l-t transition region in boundary layers at the blade surface. In turn this circumstance affects not only the boundary layer structure but also heat transfer coefficients at the blade surface and the thermal fatigue. The wakes also influence the separation behavior of boundary layers and have a large effect on the performance of airfoils as well as turbine and compressor blades.

That is why a proper numerical simulation of the transitional flows, especially under the condition of wake unsteadiness, is nowadays one of most important but also of most difficult tasks to solve. Though dramatic improvements have been achieved in the algorithms and in CFD itself, the transition modelling is still lagging behind numerical procedures. In this respect, the fundamental research is needed in two main areas. The first one is the unsteady flow in blade channels and in particular the transport of wakes in curved channels. The second important area is the unsteady interaction of wakes with boundary layer and in particular the influence of wakes and l-t transition in boundary layers on blade profiles. In fact, transition modelling still largely limits the quality of the CFD codes today, and indeed the errors in estimation of onset and extension of the transition can affect the efficiency by several percent and component life by more than an order of magnitude. Simultaneously, it is important to remember that the development of this field and physical understanding of transition can not be achieved without the experimental work. Institute of Thermal Machinery working in the TRANSPRETURB Thematic Network was one of the groups which delivered substantial amount of data for validation of wake induced transition approaches [1].

Under the turbulence modelling viewpoint [2], the non-isotropic eddy viscosity models (EVM) are considered as a fair baseline in turbomachinery simulation, as they include provisions to account for curvature and non-equilibrium effects, and to attenuate stagnation-point inconsistency. In the number of this family of turbulence closures, particular attention should be given to the non-linear EVM that adopt higher-order polynomial (e.g. third order) relationship to mimic the stress-strain relationship as a mean to recover the turbulence anisotropy. This model still undergoing validation efforts in several flow problems (such as: external aerodynamics, environmental problems, etc.), has not yet been extensively applied in the context of real turbomachinery configurations. To this end, in the workshop has been presented comparative analyses on 2D DCA compressor cascade and a 3D NFV axial ventilation fan rotor, confirmed that this non-linear model is able to provide a better base-line for non-equilibrium effects simulation with respect to the standard one. In particular the flow survey behind the fan rotor, along the whole operating line from design to throttled flow rates, shows that the non-linear solution improves the prediction of spanwise work distribution reproducing the endwall features (such as the casing under- and hub over-turning). This circumstance was related to the ability of non-linear model of adequately taking into account the stall-like behaviour on blade suction-side as well as the leakage phenomena. To this end, the sensitization of cubic stress dependence on strain and vorticity was considered as the critical modelling mean to predict severe 3D flow structure not affected by non-physical eddy diffusivity levels.

In the design of modern gas turbine combustors, a major role is played by constraints deriving from increasingly stricter pollutant emission standards (e.g. carbon monoxide CO, or nitrogen oxides NO_x) [3]. The prediction of pollutant emissions can be attempted at two levels: by using empirical correlations, and by using CFD tools. The first approach has been widely used in the design of gas turbine combustion chambers, with some degree of success. However, empirical correlations are based on global quantities, and accordingly do not allow to assess the impact of fine design changes, aimed at reducing pollutant emissions. Accordingly, nowadays CFD represents the most promising tool. Among the emissions from turbojet engines the nitric oxides are of particular environmental concern as they represent one of the main factors depleting the ozone layer. It is worth emphasizing that the fraction of NO_x emitted in the form of nitrous oxide N_2O has also an important climatic effect, since the GWP (Global Warming Potential) of nitrous oxide stands at 320, i.e., it is as many times as effective as carbon dioxide as a greenhouse gas. Despite the low amounts of nitrous oxide emitted from combustion sources, there is evidence that it contributes about 10% to global warming. A detailed model to predict pollutant emissions from gas turbines has been proposed [3], based on the stretched laminar flamelet approach. This model is able to account for the NO_x emissions from both the thermal and the nitrous oxide (N_2O) pathways; thus the emissions of N_2O itself and CO are quantified. As far as the N_2O is concerned this may turn out to be a significant source, especially in view of the fact that the increasingly higher pressure ratios adopted in modern gas turbine engines will presumably lead to an increased weight of the N_2O pathway. Present results suggest that this pathway can contribute significantly, especially at take-off conditions. The fraction of nitrous oxide not converted to NO is however substantially higher at cruise conditions. In particular, it is found that the N_2O pathway contributes 12.8% and 8% to total NO_x emissions at take-off and cruise, respectively.

Three-dimensional numerical simulation of heat transfer and film cooling is becoming a part of complex procedure of gas turbine blade design and is slowly replacing simpler two-dimensional methods mainly based on boundary layer computations and relying strongly on experimental databases. However, many fundamental and practical developments are needed before the full three-dimensional computational analysis becomes reliable. In the workshop, has been discussed the application of Reynolds-Averaged Navier-Stokes (RANS) equations and the validity of different EVMs in the treatment of the heat transfer and film cooling of gas turbines. A first critical modelling feature is the overprediction of turbulent kinetic energy effects the heat transfer predictions. To this end has been used an experimentally document test case of a film-cooled gas turbine rotor blade (VKI blade test case). The second analysis deals with an unsteady three-dimensional slot jet, in order to discuss the unsteady effects (including jet pulsations) using RANS methodology coupled with eddy-viscosity turbulence models. It was shown herein that substantially improved heat transfer and film cooling predictions can be obtained when eddy-viscosity based turbulence models are carefully used. Limiters, such as the bound on turbulence time scale T, that prevent anomalous levels of turbulent kinetic energy predictions in a turbine passage are an example of simple and efficient improvement to widely used turbulence models.

As a final illustration the improved eddy-viscosity models were used as a part of an optimal shape design algorithm used for the design of an experimental rig at Stanford University.

Acknowledgments:

Franco Rispoli and Alessandro Corsini wish to express a deep gratitude to the workshop contributors for the quality of the presented lectures, and to the Conference Organizer Prof. T. Lajos and Prof. J. Vad.

References:

[1] Contribution by W. Elsner, invited lecturer. Presentation: W. Elsner, S. Drobniak, 2003, "Modelling of Unsteady Boundary Layer Transition", Proc. Conference on Modelling Fluid Flow (CMFF'03), September 2003, Budapest, Vol. 1.

[2] Contribution by A. Corsini, invited lecture "Flow analyses in axial compressors with a non-linear eddy-viscosity closure", Dipartimento di Meccanica e Aeronautica, University of Rome "La Sapienza".

[3] Contribution by D. Lentini, invited lecturer. Presentation: D. Lentini, 2003, "Pollutant Emission Modelling in Gas Turbines", Proc. Conference on Modelling Fluid Flow (CMFF'03), September 2003, Budapest, Vol. 1.

[4] Contribution by G. Medic, P. Durbin, 2003, "Rans Simulations For Film-Cooling Analysis And Design", Proc. Conference on Modelling Fluid Flow (CMFF'03), September 2003, Budapest, Vol. 1.

Challenges in Optimisation of Axial Flow Turbomachinery Blades for 3D Flow, Including Sweep and Dihedral Effects

René A. Van den BRAEMBUSSCHE
Turbomachinery and Propulsion Department, von Kármán Institute for Fluid Dynamics
Waterloose steenweg, 72, B-1640, Sint-Genesius-Rode, Belgium
Tel.: (+32 2) 359 9609, Fax: (+32 2) 359 9600, Email: vdb@vki.ac.be

János VAD
Department of Fluid Mechanics, Budapest University of Technology and Economics
Bertalan Lajos u. 4-6. H-1111, Budapest, Hungary
Tel.: (+36 1) 463 2464, Fax: (+36 1) 463 3464, Email: vad@simba.ara.bme.hu

Summary of workshop organised at the Conference on Modelling Fluid Flow (CMFF'03), September 3 to 6, 2003, Budapest

The goals in performance that are presently put forward for advanced turbomachines can only be achieved by a careful consideration of the three-dimensional (3D) geometric effects and secondary flows. The two-dimensional (2D) cascade concept for blade design is nowadays useful only as a first guess of the optimum geometry.

Inverse methods for the design of blades for minimum loss at a given Euler work (input/output) require prescribing the optimum velocity or pressure distribution on the blade surface. This may be possible for 2D flows but is extremely difficult for flows where 3D effects are important. Inverse methods are therefore limited to 2D flows and may be used only to define a first 2D cut of the 3D blade geometry.

There is still no agreement in the literature about the 3D design rules that will lead to better performance. Optimisation techniques may give an answer to these questions only if they consider all real flow effects in the full 3D geometry with proper inlet and outlet boundary conditions [1, 2]. The use of a 3D Navier-Stokes (NS) solver is therefore unavoidable [3, 4]. However, most optimisation tools are iterative and require a large number of NS solutions. Full 3D NS solutions on fine grids in a multistage environment are very expensive in terms of computation time and it is a challenge to develop more efficient optimisers that limit the amount of NS analyses. Expert systems, based on Artificial Neural Network (ANN), Design Of Experiment (DOE) and Genetic Algorithms (GA) may provide a considerable reduction of design effort. They aim to replace the traditional trial-and-error modification of blade parameters by a faster and hence less costly procedure but further development is still needed [1, 2]. Navier-Stokes solvers still have limitations such as unreliable prediction of location of laminar to turbulent transition, and the outcome of the design depends on the accuracy of the analysis tools that are used [1].

The impact of straight lean on performance is more or less understood. It causes basically not reduction but a redistribution of losses within a given blade row. A widespread opinion is that compound lean and the resultant radial pressure gradient offer a potential for efficiency improvement by

- unloading the endwall sections, thus initiating less secondary flow,
- concentrate the flow near midspan where the blading is more efficient,
- avoiding stagnating flows near the end-walls by controlling the migration of the blade boundary layers due to centrifugal force,
- making the outflow more uniform, thus ensuring more favourable inlet conditions for the subsequent blade row.

These recommendations are not consistent and may lead to different geometries.

The impact of compound lean, stagger angle and other parameters has recently been evaluated in a 3D HP turbine stage by a systematic change of the stacking line [5]. Starting from rotor and stator blades, that have been redesigned to achieve a favourable Mach number distribution at all sections between hub and shroud, it has been shown

- how the "negative lean" (for which the suction/endwall corner is obtuse) at hub and shroud increases the loading near mid-span and unloads the near wall blade sections [1, 6]
- that the increased suction side Mach number peak and more pronounced downstream deceleration lead to higher mid-span losses [1,7].

The systematic study resulted in a modest 1 % gain in efficiency, what is less than the 2 to 3 % efficiency gain claimed by some authors. One can conclude from this that either

- the effects of compound lean are not yet completely understood, or
- lean is less effective for the low aspect ratio blading under investigation, or
- it is a phantasm that blade lean allows large performance improvements.

An explanation for the higher increase in efficiency, claimed by some authors, could be that they have started the optimisation from a non optimum blade shape and that changing the blade lean has remedied the problem during the optimisation.

Most modern turbine stator blades have some compound lean nowadays. One important reason could be that compound lean is also a sales argument and that turbomachines without a spectacular compound lean may be judged to be old-fashioned or even badly designed.

No general concept exists for prescribing the optimum spanwise blade sweep distribution for best stage efficiency [3, 6, 8]. Only a qualitative guideline, regarding the direction of optimum sweep, can be given for the near-endwall region. The loss-reducing effects of "positive sweep" (for which the blade section under consideration is upstream of the adjacent inboard section) are widely acknowledged for the near-endwall blade sections. However, even such a qualitative guideline has not yet been found in the literature regarding optimum sweep for sections further away from the endwalls.

A possible guideline is the following [3]. The blade losses away from the endwalls are governed by the blade surface friction and the streamwise adverse pressure gradient (decelerating flow downstream of the suction side velocity peak). Blade wall friction losses are proportional to the length of a given – generally 3D – fluid pathline on the blade surface and shorter streamline paths may reduce the wall friction. The 3D features of the path also tailor the distribution of streamwise adverse pressure gradient over the blade surface. This suggests the existence of an optimum pathline for minimum loss with prescribed lift. Both sweep and shed vorticity, due to spanwise changing blade circulation, introduce additional three-dimensionality in the pathlines. Therefore, spanwise distributions of sweep and blade circulation could probably be harmonised to minimise blade losses further away from the endwalls [3].

The efficiency gain due to non-radial stacking may be more significant in turbomachines with larger aspect ratios, because it may have an important impact on the radial equilibrium [8]. Sweep is applied in transonic fans of jet propulsion engines – mainly to reduce the shock losses. Sweep and lean are also used for reduction of noise and vibration in the case of ventilation fans, automotive cooling fans and ship propellers, without consideration of the possible gain in efficiency.

A "historical confusion" (aeronautics vs. turbomachinery) persists with respect to terminology on stacking line treatment (dihedral vs. lean; sweep). It is made even more confusing by introducing twist (stagger) [1, 3, 6, 8]. The terminology must be used carefully during discussion and would be beneficial to introduce a standardised terminology.

Conclusions:

1. Realistic consideration of flow phenomena and losses in axial flow turbomachinery blade design and optimisation requires the use of 3D Navier Stokes-based tools. In this respect one still faces the challenges of computer time reduction and improvement of accuracy (i.e. correct prediction of transition).
2. It would be useful if one could establish a general concept for the quantitative prescription of lean and sweep angles along the span in order to maximise the stage efficiency by improving blade row performance and creating a more uniform outlet flow. A global target needs to be defined. By lack of such concepts, the stacking line geometry is often chosen arbitrarily, on a trial-and-error basis.
3. Non-radial stacking improves the sales value of the turbomachine by providing a "modern appearance", but the related efficiency gain for low aspect ratio stages may be modest. The available efficiency gain for high aspect ratio and high performance bladings needs further investigation.
4. Non-radial stacking provides a potential for the reduction of noise and vibration for low solidity bladings of moderate pressure rise fans, blowers and hydraulic machines.
5. The terminology regarding stacking line modifications is confusing and a standardisation of definitions is desirable.

Acknowledgments:

J. Vad acknowledges the support by the Hungarian National Fund for Science and Research under contract No. OTKA T 043493, and by the Hungarian Academy of Sciences out of the István Széchenyi Fellowship under contract No. SZÖ 271/2003.

References:

[1] Contribution by R. A. Van den Braembussche, workshop leader
[2] Contribution by R. Schilling, Institute of Fluid Mechanics, Munich University of Technology
[3] Contribution by J. Vad, invited lecturer. Presentation: J. Vad, G. Constandinides, F. Peretti, M. Gutermuth, T. Régert, 2003, "Investigation on Combined Effects of Sweep and Spanwise Changing Design Circulation on Airfoil Aerodynamics," Proc. Conference on Modelling Fluid Flow (CMFF'03), September 2003, Budapest, Vol. 1, pp. 145-152.
[4] Contribution by L. Kullmann, Department of Hydraulic Machines, Budapest University of Technology and Economics

[5] "Turbine Stage Optimization by means of a full 3D Blade Design Concept", R.A. Van den Braembussche, C.H. Sieverding, L.A. Maretto, A. Sabatini; Proc. 4th European Conference on Turbomachinery, Fluid Dynamics and Thermodynamics, March 20-23, 2001.

[6] Contribution by P. Lampart, invited lecturer, Institute of Fluid Flow Machinery, Polish Academy of Sciences. Presentation: Lampart, P., 2003, "Numerical Optimisation of a High Pressure Steam Turbine Stage," Proc. Conference on Modelling Fluid Flow (CMFF'03), September 2003, Budapest, Vol. 2, pp. 1238-1245. (included also in the present volume)

[7] Contribution by S. Dykas, Institute of Fluid Mechanics, Munich University of Technology

[8] Contribution by Z. Wiercinski, Institute of Fluid Flow Machinery, Polish Academy of Sciences

FLOWS WITH CHEMICAL REACTIONS; CHEMICAL AND PROCESS ENGINEERING; ENVIRONMENT; EXTERNAL FLUID DYNAMICS

Flame Stabilization of Highly Diffusive Gas Mixtures in Porous Inert Media

Dr.-Ing. Klemens WAWRZINEK
Gustav-Meyer-Allee 25, D-13355 Berlin
Tel.: (+49 30) 46307-546
Email: klemens.warzinek@enginion.com

Dimosthenis TRIMIS,
Institute of Fluid Mechanics,
Friedrich-Alexander-University of Erlangen-Nuremberg
Cauerstr. 4, D-91058 Erlangen, Germany
Tel : (+49 9131) 852 9490, Fax : (+49 9131) 852 9503
E-mail: dimos@lstm.uni-erlangen.de

Nomenclature

a	[m^2/s]	Thermal Diffusivity
l	[m]	Length
t	[s]	Time
D	[m]	Diameter
D	[m^2/s]	Diffusion Coefficient
S_L	[m/s]	Laminar Flame Speed
λ_f	[$W/(m\ K)$]	Thermal Conductivity of the Fluid
ρ_f	[kg/m^3]	Density of the Fluid
τ	[s]	Characteristic Reaction Time

Subscripts

c	Component
crit	Critical
eff	Effective
f	Fluid, fluidic
p	Pore
L	Laminar

Abbreviations:

HVR	High Velocity Regime
LVD	Low Velocity Detonation
LVR	Low Velocity Regime
MFC	Mass Flow Controller
ND	Normal Detonation
RCR	Rapid Combustion Regime
SVR	Sound Velocity Regime

Dimensionless Numbers:

$Le_c \equiv \frac{a_f}{D_c}$ Lewis-number

$Pe \equiv \frac{S_L d_{p,eff}}{a_f}$ Modified Péclet-number

1. Introduction

Flame propagation in porous inert media depends on structure and physical properties of the solid matrix and on the properties of the combustible gas. The resulting flame propagation modes can be classified into different regimes, of which some important parameters are given in **Table 1**.

Table 1. Flame propagation regimes in porous media [1], [2], [6], [7]

Regime	Speed of combustion wave (m/s)	Mechanism of flame propagation
Low velocity (LVR)	$0 - 10^{-4}$	Heat conduction and interphase heat exchange
High velocity (HVR)	0,1 - 10	High convection
Rapid Combustion (RCR)	10 - 100	Convection, low pressure gradient
Sound Velocity (SVR)	100 - 300	Convection with significant pressure gradient
Low velocity detonation (LVD)	500 - 1000	Self ignition with shock wave
Normal detonation (ND)	1500 - 2000	Detonation with momentum and heat loss

In this paper only such regimes are discussed in which no pressure gradient occurs in the reaction zone, i.e. the low velocity regime (LVR) and the high velocity regime (HVR). The solid matrix strongly influences the reaction conditions of the combustion, mainly because the heat transport in the burner is dominated by the properties of the solid material. Ceramic spheres and other packing material can be used in this application. In [3] and [4] an overview about applicable porous media for porous burners is given.

Since the major influence of the solid matrix derives from the enhanced heat transport, the effects of the matrix properties on flame stabilization in porous media are described. Furthermore, the diffusive mass transfer and its effects on flame stabilization are discussed and experimentally investigated.

2. Influence of Heat Transport

Concerning the flame stabilization in porous inert media significant differences occur in comparison with free flames. The stabilization depends mainly on the heat transport properties of the solid matrix. The heat transport inside of a porous medium is often described by an effective heat conductivity which comprises radiation and heat conduction of both, solid and gas phase and additionally gas convection and dispersive mechanisms [5]. The effective heat transport inside of a porous medium is 2 – 3 orders of magnitude higher than in free flames and can be considered the dominant parameter for the flame propagation for most cases. Compared with free flames the higher heat transfer leads to a faster flame propagation, which hampers the flame stabilization in porous media. For solving the task of flame stabilization in porous media different approaches have been developed, which are summarized in **Figure 1**.

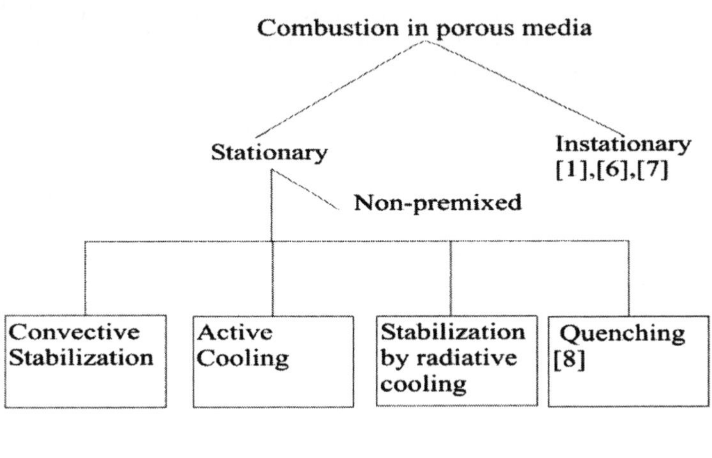

Fig. 1. Overview flame stabilization in porous media

In the most relevant for burner applications LVR and HVR regimes flame propagation or extinction can be described by a modified Péclet-number Pe, which is defined by

$$Pe = \frac{S_L \, d_{p,eff} \, \rho_f \, c_{p,f}}{\lambda_f} = \frac{S_L \, d_{p,eff}}{a_f} \tag{1}$$

Pe is formed with the laminar flame speed S_L instead of a flow velocity and describes the ratio between the heat release due to combustion in a pore and the heat removal on the walls of a pore. This ratio must exceed a critical value for flame propagation in a cold porous medium, which is given by [1] with 65 ± 45. Flame propagation in this regime is dominated by convection but also the high conductivity and radiative properties of the solid matrix influence the flame speed. Further parameters of Pe are the effective pore diameter $d_{p,eff}$ and the thermal diffusivity a_f.. $d_{p,eff}$ represents an equivalent length scale for the heat transport. The decision, whether or not flame propagation in a porous inert medium will occur, can be made by the choice of the pore size of the solid matrix. Thus, a critical pore size exists above which flame propagation and below which flame quenching occur. Nevertheless, flame propagation may also occur in subcritical cavities, if the temperature of the matrix is high enough, so that reactions are not quenched by the low wall temperature. This kind of operation is often called filtration combustion.

2.1. Stabilization by Unsteady Operation

Combustion in subcritical porous media will only take place, if the temperature of the matrix is high enough to ignite the mixture. Flame propagation is driven mainly by the heat conductivity of the solid material and the interphase heat exchange where the heat transport in the gas phase is negligible. The so called combustion wave travels very slowly through the porous matrix with a speed of 10^{-5} - 10^{-4} *m/s* [9], [10], [11]. This wave travels either in or against the flow direction. In the former case superadiabatic and in the latter subadiabatic combustion temperatures occur. The direction of the wave is mainly dependent on the heat capacities of the solid and the gas phase and on the interphase heat exchange. For a standing wave only one single operational point is possible for a certain gas mixture. However, the very low speed of the combustion wave allows operation with varying gas velocities by changing the flow direction when the combustion wave is reaching the end of the reactor. In Figure 2 the temperature profiles inside of a reactor with a subcritical porous matrix are shown.

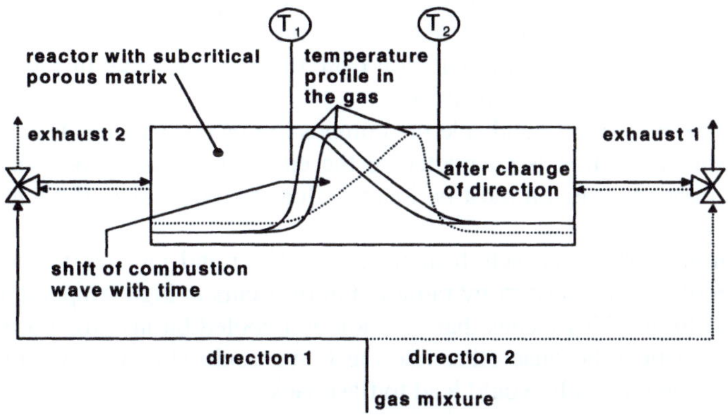

Fig. 2. Scheme of a subcritical porous reactor in alternating operation

In this case the combustion wave travels in flow direction. The heat of the combustion is heating up the porous matrix downstream of the combustion wave.

When the flow direction is reversed the hot porous matrix preheats the fresh gas mixture and superadiabatic combustion occurs [10], [11]. With this principle gas mixtures with very low heat content can be burned. The flame speeds in such reactors are about 2 - 4 times higher than the laminar flame speed [12], [13].

2.2. Stabilization in Steady Operation

In supercritical porous media in principle the same mechanisms of flame propagation act as in free flames, but the higher effective heat conductivity must be taken into consideration. The effective heat conductivity as well as the thermal diffusivity are 2 - 3 orders of magnitude higher than in a gas. Following the simplified theory of flame propagation [14] the flame speed S is proportional to the square root of the temperature diffusivity:

$$S = \sqrt{\frac{a_f}{\tau}} \tag{2}$$

with the thermal diffusivity a_f and a characteristic time scale of the reaction τ. From this results that the flame speed in supercritical porous media is 10 - 30 times higher than the laminar flame speed. One possibility to stabilize the combustion in supercritical porous media is to induce a change of the flow speed by a stepwise or continuous change of the cross sectional area. For methane/air mixtures in porous media the required flow speed to avoid flash back lies in the range of 2 – 5 *m/s*, which is about ten times higher than in free flames.

Another possibility for flame stabilization is to apply cooling of the reaction zone, e.g. by embedded water cooled tubes in the main reaction zone. The cooling can also be realized by intense radiation from the reaction zone. In **Figure 3** these two principles are schematically shown.

In the case on the left hand side the combustion takes place inside of the pores, and the heat of reaction is conducted to the embedded water tubes. In the other case the combustion region must be located close to the surface of the porous matrix.

The porous surface extracts heat from the flame and radiates to the environment. Effective heat transport by radiation only occurs at high temperatures above the ignition limits. This means that for radiation cooled burners the porous matrix must be subcritical, because otherwise the flame could slowly move into the porous matrix, which finally would lead to flashback.

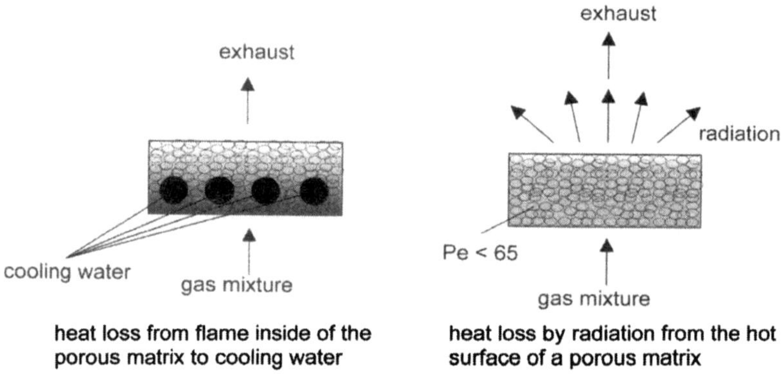

Fig. 3. Flame stabilization in porous media by cooling

In [8] another principle for the flame stabilization was introduced, which made use of thermal flame quenching. The combination of two regions with different pore sizes was carried out, which result in supercritical and subcritical Péclet-numbers, respectively. This principle is shown in **Figure 4**.

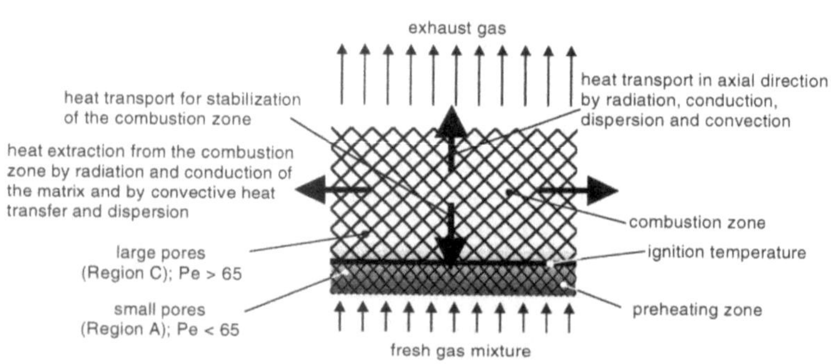

Fig. 4. Heat transport and flame stabilization in a two layer porous burner [8]

In region A, through which the unreacted cold gas mixture flows, combustion is not possible due to the subcritical Péclet-number, so combustion only occurs in the supercritical region C with larger pores. Thus the flame stabilizes at the interface between the two regions. This arrangement allows flame stabilization at the same position over a wide range of flow velocities. In order to prevent flash-back the amount of heat which is transported against the flow direction must not be higher than the amount which is carried convectively by the fresh gas mixture into the combustion region, because otherwise a combustion wave can develop starting from the hot interface between region A and C, which possibly travels against the flow direction. This can be satisfied by means of a low conductive region A, which allows only a small amount of heat transport against the flow direction. In contrast region C should feature a high effective heat conductivity, because this allows to operate at high flow rates without the danger of blow-off. Typically, thermal heat loads of 3000 kW/m^2 in steady operation and up to 4500 kW/m^2 for peak loads are feasible. In [15] a power modulation ratio of 20:1 was realized with very low emissions over the whole modulation range and high flame stability in a wide equivalence ratio range. In [16] a flat porous burner was designed, which is applied in industrial applications and makes use of the high emissivity of a SiC foam. The heat is radiated out of the volume of the SiC foam and enables to achieve high radiation efficiency compared with standard surface radiating burners.

The principle of flame stabilization by changing the Péclet-number may be used advantageous in many application fields. However, in the Péclet-number criterion mainly heat transport processes are considered, and if diffusive mass transport becomes dominant, wrong critical pore diameters are predicted.

3. Influence of Diffusive Mass Transport

Up to now the influence of the diffusive mass transport on the stabilization of a flame in a porous medium was not investigated intensively. In [1] and [17] flame instabilities were observed for certain gas mixtures. These instabilities were linked to the increased influence of the diffusive mass transport, which decreased the dominance of the heat transport on the flame stabilization. The ratio between diffusive mass transport and heat transport can be described by the Lewis-number Lec of a component c of the gas mixture

$$Le_c = \frac{\lambda_f}{D_c \rho_f c_{P,f}} = \frac{a_f}{D_c} \tag{3}$$

which is the ratio between the temperature diffusivity a_f and the diffusion coefficient of the component D_c. According to [1] and [18] the flame structure changes for Lewis-numbers smaller than unity. For many widely used gases, e.g. methane/air mixtures, Le is close to unity and therefore its influence is often neglected. However, mixtures with Lewis-numbers far below one require the consideration of mass diffusion. For example hydrogen/air and hydrogen/chlorine mixtures feature Lewis-numbers of about 0.4. For such low values one can expect a strong influence of the diffusion on the flame stabilization in porous media. This influence is neglected in the Péclet-number criterion, because only the laminar flame speed is considered, which does not account for deformations of the flame front. This means that the constant value of the critical Péclet-number is not sufficient for the design of a porous burner flame trap.

In this study experiments concerning flame propagation in porous media were performed with gas mixtures of different Lewis-number. The aim of these experiments was to gain a better knowledge for the design of the porous burner flame trap.

3.1. Experimental determination of critical Péclet-number at various Lewis-numbers

In the following, experiments concerning the flame propagation of gases at different Lewis-numbers are described, with the aim of finding the critical Péclet-number for each mixture. Packings from steel spheres were used as the porous medium. In **Table 2** the gases and the sphere packings, which were investigated are summarized.

Table 2. Gas mixtures with different Lewis-number

Composition	Sphere sizes d_{sphere} in mm	Lewis-number
CH_4/air	10.0, 11.9, 14.3	≈ 0.96 in fuel rich flames
C_2H_6/air	10.0, 11.9, 14.3	≈ 1.42 in fuel lean flames
C_3H_8/air	10.0, 11.9, 14.3	≈ 1.8 in fuel lean flames ≈ 0.8 in fuel rich flames
H_2/air	2.5, 3.5, 4.0	≈ 0.4 in fuel lean flames ≈ 2.0 in fuel rich flames
$H_2/Cl_2/HCl$	3	≈ 0.33

It is important to note, that the limiting Lewis-number is defined by the lacking component, i.e. oxygen in fuel rich mixtures and the fuel component in fuel lean mixtures. The equivalent pore diameter $d_{p\,eff}$ of the sphere packing can be estimated according to [2] with $d_{sphere} = 2.77\, d_{p\,eff}$. The gases in Table 1 cover a wide range of Lewis-numbers and by the investigation of the flame propagation limits in sphere packings a relation between critical Péclet-number and Lewis-number can be given.

3.1.1. Experimental Set-up for Detection of Flame Propagation

In order to investigate the critical Péclet-number, the critical pore diameter for the flame propagation of a certain mixture must be determined. The experiments were carried out in cold sphere packings. The gas mixture was ignited at the reactor exit. The detection whether or not the flame propagated through the matrix was carried out by two photo diodes, which gave the opportunity also to measure the propagation velocity of the flame inside of the porous medium. In **Figure 5** the experimental set-up is shown schematically.

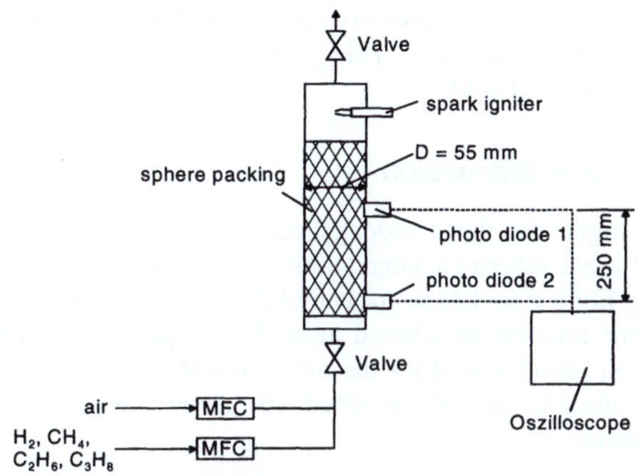

Fig. 5. Experimental set-up for the investigation of the critical Péclet-number

The flow rates of the gas components were adjusted with mass flow controllers (MFC) in order to reach the required composition. In order to bring the gas to rest, the valves were closed, when the correct composition was reached inside the packing. Just before ignition the upper valve was opened. The flame was detected by the two photo diodes. **Figure 6** presents a typical signal sequence.

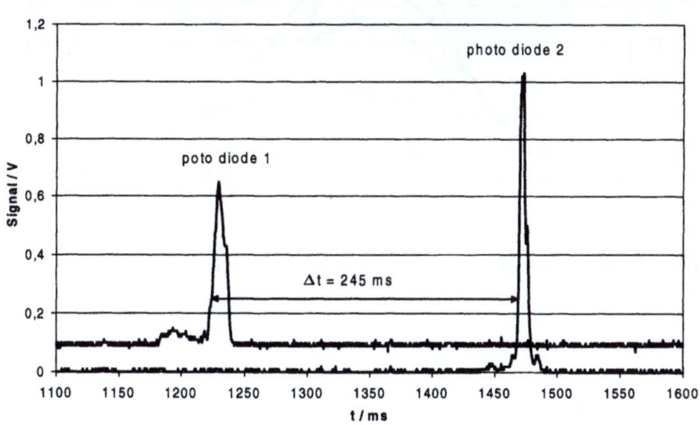

Fig. 6. Signal sequence with flame propagation

For each gas and sphere size the flame velocity in the packing and both the rich and the lean limit of flame propagation was determined. From the laminar flame speed and the physical properties of the mixture, the actual Péclet-number was determined.

For the last gas mixture of Table 1, the $H_2/Cl_2/HCl$ mixture, a different experimental procedure was applied. A closed system was used, because of the toxic gas mixture, with a two layer porous burner in which the flame trap consisted of 3 *mm* Al_2O_3 spheres. The temperature inside of this region was measured as an indication whether or not the flame was quenched. In these experiments the H_2/Cl_2 ratio was kept constant close to stoichiometry and the HCl fraction was varied until the limit point of flash back was found. This gas mixture was of interest because of its low Lewis-number around 0.33.

3.1.2. Results and Discussion

In order to investigate the flame propagation behavior, the equivalence ratio was varied in the different sphere packings, until no flame propagation was registered. In the following Figs. 7 to 10 the measured flame speeds and the laminar flame speed of the same mixture are plotted against the equivalence ratio. The critical Péclet-number can be calculated for the minimal and the maximal equivalence ratio, the corresponding laminar flame speed, the gas mixture properties and the diameter of the spheres.

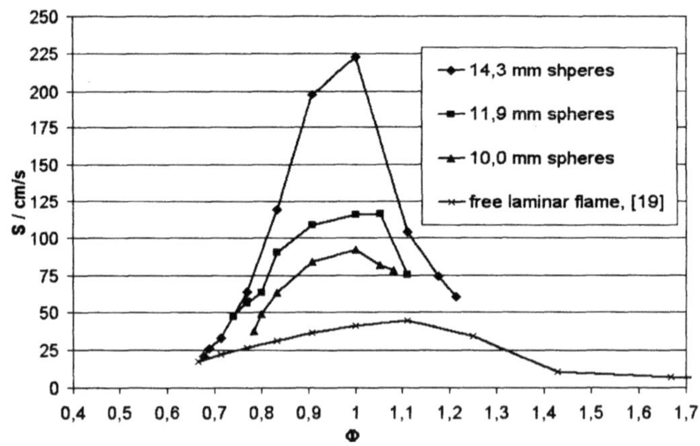

Fig. 7. Flame speeds of methane/air mixtures in sphere packings

In **Figure 7** the flame speed of methane/air mixtures in different sphere packings is shown. The flame propagation is 2 to 6 times higher than in a free laminar flame. Similar behavior is observed for ethane/air and propane/air mixtures, for which the measured flame speeds are presented in **Figures 8 and 9**. In both cases flame propagation in the packings was up to 10 times higher than in the laminar free flame. The maximum value of the flame speed occurred at the same equivalence ratio as for the laminar flame.

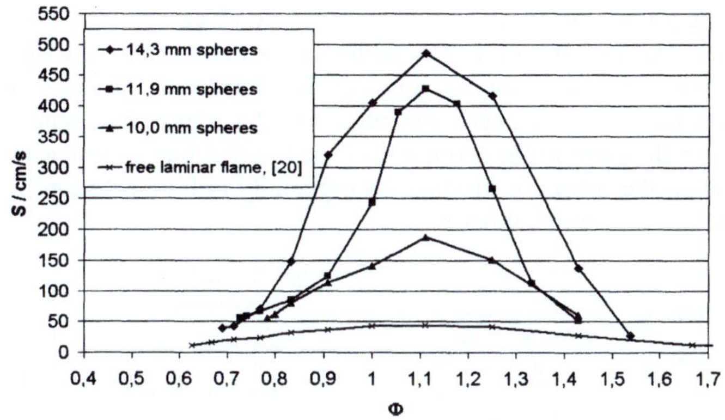

Fig. 8. Flame speeds of ethane/air mixtures in sphere packings

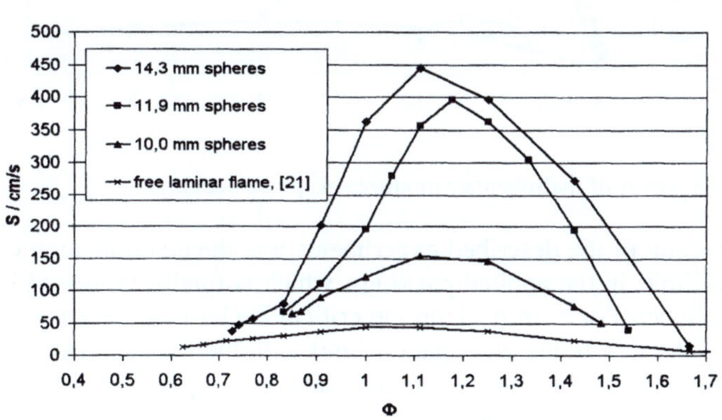

Fig. 9. Flame speeds of propane/air mixtures in sphere packings

This does not apply for the hydrogen/air mixtures (**Figure 10**), for which the maximum flame propagation speed in the packings were measured at equivalence ratios just lower than one, whereas the maximum flame speeds in a laminar hydrogen/air flame occurs at fuel rich conditions. This effect can be explained by the physical properties of the gas mixture. In hydrogen/air flames the heat conductivity of the mixture increases significantly with increasing equivalence ratio and the high heat conductivity of hydrogen becomes dominant in the mixture. This leads to an acceleration of the flame propagation due to the high heat conductivity of hydrogen. In porous media the effective heat conductivity is dominated by the transport properties of the solid phase almost regardless of the equivalence ratio of the gas mixture. Therefore, the influence of the increase of the gas conductivity at high equivalence ratios is negligible regarding the flame propagation. With increasing equivalence ratio the influence of the thermal flame propagation gains influence against the diffusive mechanism especially at high reaction temperatures, which occur close to stoichiometry.

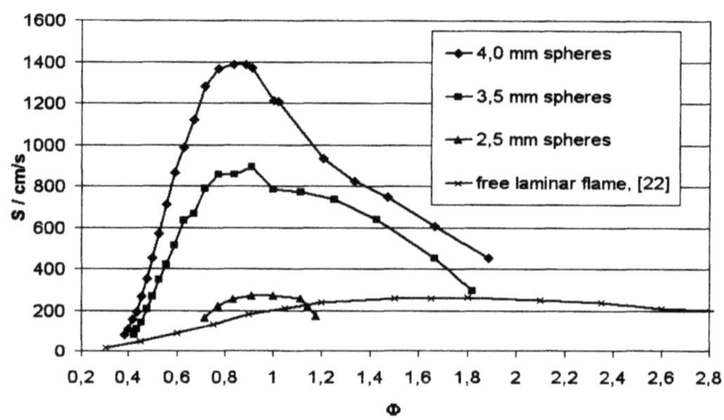

Fig. 10. Flame speeds of hydrogen/air mixtures in sphere packings

A major result of the described experiments was the determination of the flame propagation limits in the applied packings, which is finally expressed by the critical Péclet-number. For each packing the critical Péclet-number was calculated for the lean and the rich flame propagation limit with the laminar flame speed at the corresponding equivalence ratio. The procedure was repeated for different diameters of the spheres. The average of the critical Péclet-numbers is plotted against the Lewis-number in **Figure 11**. For Lewis-numbers close to and larger than one the criterion of [1] of $Pe_{crit} \approx 65$ could be confirmed.

However, for mixtures of Lewis-numbers below one the critical Péclet-number was much lower than this value. For rich C_3H_8/air mixtures with a Lewis-number of 0.8 the critical Péclet-number reached 27. With 17 the critical Péclet-number was even lower for lean hydrogen/air-mixtures with a Lewis-number of about 0.4.

The experiment with the $H_2/Cl_2/HCl$-mixture also led to an interesting result. For mixtures with a molar HCl/Cl_2-ratio less than 3, the temperature in the flame trap region was rising significantly fast. Thus, at these conditions the critical Peclet number was reached. The laminar flame speed for the $H_2/Cl_2/HCl$-mixture upon which the critical Péclet-number was determined was calculated with the CHEMKIN 3.6 software applying kinetic data of [23]. The resulting critical Péclet-number was with 6.5 the lowest reached in this study.

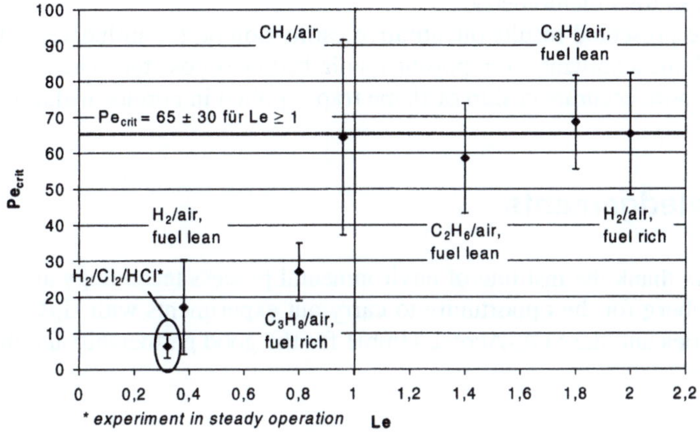

Fig. 11. Critical Péclet-number over Lewis-number

The uncertainties in Fig. 11 derive mainly from the uncertainty on the determination of the laminar flame speed, with which the critical Péclet-number was calculated. The laminar flame speed was determined at the corresponding equivalence ratio, which could be adjusted by means of the mass flow controllers with an accuracy of ± 0.05. The relative uncertainty of the calculation of the effective pore diameter applying the formula of [2] was estimated with 10 %. The same uncertainty was also estimated for the value of the thermal diffusivity. The spreading of the critical Péclet-numbers for the different sphere sizes fit all well within these calculated limits except for methane/air mixtures. The spreading of the values of the critical Péclet-number for methane/air-mixtures for the different sphere sizes was almost doubled than it would be expected from the above described calculation. A possible explanation for this is the fact that the Lewis-number of these mixtures is almost unity, which means that both, heat transport and mass diffusivity play an important role in flame propagation. Therefore, both effects are superposed, which would lead to an increased sensitivity to small variations. However, this effect cannot be calculated exactly. The uncertainty bar of the methane/air-mixture corresponds to the spreading of the critical Péclet-numbers for the different sphere sizes.

4. Conclusion

The results of this experimental study show, that gas mixtures with a high diffusivity compared with the thermal diffusivity of the mixture, i.e. Lewis-numbers less than unity, show lower critical pore diameters or critical Péclet-numbers than mixtures with Lewis-number equal or larger then unity, concerning flame propagation. The Lewis-number is calculated for the reaction component, of which the mixture lacks for stoichiometry.

With the presented results quantitative estimation on the influence of the diffusivity on flame stabilization in porous media becomes possible. The results can be used for a more accurate design of flame traps applied in porous media burners.

Acknowledgements

The authors thank the institute of environmental process technology at the University of Freiberg for the opportunity to carry out experiments with H_2/Cl_2-flames in their facilities and the SGL-Acotec GmbH for the good partnership and the fruitful discussions.

References

[1] Babkin VS, Korzhavin AA, Bunev VA (1991) Propagation of premixed gaseous explosion flames in porous media. Combustion and Flame Vol. 87, S. 182-190

[2] Pinaev AV, Lyamin GA (1989) Fundamental Laws Governing Subsonic and Detonating Gas Combustion in Inert Porous Media. Combustion, Explosion and Shock Waves USSR, Vol. 25, No. 4, S.448-458, Translated from Fizika Goreniya i Vrzyva, Vol. 25, No. 4, pp 75-85

[3] Pickenäcker O, Pickenäcker K, Wawrzinek K, Trimis D, Pritzkow WEC, Müller C, Goedtke P, Papenburg U, Adler J, Standke G, Heymer H, Tauscher W, Jansen F (1999) Innovative Ceramic Materials for Porous-Medium Burners. Interceram, Vol. 48, No. 5 & 6, pp 326-330 & 424-434

[4] Pickenäcker O, Pickenäcker K (2001) Eigenschaften von keramischen porösen Strukturen für die Porenbrennertechnik. Keramische Zeitschrift, Vol. 53, No. 9, pp 780-787.

[5] Schlünder E-U, Tsotsas E (1988) Wärmeübertragung in Festbetten, durchmischten Schüttgütern und Wirbelschichten. Georg Thieme Verlag, Stuttgart

[6] Lyamin GA, Pinaev AV (1986) Combustion regimes in an inert porous material. Combustion, Explosion and Shockwaves, USSR, Vol. 22, No. 5, S. 553-558, translated from Fizika Goreniya I Vrzyva, Vol. 22, No. 5, pp 64-70

[7] Lyamin GA, Pinaev AV (1987) Fast subsonic combustion of gases in an inert porous medium with a smooth rise in the pressure in the wave. Combustion, Explosion & Shockwaves, USSR, Vol. 23, No. 4, pp 399-402, from Fizika Goreniya I Vrzyva, Vol. 23, No. 4, pp 27-30

[8] Trimis D, Durst F (1996) Combustion in a porous medium - advances and applications. Combustion Science and Technology, Vol. 121, pp 153-168.
[9] Bingue JP, Saveliev VA, Fridman AA, Kennedy LA (2002) Hydrogen Production in Ultra-rich Filtration Combustion of Methane and Hydrogen Sulfide. International Journal of Hydrogen Energy 27, pp 643-649
[10] Hanamura K, Echigo R, Zhdanok SA (1993) Superadiabatic Combustion in a Porous Medium. Int. J. Heat Mass Transfer, Vol. 36, No. 13, pp 3201-3209
[11] Gavrilyuk VV, Dmitrienko YM, Zhdanok SA, Minkina VG, Shabunya SI, Yadrevskaya NL, Yakimovich AD (2001) Conversion of Methane to Hydrogen under Superadiabatic Filtration Combustion. Theoretical Foundations of Chemical Engineering, Vol. 35, No. 6, pp 589-596

A Hybrid Approach to Model Combustion Instabilities

Christian Oliver PASCHEREIT
Hermann-Föttinger-Institute
Technical University Berlin
Germany
oliver.paschereit@tu-berlin.de

Bruno SCHUERMANS, Valter BELLUCCI
ALSTOM (Schweiz) AG
CH-5405 Baden-Dattwil
Switzerland
bruno.schuermans@power.alstom.com
valter.bellucci@power.alstom.com

Abstract An acoustic network approach was used to model and control thermoacoustic interaction in gas turbines. A hybrid approach was chosen: numerical, experimental and analytical techniques are combined to describe the system. The network modules are represented as state-space realizations.

A modal expansion technique is used to obtain a state-space representation of the acoustic propagation through complex 3-dimensional geometries. The modal expansion can be based on analytic models or finite element analysis. Complex modules, as the thermoacoustic burner properties, are modeled using a combined experimental and analytic approach. The method is not restricted to symmetries: configurations with geometrically or operationally different burners are simulated in time or frequency domain.
Key Words: combustion instabilities, thermoacoustic non-linear low order modeling

Nomenclature

$\overline{(\cdot)}$ Mean flow quantities.

$(\cdot)'$ Acoustic perturbations.

$\hat{(\cdot)}$ Laplace transform of acoustic perturbations.

$\hat{(\cdot)}$	Modal amplitude.
c	Speed of sound.
ρ	Density.
p	Pressure.
u	Velocity
x	State vector.
$\mathbf{x}$	Spatial coordinate.
ω	Circular frequency $2\pi f$.
s	Laplace coefficient $s = i\omega$
δ_{kron}	Kronecker delta $\delta_{kron}(0) = 1$, $\delta_{kron}(n) = 0$ if $n \neq 0$

1. Introduction

Thermoacoustic interactions in industrial combustion systems are difficult to model because they involve complex interactions between several physical mechanisms. An acoustic network approach has been used to model and control thermoacoustic interaction in gas turbines. There are an enormous number of papers dealing with modeling and control of combustion instabilities. An overview will not be given here, the interested reader may consult the references.

Lumping the combustion system into several subsystems and combining the subsystems in a network of acoustic elements allows for a combination of different modeling techniques. The idea of such a lumped-element representation is not new (Lang et. al., 1987; Stow & Dowling, 2001; Schuermans et. al., 2000; Pankiewitz & Sattelmayer, 2002) just to cite a few. However, in this new approach, a methodology is developed that includes geometries of any complexity while the resulting systems can be analyzed in a time-efficient, straightforward manner. Because of the lumped-element representation, a compactness assumption is made for the burners and the combustion process: it is assumed that the geometrical extent of these modules is small compared to the wavelength. Therefore, this method is restricted to the low frequency regime. The method has been developed for modeling the acoustic interaction in lean-premix gas turbines. Such combustion systems are characterized by low Mach number flow in the combustion chamber (the Mach number being typically less than one order of magnitude larger than the laminar flame speed Mach number). Therefore, at least in a first approach, the Mach number is neglected in the wave propagation. However, the Mach number is being taken into account on the boundaries in the flame and burner modules. This paper will first demonstrate how a state-space representation of geometries (without combustion) can be obtained. As an example, the state-space representation of an annular duct will be derived. A comparison with results obtained from Finite Element Analysis is made. Interconnection of several systems into a network of acoustic systems is done using *Linear Fractional Transforms*. A stability analysis is then made by evaluating the eigenfrequencies of the interconnected system. The validity of this approach is demonstrated on a very simple, one-dimensional thermoacoustic system. The eigenfrequencies are solved for analytically and compared with the results obtained from modal expansion and Linear Fractional Transforms. A second validation is performed on a system consisting of two annular ducts interconnected by 1-dimensional tubes, the eigenfrequencies are compared against results obtained from Finite Element Analysis. A network model of an annular, multi burner, gas turbine combustion chamber is then derived. The models for the burner and combustion process are not explained in detail in this paper, but reference is made to other publications. The interconnected network model is then used to synthesize an active controller, using $H\infty$ optimization. A modal decomposition technique has been used to facilitate the design procedure of the active controller. A set of controllers is then obtained to control each mode individually. Results are presented that demonstrate the effectiveness of the controller, even in the presence of non-linearities and (parametric) noise.

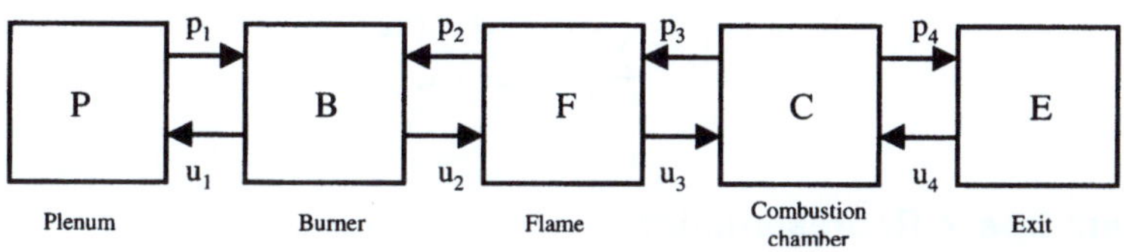

Fig. 1. Interconnection of subsystems of the combustion system. Note that all arrows represents vectors of input or output signals.

2. Modal Expansion

In this section acoustic transfer functions are derived for geometries with multiple inputs and multiple outputs (MIMO). Starting from the wave equation with sources on the surface (*f*), but without sources in the volume:

$$c^2\nabla^2 p' - \frac{\partial^2 p'}{\partial t^2} = 0 \tag{1}$$

$$\hat{\mathbf{n}} \cdot \nabla p' = -f \tag{2}$$

A solution can be obtained by making use of Green's functions. As shown in (Culick, 2001), the acoustic pressure at any point of the volume can be written as a function of the source, model eigenvalues (ω_n), speed of sound (c), and the eigenvectors (ψ):

$$\hat{p}(\mathbf{x}) = \sum_{n=0}^{\infty} \frac{c^2 \ \psi_n(\mathbf{x})}{\Lambda(\omega^2 - \omega_n^2)} \oint_S \psi(\mathbf{x}_s)\hat{f}(\mathbf{x}_s)dS. \tag{3}$$

where $\Lambda = \int \psi^2 dV$. The eigenfunctions (ψ) and eigenfrequencies (ω_n) can be obtained analytically for simple geometries or numerically (e.g., Finite Element Method) for more complex systems. If the source function is a source of acoustic velocity on the boundary, it can be written as:

$$\hat{f}(\omega, \mathbf{x_s}) = -i\omega\rho\hat{u}_s(\omega, \mathbf{x_s})$$

An acoustic transfer function can be defined as the ratio between the acoustic pressure at a certain position (x) in the volume to the acoustic velocity acting as an input on an area A_s centered on the boundary at x_s. If the extent of the area A_s is small compared to the wavelength, then Eq. 3. can be re-written to obtain the transfer function $H(\omega)$ between $\hat{p}$ (x) and $\hat{u}_s(x_0)$.

$$H(\omega) = \frac{\hat{p}(\mathbf{x})}{\hat{u}_s(\mathbf{x_s})} = -i\omega\rho A_s c^2 \sum_{n=0}^{\infty} \frac{\psi_n(\mathbf{x})\psi_n(\mathbf{x_s})}{\Lambda(\omega^2 - \omega_n^2)} \tag{4}$$

Equation 4. relates the acoustic pressure at one location to the acoustic velocity at one other location. This Single Input Single Output (SISO) representation (which corresponds to the acoustic impedance) can easily be extended to the general MIMO case. The *K* velocities at $\vec{x}_{in}$ are then related to the *J* pressures at $\vec{x}_{out}$ by a JxK *transfer matrix* **H**: $\mathbf{p}(\vec{x}_{out}) = \mathbf{H}\,\mathbf{u}(\vec{x}_{in})$,in which the elements of **H** are given by:

$$\mathbf{H}_{jk} = -i\omega\rho A_k c^2 \sum_{n=0}^{\infty} \frac{\psi_n(\mathbf{x_j})\psi_n(\mathbf{x_k})}{\Lambda(\omega^2 - \omega_n^2)} \tag{5}$$

2.1. State Space Representation

Because all elements of the transfer matrix **H** have the same eigenvalues (ω_n) the transfer matrix can be expressed more conveniently by a *state-space representation*. One mode of the SISO system of 4 can be represented as:

$$\begin{aligned} \dot{x}(t) &= \mathbf{A}_n x(t) + \mathbf{B}_n u(t) \\ \frac{p_n(t)}{\rho c} &= \mathbf{C}_n x(t) + \mathbf{D}_n u(t) \\ \mathbf{A}_n = \begin{bmatrix} -\alpha_n & -\omega_n \\ \omega_n & -\alpha_n \end{bmatrix}&, \quad \mathbf{B}_{nj} = \begin{bmatrix} 0 \\ \psi_n(\mathbf{x}_j) \end{bmatrix} \\ \mathbf{C}_{nk} = \begin{bmatrix} 0 & \frac{cA_k}{\Lambda}\psi_n(\mathbf{x}_k) \end{bmatrix}&, \quad \mathbf{D} = [0] \end{aligned} \tag{6}$$

Note that α, the modal damping has been introduced here. It is assumed that the value of α is small compared to ω_n. There are many different possible state space representations of a system. The representation chosen here has the advantage that it can easily be extended to the MIMO case. The state of the system is represented by the 2×1 vector x_n (this notation is chosen to be consistent with notation used in control theory and should not be confused with geometrical position **x**). The structure of the equations for the general case, having *N* modes, *J* inputs and *K* outputs is the same as Eq. 6. However, the matrix **A** becomes a $2N \times 2N$ block-diagonal matrix and the matrices **B** and **C** become $2N \times J$ and $K \times 2N$ matrices respectively:

$$\begin{bmatrix} \dot{x}_1 \\ \vdots \\ \dot{x}_{2N} \end{bmatrix} = \begin{bmatrix} \mathbf{A}_1 & & \\ & \ddots & \\ & & \mathbf{A}_N \end{bmatrix} \begin{bmatrix} x_1 \\ \vdots \\ x_{2N} \end{bmatrix} + \begin{bmatrix} \mathbf{B}_{11} & \dots & \mathbf{B}_{1J} \\ \vdots & \ddots & \vdots \\ \mathbf{B}_{N1} & \dots & \mathbf{B}_{NJ} \end{bmatrix} \begin{bmatrix} u_1 \\ \vdots \\ u_J \end{bmatrix}$$

$$\frac{1}{\rho c}\begin{bmatrix} p_1 \\ \vdots \\ p_K \end{bmatrix} = \begin{bmatrix} \mathbf{C}_{11} & \dots & \mathbf{C}_{1N} \\ \vdots & \ddots & \vdots \\ \mathbf{C}_{K1} & \dots & \mathbf{C}_{KN} \end{bmatrix} \begin{bmatrix} x_1 \\ \vdots \\ x_{2N} \end{bmatrix} + \begin{bmatrix} 0 & \dots & 0 \\ \vdots & \ddots & \vdots \\ 0 & \dots & 0 \end{bmatrix} \begin{bmatrix} u_1 \\ \vdots \\ u_J \end{bmatrix} \tag{7}$$

2.2. State Space Representation of an Annular Duct.

In order to obtain an acoustic transfer function or matrix of some geometry, the eigenfrequencies (ω_n) the values of the eigenvectors at the interface locations ($\psi_n(x)$) need to be known. For practical (often very complicated) systems the eigenfrequencies and vectors can be obtained from a finite element analysis. Note that in the Finite Element Analysis, only a modal analysis is required (which is very computationally efficient). Moreover, only the modal values at one position on the interface locations are required, thus very little output is needed. For more simple geometries the eigenfrequencies and vectors can be obtained analytically. As an example, the transfer function of an annular duct is derived. A combustion chamber of a gas turbine can be represented as an annular duct with *J* input and output ports (Fig. 2), *J* being the number of burners.

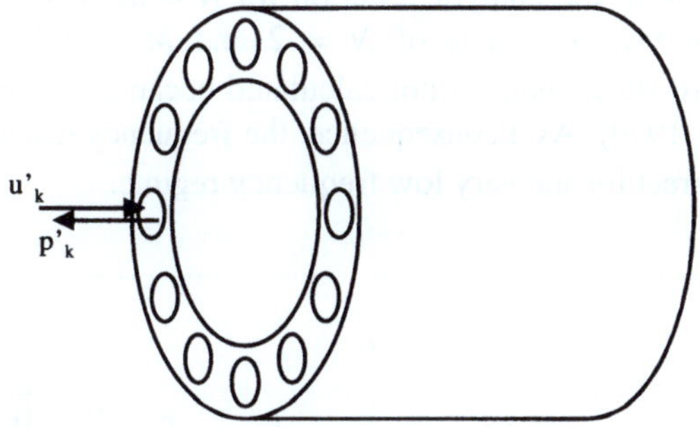

Fig. 2. Annular duct geometry

The input–output relation is given by a transfer matrix relating *J* inputs to *J* outputs. The required eigenvalues and vectors for a thin annular duct of length (*L*) and mean diameter (*D*) and height (*h*) are given by:

$$\omega_{n,m} = \sqrt{(\frac{2cm}{D})^2 + (\frac{\pi cn}{L})^2},$$

$$\psi_{n,m} = cos(\frac{\pi nx}{L})\begin{cases} cos(m\phi), \\ sin(m\phi) \end{cases}$$

$$\Lambda_{n,m} = \begin{cases} \frac{LD\pi h}{(2-\delta_{kron}(n))(2-\delta_{kron}(m))} \\ \frac{LD\pi h}{2(1+\delta_{kron}(n))\delta_{kron}(m)} \end{cases} \tag{8}$$

In which *n* and *m* are the numbers of the longitudinal and azimuthal modes respectively. Thus, the mode is notated here as (*n, m*).

Because of the rotational symmetry of the annular duct, all eigenvalues are two-fold degenerate (except for n=0) and have two orthogonal eigenmodes. Substitution of Eq. 8. into Eq. 6. and Eq. 7. yields the state-space representation of a thin annular duct with collocated inputs and outputs at one side of the duct:

$$A_{n,m} = \begin{bmatrix} -\alpha & \omega_{n,m} & & \\ \omega_{n,m} & -\alpha & & \\ & & -\alpha & \omega_{n,m} \\ & & \omega_{n,m} & -\alpha \end{bmatrix},$$

$$B_{n,m} = \begin{bmatrix} 0 & \dots & 0 \\ \cos(m\phi_1) & \dots & \cos(m\phi_J) \\ 0 & \dots & 0 \\ \sin(m\phi_1) & \dots & \sin(m\phi_J) \end{bmatrix},$$

$$C_{n,m} = \frac{(2-\delta_{kron}(n))(2-\delta_{kron}(m))}{JL} B^T_{n,m}. \tag{9}$$

The D matrix being empty. This modal expansion representation of the transfer function (or impedance) of the annular duct has been compared with an analytic solution and with a solution obtained from the commercial Finite Element Method package Sysnoiser® (Fig. 3). Note that the modal expansion representation is mathematically equal to the analytic solution for $N = M = \infty$. However, for the result presented in Fig. 3. values of $N = 2$ and $M = 4$ have been used. In Sysnoiser®, the zero Hertz mode is not calculated because it causes numerical difficulties (Sysnoise, 1999). As a consequence, the frequency response calculated by Sysnoiser® is incorrect for the very low frequency regime.

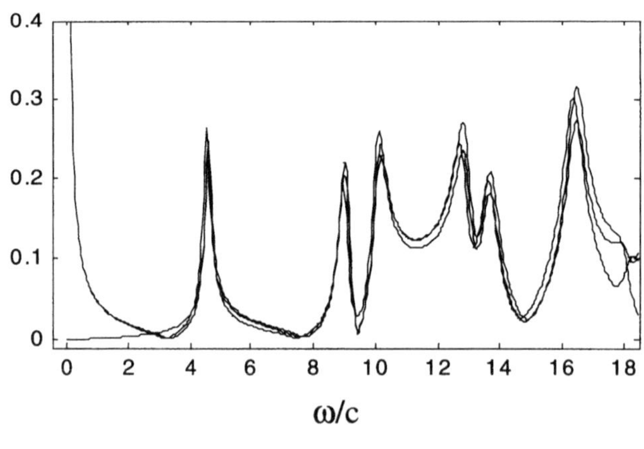

Fig. 3. Frequency response of annular duct, calculated with Sysnoiser® (solid), analytic solution (dotted), and modal expansion (dashed).

This representation can easily be extended to the more general case with inputs and outputs on both sides of the duct (at $x = 0$ and $x = L$). When doing so, it is helpful to apply the following partioning of the B,C and D matrices:

$$
\begin{aligned}
\dot{\mathbf{x}} &= \mathbf{A}\mathbf{x} + \mathbf{B}_\mathbf{l}\mathbf{u}_\mathbf{l} + \mathbf{B}_\mathbf{r}\mathbf{u}_\mathbf{r} \\
\frac{\mathbf{p}_\mathbf{l}}{\rho c} &= \mathbf{C}_\mathbf{l}\mathbf{x} + \mathbf{D}_\mathbf{ll}\mathbf{u}_\mathbf{l} + \mathbf{D}_\mathbf{lr}\mathbf{u}_\mathbf{r} \\
\frac{\mathbf{p}_\mathbf{r}}{\rho c} &= \mathbf{C}_\mathbf{r}\mathbf{x} + \mathbf{D}_\mathbf{rl}\mathbf{u}_\mathbf{l} + \mathbf{D}_\mathbf{rr}\mathbf{u}_\mathbf{r}
\end{aligned} \tag{10}
$$

In which l and r refer to the left or right hand side inputs and outputs. The matrix $\mathbf{A}$ is the same as for the single sided duct. The matrices $\mathbf{B_l}$ and $\mathbf{C_l}$ contain the values of the eigenvector on the left side of the duct and are identical to the matrices $\mathbf{B}$ and $\mathbf{C}$ in Eq. 9. The matrices $\mathbf{B_r}$ and $\mathbf{C_r}$ contain the values of the eigenvectors at $x = L$. Because $cos(\frac{\pi n x}{L}) = (-1)^n$ if $x = L$ the following expressions are obtained for $\mathbf{B_r}$ and $\mathbf{C_r}$ in annular ducts:

$$
\begin{aligned}
\mathbf{B}_{\mathbf{r}n,m} &= (-1)^n\,\mathbf{B}_{\mathbf{l}n,m} \\
\mathbf{C}_{\mathbf{r}n,m} &= (-1)^n\,\mathbf{C}_{\mathbf{l}n,m}
\end{aligned}
$$

Although the $\mathbf{D}$ matrices are empty again, they are shown here in order to be more consistent with a more general notation of partitioned state-space systems:

$$
H = \begin{bmatrix} H_{11} & H_{12} \\ & \\ H_{21} & H_{22} \end{bmatrix} = \left[\begin{array}{c|cc} A & B_l & B_r \\ \hline C_l & D_{ll} & D_{lr} \\ C_r & D_{rl} & D_{rr} \end{array}\right]
$$

3. Network Interconnection

In order to obtain a model of the acoustic behavior of a gas turbine combustion system, acoustic transfer matrices can be combined to form a network of acoustic elements. Such a network representation has the advantage that different modelling techniques can be used for different elements. The acoustic wave propagation can then be modelled using the modal expansion technique, while the burner and flame sub-systems are modelled using a combined analytical and experimental technique.

For a realistic acoustic model of a combustion system, it is essential to take into account the acoustic losses, or damping. The most important acoustic losses are caused by dissipation on the boundaries of the system (e.g., air supply system, high Mach number combustor exit) and due to the process of converting acoustic energy into vorticity. This later mechanism is very important in the model of the burner and is directly associated with the mean flow loss coefficient of the burner. Acoustic losses purely associated with wave propagation through the combustion chamber (taken into account by the parameter α in Eq. 6) are generally very small compared with the losses on the boundaries and in the burner element. The acoustic losses are thus explicitly accounted for in the network model. This is an advantage compared to the approach used e.g., by (Annaswamy *et. al.*), where acoustic losses are not taken into account in the model.

Every subsystem will be represented as a state-space system having the shape of Eq's.6 or 11. These subsystems can be inter-connected in a very convenient way by making use of the *Redheffer Star Product*. The Redheffer Star Product is a matrix operation based on a *Linear Fractional Transform* (Zhou & Doyle, 1998). It is often used in control theory to model uncertainty in systems, but can be used to interconnect any network of state space systems. The interconnection of two ducts (or any other systems) H and G is then simply given by: $H \star G$ in which $\star$ denotes the Redheffer Star Product, and is defined by:

$$H \star G = \begin{bmatrix} \mathcal{F}_l(H, G_{11}) & H_{12}(I - G_{11}H_{22})^{-1}G_{12} \\ G_{21}(I - H_{22}G_{11})^{-1}H_{21} & \mathcal{F}_u(G, H_{22}) \end{bmatrix},$$

In which $\mathcal{F}_u()$ and $\mathcal{F}_l()$ denote the upper and lower Linear Fractional Transform, defined as: $\mathcal{F}_l(M, g) = M_{11} + M_{12}g(I - M_{22}g)^{-1} M_{21}$. The interconnection is illustrated in Fig. 4.

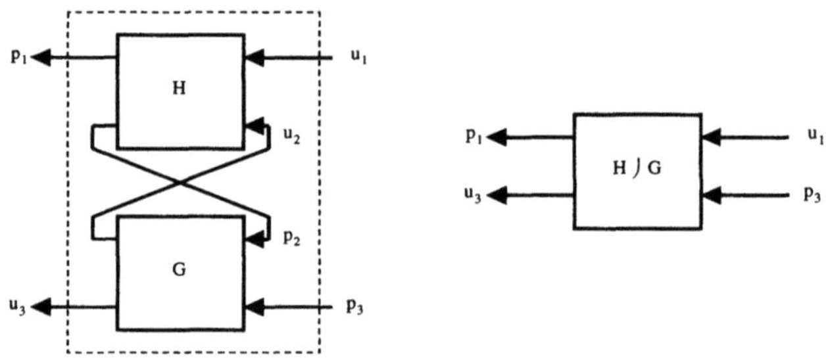

Fig. 4. Interconnection of two MIMO systems H and G, using the Redheffer Star Product. Note that all arrows represent vectors of input or output signals

As a first example, the 1-dimensional thermoacoustic system described in (Lang et. al., 1987) will be analysed. This system consists of a straight duct, closed on one side, open on the other side, with a flame stabilized in the middle of the duct. The pressure drop across the flame sheet is assumed to be negligible. The acoustic velocity jump is modelled by the so-called $n - \tau$ model: $u_2(t) = u_1(t) + nu_1(t - \tau)$. In which τ is a delay time and n is referred to as the interaction coefficient. The impedance of the open end is simply: $Z_3 = 0$. A sketch of the system and its lumped element block diagram is given in Fig. 5.

Note that the $n - \tau$ model contains a delay and is thus of infinite order. In order to avoid systems of infinite order, the time delay is approximated by a *Pad'e approximation*, a technique commonly used in control theory. The upstream duct is represented by a transfer function similar to Eq. 4, but becomes more simple because in the one-dimensional case $m = 0$. Using the Redheffer Star Product, the system can be represented as: $\mathbf{S} = \mathbf{P} \star \mathbf{F} \star \mathbf{C} \star \mathbf{Z}$. Note that this system has no inputs or outputs, but these could of course be added. However, in order to investigate internal stability it is sufficient to analyze the homogeneous system. The eigenfrequencies or poles of the system are the complex eigenvalues of the matrix $\mathbf{S}$. Note that $\mathbf{S} \in \mathbb{R}$, is time invariant and does not depend on ω. Computing the eigenvalues can be done using standard methods available in linear algebra. The eigenvalues of $\mathbf{S}$ corresponding to the first resonant mode have been calculated for several values of τ. According to (Lang et. al., 1987), the eigenvalues of the system are shown to be the roots of:

$$\cos(2\frac{\omega L}{c}) - \sin^2(\frac{\omega L}{c})\, n\, e^{-i\omega\tau} = 0, \tag{11}$$

which requires a numerical search. Both the results obtained via modal expansion and the roots of Eq. 11 are plotted in Fig. 6., in which ω_0 is the resonance frequency in the case $n = 0$. For large values of $\omega_0\tau$ both curves deviate, this is because the order of the Pad´e approximation was relatively low (six). However, a perfect match can be obtained by increasing the order of the Pad´e approximation.

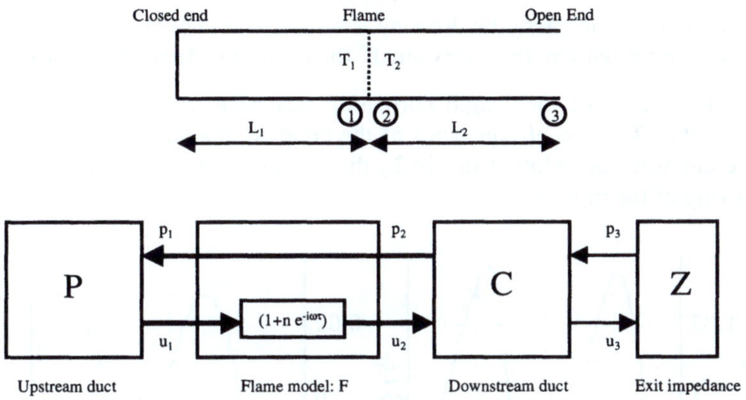

Fig. 5. Sketch of 1-dimensional combustion system, and block diagram representation

Using the state-space representation and the Linear Fractional Transforms, the interconnection of complex MIMO systems can be done in a straightforward manner. As an example, two annular ducts are interconnected by 24 smaller 1-dimensional ducts, very similar as in (Evesque & Polifke, 2002). This represents in essence a gas turbine combustion chamber geometry: the first annular duct corresponds to the plenum chamber, the smaller ducts represent the burners and the second annular duct represents the combustion chamber. The temperature in the second annular duct is different from the temperature in the first duct. This geometry has been modelled in Sysnoiser® and was also evaluated by calculating $P \star B \star C$, in which P represents the upstream, cold, annular duct, B represents 24 parallel 1-D ducts and C is the downstream, hot, annular duct. Thus, P and C have 24 inputs and 24 outputs, while B has 2 × 24 inputs and 2 × 24 outputs. Another possibility is to calculate the eigenmodes of the upstream geometry together with the 1-D ducts, and the downstream geometry (consisting of the hot annular duct only). These two geometries can be calculated separately in Sysnoiser®, and coupled after. These modes can now directly be used to apply modal expansion and represent the two subsystems in state-space. If PB is the upstream system and C the downstream system the inter-connected system can be represented as: $PB \star C$. Note that once the eigenfrequencies $\omega_{n,ref}$ obtained for the geometry at a specific temperature, the eigenfrequencies at different temperatures are easily obtained from: $\omega_n = \frac{c}{c_{ref}}\omega_{n,ref}$. The scaled eigenvectors: $\psi/\sqrt{\Lambda}$ remain unchanged. Although the system interconnected in this way approximates the numerical solution, a perfect match is not obtained - even for a system of very high order. This is because the interconnection at each interface position is 1-dimensional, and some important 3- dimensional effects close to the interface are neglected. This phenomenon is well known from Helmholtz resonator theory: a length correction factor (virtual length) has to be applied to compensate for the local deformation of the potential field. This can easily be done by adding a 1-D duct element at the interconnection. The virtual length is very small, thus a zero or first order expansion is sufficient. The value of the length correction has been set to: $\sqrt{\frac{A_i}{\pi}}$, based on (Rienstra, 1999). The eigenfrequencies of the coupled annular ducts calculated in three different ways are plotted in Fig.2., the values on the x-axis correspond to the numbering of the modes.

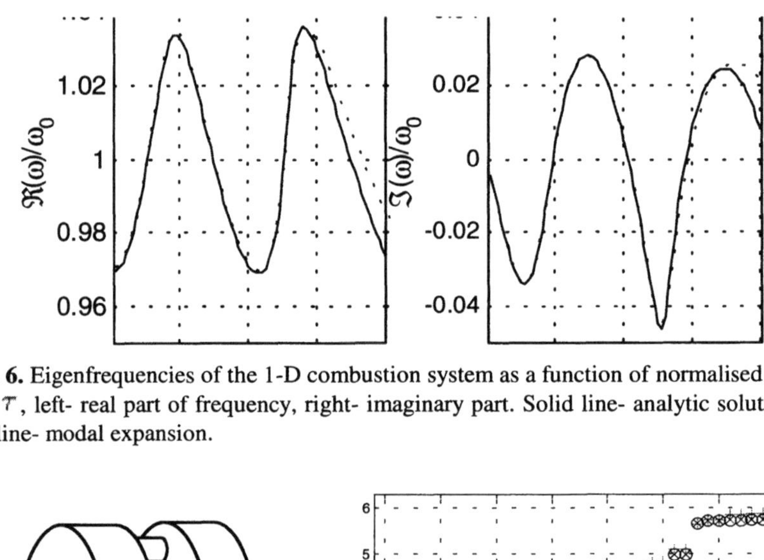

Fig. 6. Eigenfrequencies of the 1-D combustion system as a function of normalised time delay, τ, left- real part of frequency, right- imaginary part. Solid line- analytic solution, dotted line- modal expansion.

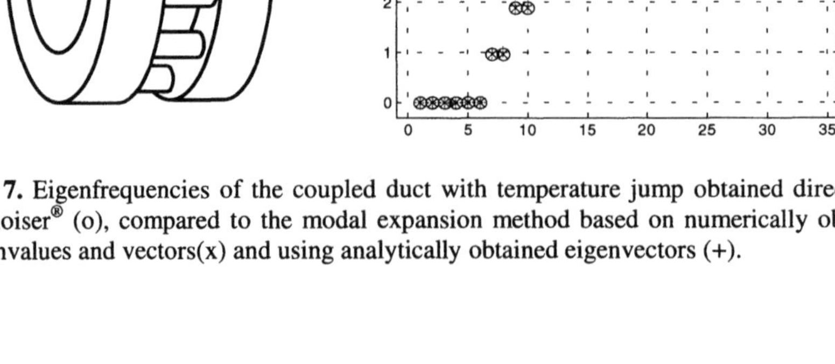

Fig. 7. Eigenfrequencies of the coupled duct with temperature jump obtained directly by Sysnoiser® (o), compared to the modal expansion method based on numerically obtained eigenvalues and vectors(x) and using analytically obtained eigenvectors (+).

3.1 Annular Combustion System

From the previous two examples it is only a small step to a representation of an annular gas turbine combustion chamber. The lumped element representation of the annular combustion system is represented by the block diagram in Fig. 1.

The burner will be modeled as an $L - \zeta$ model as described in (Schuermans et. al., 1999). This model is derived from the unsteady Bernoulli equation. The parameter L is a measure of the amount of air fluctuating in the burner nozzle, the parameter ζ represents the effect of dissipation of acoustic energy to the mean flow. The block diagram of the L - ζ model is given in Fig. 8, in which: $L\zeta(s) = \left[-\frac{L}{c}s + M(1 - \zeta - (\frac{A_1}{A_2})^2\right]^{-1}$. The values of L and ζ are obtained from a fit to measured transfer functions (Schuermans *et. al.*, 2000). For one burner, this element (denoted by B) has two inputs (p_1 and p_2) and two outputs (u_1 and u_2). In a multi burner configuration with J burners, a block diagonal matrix has to be formed: B= $diag[B_1, B_2, \ldots B_J]$. Note that if geometrically different burners are used, the elements B_j will be different.

The flame module, in which the interaction of the combustion process with the acoustic field is modeled, is again represented as an $n - \tau$ model. However, rather than having one delay, a distribution of time delays is assumed here. The block diagram is shown in Fig. 8. The relation between the acoustic velocities across the flame is then given by: $\hat{u}_2(s) = (1 - n \int_0^\infty \xi(\tau)\, e^{-s\tau} d\tau)\hat{u}_1(s)$. In which τ is the convective time delay between fuel injection and consumption, and $\xi(\tau)$ is the probability density distribution function of time delays. The interaction coefficient is given by $n = 1 - \frac{T_2}{T1}$. The distribution of time delays can either be obtained numerically (Flohr *et. al.*, 2001; Polifke *et. al.*, 2001) or from experimental fits (Belluci *et. al.*, 2001).

If a Gaussian distribution of time delays is assumed with mean value $\bar{\tau}$ and standard deviation σ_τ, then, after carrying out the integration, the flame model can be written as: $\hat{u}_2(s) = (1 - ne^{s^2\frac{1}{2}\sigma_\tau^2}\, e^{-s\bar{\tau}})\hat{u}_1(s)$. In this case, the values $\bar{\tau}$ and σ_τ have been generated from fits to experiment tally obtained frequency responses of the flame transfer function. The interconnection of the flame block is given in Fig. 8. The flame subsystem will be denoted by F, the diagonal system containing J flame transfer function is then denoted by **F.**

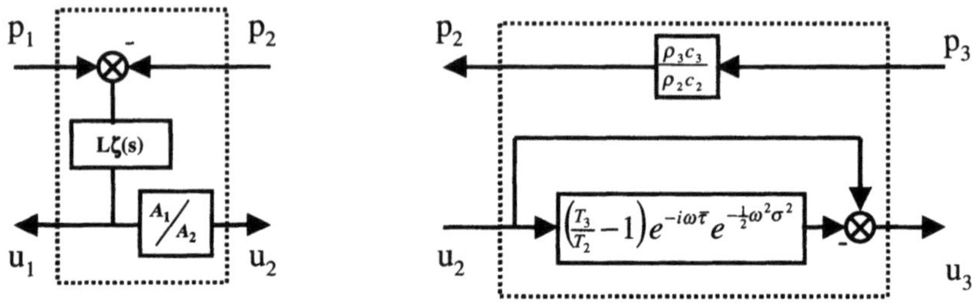

Fig. 8. Block diagrams of the burner model (left), and the flame model (right).

With the plenum chamber represented by annular duct **P**, and the combustion chamber as C , all the sub-modules can be combined as: **S** = **P** $\star$ **B** $\star$ **F** $\star$ **C**.

3.2 Inputs and Non-Linearities

The system is extended to take into account inputs to the system. Two types of system inputs are considered:

1. Sources inherent to the combustion process: they consist of sound created by turbulence, that propagates to the acoustic far-field, but of which the generation itself is not influenced by the acoustic field.
2. External excitation with fuel flow actuators, necessary for active control.

The frequency spectra of the combustion source terms have been determined experimentally (Schuermans et. al., 2000). A transfer function H_{source} was then fitted to the magnitude of the frequency spectra. A time domain source signal can then be obtained by filtering a white noise signal ($\omega(t)$) with the transfer function H_{source} as described in (Campos-Delgado et. al., 2001). In a multi burner configuration, the source terms of the individual burners are, by definition, linearly independent. Thus different white noise sequences have to be generated for each burner. A modulation of the fuel supply with fuel flow actuators will result in a modulated heat release in the flame, which causes a periodic volume expansion in the flame zone. The transfer function, H_{act} describing the dynamics of this valve, the effect on heat release and volume expansion has also been determined experimentally in (Campos-Delgado *et. al.*, 2001). So far, the entire system is considered to be linear. It is very likely that the actual system is not linear, especially when the (linearized) system is unstable. Therefore, a non-linear saturation of the heat release signal was included, similar to the approach used in (Pankiewitz & Sattelmayer, 2002). Since the command of the actuation valve is bounded (Campos-Delgado et. al., 2001), a saturation function is included here as well.

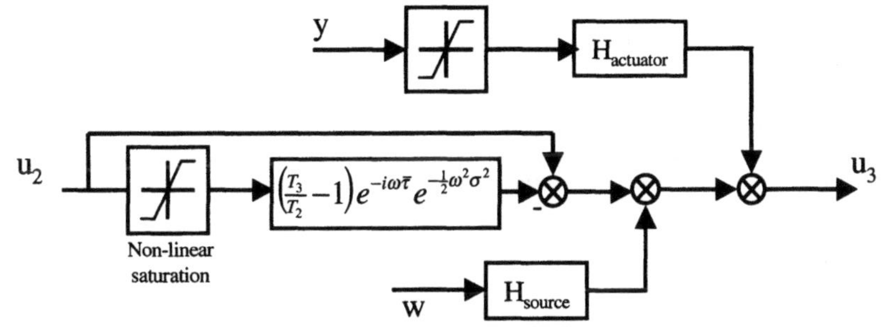

Fig. 9. Flame model with sources and non-linear saturation.

4. Application to a Gas-Turbine combustor

The method was applied to a silo gas turbine combustor. The network model included burner asymmetries and asymmetric arrangement of damper elements. The Helmholtz resonators applied for advanced damping were modeled by an analytical non linear model. The flame transfer functions were based on measurements combined in an analytic model with non-linearities. The acoustic eigenfunctions and eigenmodes of the combustor and the hood were derived from finite element calculations.

Two different setups were considered: (1) Helmholtz dampers designed to suppress two unstable modes. A strong instability at $St \approx 0.7$ was observed. (2) Additional Helmholtz dampers were designed to suppress the third unstable mode. Only a limited volume was available to mount the resonators. The additional damping power was thus limited as well. The design task had thus the extended goal to suppress the instability by most efficient placement of the damper elements. The results are displayed in Fig. 10. Good agreement was found between modeling and measurement. The third unstable mode was effectively suppressed by an additional Helmholtz damper.

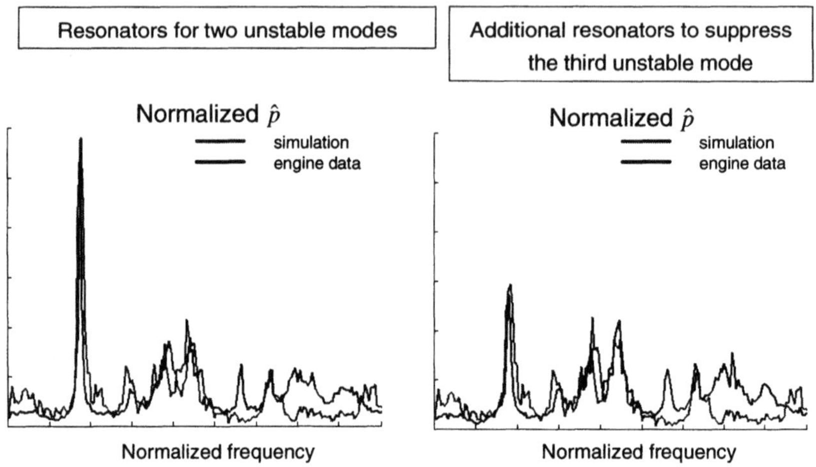

Fig. 10. Application of the state space modeling to a silo gas turbine combustor. Left: two unstable modes controlled by Helmholtz dampers. Right: suppression of the third unstable modes by additional damper elements

5. Summary

The state-space representation of acoustic systems via modal expansions, provides a straightforward, fast, efficient way to model complex (thermoacoustic) systems. Very complex geometries can be modelled by making use of input from finite element codes. For stability analysis, all eigenvalues of a system can be obtained without the need of iterative search. All 322 eigenvalues of the entire system were calculated in less than a second on a normal PC. Making use of state-space representations and linear fractional transforms enables one to setup a network of acoustic elements with multiple inputs and outputs in a convenient manner. This method has been validated by comparing the results with analytic solutions and numerical solutions. Good agreement has been obtained, even for relatively low order modal expansions. Time simulations of the system, including non-linear modules, are easily made using the Matlab/Simulink environment. This approach allows for the acoustic analysis of multi-burner gas turbine combustion chambers. The state space approach was applied to a real silo gas turbine combustor. The flame transfer functions were based on measurements combined in an analytic models with non-linearities. Helmholtz dampers were modeled by a non-linear model. The acoustic eigenfunctions and eigenmodes of the combustor and the hood were derived from finite element calculations. Good agreement was found between simulation and measurement. The state-space representation is also a perfect basis for active control design.

References

[1] Rienstra, S.W.,Hirschberg, A. An Introduction to Acoustics, Report IWDE99-02, TU Eindhoven, 1999.

[2] Zhou, K., Doyle, C.D., Essentials of Robust Control, Prentice-Hall, 1998.

[3] LMS International SYSNOISE Revision 5.4 documentation Version 1.0, (May 1999) LMS Intrenational, Leuven, Belgium

[4] Schuermans, B.B.H., Polifke, W., Paschereit, C.O., and van der Linden, J.H. Prediction of Acoustic Pressure Spectra in Combustion Systems Using Swirl Stabilized Gas Turbine Burners, ASME 2000-GT-0105 Munich, Germany, May 8-11, 2000.

[5] Stow, S.R., and Dowling, A.P. Thermoacoustic Oscillations in an Annular combustor, ASME 2001-GT-0037, New Orleans, LA, June 8-11, 2001.

[6] Flohr, P., Paschereit, C.O., Van Roon, B., and Schuermans, B.B.H. Using CFD for Time-Delay Modeling of Premixed Flames, ASME 2001-GT-0376, New Orleans, LA, June 8-11, 2001

[7] Schuermans, B.B.H., Polifke, W., and Paschereit, C.O.,Modeling transfer matrices of premixed flames and comparison with experimental results, ASME 1999-GT- 0132, Indianapolis, USA, June 7-10, 1999

[8] Culick, F.E.C., Dynamics of Combustion Systems: Fundamentals, Acoustics, and Control, Short course, Von Karman Institute of Fluid Dynamics, Active Control of Engine Dynamics, 14-18 May 2001.

[7] Lang, W., Poinsot, T., Candel, S., Active Control of Combustion Instability, Combustion and Flame 70: 281-289, 1987.

[8] Evesque, S., Polifke, W. Low-order Acoustic Modelling for Annular Combustors: Validation and Inclusion of Modal Coupling., Paper GT-2002-30064, Proceedings of ASME TURBO EXPO 2002, June 3-6,2002,Amsterdam, The Netherlands.

[9] Daniel U. Campos-Delgado, Bruno Schuermans, Kemin Zhou, Christian Pachereit, Eduardo Gallestey and Andreas Poncet., Thermoacoustic Instabilities: Modeling and Control, Accepted for publication at IEEE Transactions on Control Systems Technology, March 2001.

[10] Polifke, W., Kopitz, J., Serbanovic, A., Impact of the Fuel Time Lag Distribution in Elliptical Premix Nozzles on Combustion Stability, 7th AIAA/CEAS Aeroacoustics Conference May 28-30, 2001, Maastricht, The Netherlands.

[11] Sattelmayer, T., Influence of the Combustor Aerodynamics on Combustion Instabilities from Equivalence ratio Fluctuations , ASME 2000-GT-0082, Proc. ASME Turbo Expo 2000, Munich, Germany, May 8-11, 2000.

[12] Pankiewitz, C., Sattelmayer, T., Time Domain Simulation of Combustion Instabilities in Annular Combustors, Paper GT-2002-30063, Proceedings of ASME TURBO EXPO 2002, June 3-6,2002,Amsterdam, The Netherlands.

[13] Bellucci, V., Schuermans,B., Paschereit, C., Flohr, P., Thermoacoustic Simulation of Lean Premixed Flames Using an Enhanced Time-Lag Model, AIAA Paper 2001-2974, Anaheim, CA, 11-14 June 2001.

[14] Annaswamy, A.M, Fleifil, M., Rumsey, J.W., Prasanth, R., Hathout, J.P., Ghoniem, A.F., Thermoacoustic Instability: Modelbased Optimal Control Design and Experimental Validation, IEEE Transactions Control Systems Technology, vol 8,no 6, 2000.

MODELING OF REACTION AND TRANSPORT IN CVD PROCESS: PREDICTIONS OF GROWTH AND UNIFORMITY OF TIN OXIDE FILMS

Theodora C. Xenidou, Andreas G. Boudouvis, Nicolas C. Markatos

School of Chemical Engineering, National Technical University of Athens
9 Heroon Polytechniou Str., Zographou Campus, GR 15780, Athens
Greece

Abstract. The design of a chemical vapor deposition process is performed by combining a detailed computational fluid dynamics (CFD) model and a Langmuir-Hinshelwood type kinetic scheme. This methodology is implemented in a cold-wall reactor, where tin oxide thin films are grown under atmospheric pressure conditions. A set of available experimental data is used to estimate the parameters required by the kinetic model. The coupled CFD/kinetic model is then used for a systematic analysis of the reactor performance. Simulation results are suggestive of modifications in the operating parameters that could enhance the uniformity of the layer thickness.

Nomenclature

$\boldsymbol{g}$	[m/s^2]	gravity acceleration
$\boldsymbol{j}$	[kg/s]	diffusive mass flux
$\boldsymbol{u}$	[m/s]	velocity vector
C^s	[mol/m^3]	concentration at the wafer surface
C_p	[J/(kg K)]	specific heat
D^T	[kg/(m s)]	thermal diffusion coefficient
D^{eff}	[m^2/s]	effective diffusion coefficient
D	[m^2/s]	binary diffusion coefficient
F_d	[m^3]	flow rate of the diluent gas
F_o	[m^3]	flow rate of O_2
G	[Å/min]	growth rate
H	[J/mol]	enthalpy of formation
$\boldsymbol{I}$	[-]	unity tensor
K_a	[m/s]	adsorption rate coefficient
K_d	[m/s]	desorption rate coefficient
M	[kg/mol]	molecular weight
N	[-]	number of gas species

Nomenclature (cont.)

P	[Pa]	pressure
R_d	[mol/(m^{2}s)]	deposition rate
T	[K]	temperature
λ	[W/(m K)]	thermal conductivity
μ	[Pa s]	viscosity
ρ	[kg/m^3]	density
ω	[-]	mass fraction

Subscripts and Superscripts

i	with respect to ith gas species
(g)	gas phase species
(a)	adsorbed species
(s)	solid species
max	maximum value
min	minimum value
ave	average value
T	due to temperature gradients
C	due to concentration gradients

1 Introduction

Thin solid films are grown on heated substrates through Chemical Vapor Deposition (CVD) from chemically reactive gases. The use of these thin films is widespread in many industrial applications for producing semiconductors, protective coatings, gas sensors, electrodes, transducers and other opto-electronic devices. The great advantage of the CVD process compared to other deposition techniques is its capability to produce thin solid films of a wide variety of materials with precisely defined and highly reproducible properties [1].

The CVD process involves the simultaneous transport of mass, momentum and energy coupled with complex chemical reactions in the gas phase and on heated surfaces. Hence, the quality of the produced films is determined by the interactions of transport processes and chemical reactions, which depend on reactor design and process variables such as pressure, temperature, inlet flow rates and gas mixture composition [2]. Many computational fluid dynamics codes have been developed for solving such a coupled set of processes in different configurations of CVD reactors [3,4]. However, due to the lack of reaction-rate data for many chemical systems, the design of a new CVD process requires substantial time in experiments to empirically determine reactor configuration and operating parameters for optimizing the production efficiency. Clearly, it is difficult to apply such a method to the design of a CVD reactor with complex geometry. Therefore, a systematic design approach for CVD reactors is required.

In our days, the microelectronics industry imposes very strict requirements on the quality of the produced films, such as high deposition rates, good thickness uniformity across the wafer (less than 2% thickness non-uniformity) and economic use of the reactants. In this aspect, the purpose of the present work is to examine t he possibility of coupling a detailed three-dimensional CFD model [5] with a relatively simple chemical kinetic scheme, aiming at the efficient design and optimization of a CVD process. The proposed design methodology is implemented in a horizontal, single-wafer, cold-wall CVD reactor, where tin oxide (SnO_2) thin films are grown from tin tetrachloride ($SnCl_4$) under atmospheric pressure conditions.

Tin oxide thin films are attractive due to the stability of their electrical, optical, chemical and mechanical properties. In particular, they remain stable up to high temperatures and they have excellent resistance at room temperatures as well as good adhesion to many substrates. Tin oxide deposition by $SnCl_4$ oxidation has been experimentally studied by several groups [6-8], including the original work of Ghoshtagore [9]. Most of these studies have focused on the optical and electrical properties as well as the surface morphologies of the produced films. Although Ghoshtagore [9] addressed the issue of tin oxide growth kinetics in part, limited work has been devoted to the investigation of the chemical mechanisms involved in the gas phase and on surfaces during the growth of tin oxide CVD from $SnCl_4$.

Recently, a Langmuir-Hinshelwood type kinetic mechanism for this specific chemical system has been proposed [10]. In the present work, that kinetic model describing heterogeneous phenomena coupled to a CFD model is used to further analyze the influence of some key operating parameters on the process performance. In particular, the uniformity of the layer thickness is investigated, with special attention paid to the origins of the axial and transversal heterogeneities on the wafer for each of the parameters examined, thus opening possible ways of process improvement.

2 The CVD Reactor

Tin oxide deposition has been experimentally investigated through the following oxidation reaction [10]:

$$SnCl_4 + O_2 \rightarrow SnO_2 + 2Cl_2 \tag{1}$$

The main geometrical parameters of the horizontal, cold-wall, single wafer CVD reactor are summarized in Table 1 while Fig. 1 shows a schematic view of the experimental setup used in that study.

Table 1. Geometrical features of the CVD reactor [10]

Tube diameter [cm]	Tube length [cm]	Wafer diameter [cm]	Inlet holes diameter [cm]	Outlet hole diameter [cm]	Susceptor size [cm]
11.00	40.00	7.50	0.38	1.62	8.98x12.8x1.10

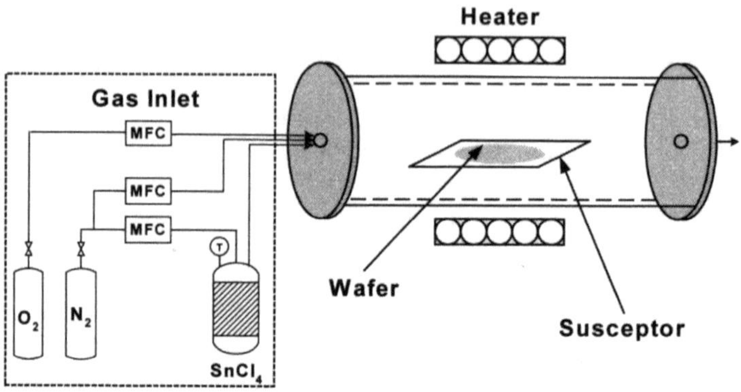

Fig. 1. Schematic view of the cold-wall CVD reactor

The liquid precursor ($SnCl_4$) is delivered from the bubbler to the reactor chamber using nitrogen as carrier gas. The reactant gases are injected through five different inlet holes in the reactor, which contains a silicon wafer (substrate) placed on a graphite susceptor. The wafer is heated radiatively by lamp arrays, placed outside the quartz reactor, at a constant temperature (743K is a typical wafer temperature). Finally, the reactant and the product gases leave the reactor through the outlet, which is fixed at atmospheric pressure. The exterior surfaces of the reactor are cooled with air fans.

3 The Mathematical Model

3.1 Transport Equations

The gas mixture in the CVD reactor is treated as an ideal gas and the flow as laminar. The composition of the N-component gas mixture is described in terms of the dimensionless mass fractions ω_i (i=1…N) of its constituents, which sum up to unity. The transport of mass, momentum and heat are described by the continuity equation, the Navier-Stokes equations and the transport equation of thermal energy, expressed in terms of temperature [11]:

Mass balance:

$$\nabla \cdot (\rho \boldsymbol{u}) = 0 \tag{2}$$

Momentum balance:

$$\nabla \cdot (\rho \boldsymbol{u}\boldsymbol{u}) = -\nabla P + \rho \boldsymbol{g} + \nabla \cdot \left[\mu \left(\nabla \boldsymbol{u} + (\nabla \boldsymbol{u})^T \right) - \frac{2}{3} \mu (\nabla \cdot \boldsymbol{u}) I \right] \tag{3}$$

Energy balance:

$$C_p \nabla \cdot (\rho \boldsymbol{u} T) = \nabla \cdot (\lambda \nabla T) - \sum_{i=1}^{N} \boldsymbol{j}_i \cdot \frac{\nabla H_i}{M_i} \tag{4}$$

Species balance:

$$\nabla \cdot (\rho \boldsymbol{u} \omega_i) = -\nabla \cdot \boldsymbol{j}_i \tag{5}$$

The diffusion of gas species in the multicomponent gas mixture is expressed through their diffusive mass fluxes, which sum up to zero. According to the kinetic theory, the driving forces for diffusion in thermal CVD processes are the concentration and the temperature gradients [12]. Consequently, the diffusive mass fluxes are composed of two terms:

$$\boldsymbol{j}_i = \boldsymbol{j}_i^C + \boldsymbol{j}_i^T \tag{6}$$

The first term in Eq. (6) is provided by the Stefan-Maxwell equations, while the second one represents thermal diffusion:

$$\boldsymbol{j}_i^C = -\rho D_i^{eff} \nabla \omega_i - \rho \omega_i D_i^{eff} \nabla (\ln M) + M \omega_i D_i^{eff} \sum_{j=1, j \neq i}^{N} \frac{\boldsymbol{j}_j^C}{M_j D_{ij}} \tag{7}$$

$$\boldsymbol{j}_i^T = -D_i^T \nabla (\ln T) \tag{8}$$

The transport equations are supplemented with the appropriate boundary conditions, which include prescribed velocity, temperature and concentration at each of the five inlet holes and no-slip on walls. The operating pressure is specified at the outlet, where zero normal derivatives are assumed for all other variables. The surface reaction takes place only on the heated substrate. Experimentally measured values of temperature on the substrate and on the reactor walls are used as boundary conditions for the energy equation.

The transport and thermochemical properties of the individual gas species and the binary diffusion coefficients are estimated using the kinetic gas theory. Mixing rules are used for calculating the properties of the gas mixture in terms of pressure, temperature and composition [12,13].

3.2 Kinetic Model

The control of thickness and chemical composition of the deposited film is of crucial importance in all CVD processes. Both of them are strongly affected by the reactions taking place in the gas phase and on the deposition surface as well as by the flow dynamics inside the reactor chamber.

Regarding the tin oxide deposition kinetics, the oxidation reaction in Eq. (1) summarizes much more complex phenomena. To account for the surface reactions, a Langmuir-Hinshelwood type kinetic mechanism was determined [10]. According to the proposed mechanism, the formation of the tin oxide film consists of the two following steps:

$$SnCl_{4(g)} \rightarrow SnCl_{2(a)} + Cl_{2(g)} \tag{9}$$

$$SnCl_{2(a)} + O_{2(g)} \rightarrow SnO_{2(s)} + Cl_{2(g)} \tag{10}$$

At equilibrium, the deposition rate of the SnO_2 film is given by the equation:

$$R_d = \frac{K_a C^s_{SnCl_4} K_d C^s_{O_2}}{K_a C^s_{SnCl_4} + K_d C^s_{O_2}} \tag{11}$$

Coupling the kinetic model of tin oxide deposition with the CFD model requires determining the temperature dependent kinetic coefficients K_a and K_d. Their values are evaluated by a trial and error procedure so that a good agreement between experimental data and simulation results, in terms of average deposition rate, is obtained. The values of the kinetic coefficients K_a and K_d at three different substrate temperatures are shown in Table 2. It is evident that the tin oxide deposition rate is more sensitive to desorption rate coefficient K_d compared to the value of the adsorption rate coefficient K_a.

Table 2. Kinetic coefficients at different substrate temperatures

Temperature [K]	K_a [m/s]	K_d [m/s]
703	1.000E-03	7.994E-06
743	1.000E-03	1.241E-05
763	1.000E-03	1.284E-05

3.3 Layer Thickness Uniformity

The deposition rate R_d (mol/(m^{2}s)) is converted to the growth rate G (Å/min) as

$$G = 6 \cdot 10^{11} R_d \frac{M_{(s)}}{\rho_{(s)}} \tag{12}$$

The layer thickness of the tin oxide film is estimated as growth rate multiplied by the deposition time, which is, according to the experiments, equal to 10min.

The uniformity of the layer thickness is defined through the maximum, minimum and average growth rates i.e., by $(G_{max} - G_{min})/(2 \cdot G_{ave})$.

3.4 Computational Details

The resulting system of coupled nonlinear partial differential equations, along with the boundary conditions are spatially discretized with the finite volume method. The momentum and continuity equations are coupled through the SIMPLEST pressure correction scheme and the upwind differencing scheme is

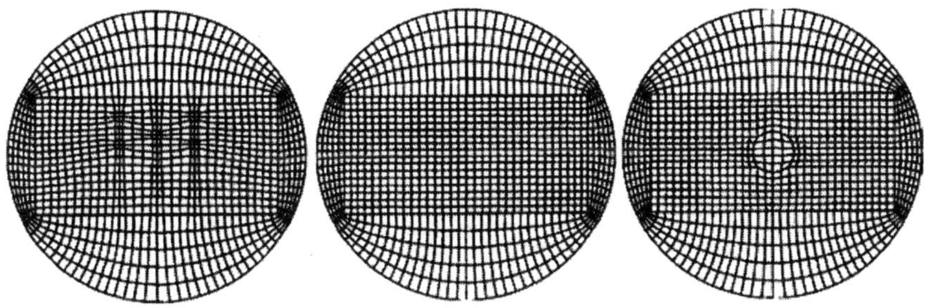

(a) (b) (c)

Fig. 2. Domain cross-sectional discretization at (a) inlet (b) wafer and (c) outlet

used for the convective terms [11,14]. A 3D non-orthogonal, multi-domain grid is used for the reactor geometry discretization, as shown in Fig. 2.

The total number of cells in the domain is equal to 10857, as further refinement gave no substantial change in the computed results. A Pentium IV/2.4 GHz is used for the computations, which consumed approximately 2h CPU time to obtain a single CVD reactor state.

4 Results and Discussion

All simulations are performed by first selecting a set of reference conditions and then by varying the value of each one of the operating parameters. The reference operating conditions selected in our simulation study are summarised in Table 3.

In Figure 3, the species mass fractions across the wafer surface, at the reference operating conditions, are shown. The simulation results predict a relatively high non-uniformity of the distribution of the reaction product. In particular, the Cl_2 mass fraction was found to be smaller in the wafer center compared to its periphery. This result actually represents the distribution of the surface reaction, thus the formation of the tin oxide thin layer.

Table 3. Reference operating conditions considered for tin oxide deposition

Simulation variable	Value
N_2 diluent flow rate [ml/min][a]	400.0
N_2 carrier flow rate[ml/min][a]	200.0
O_2 flow rate [ml/min][a]	150.0
$SnCl_4$ bubbler temperature [K]	297.2
Substrate temperature [K]	743.0
Inlet temperature [K[	323.0
Outlet temperature [K]	378.0
Wall temperature [K]	413.0

[a]Values refer to standard conditions [20°C, 1atm].

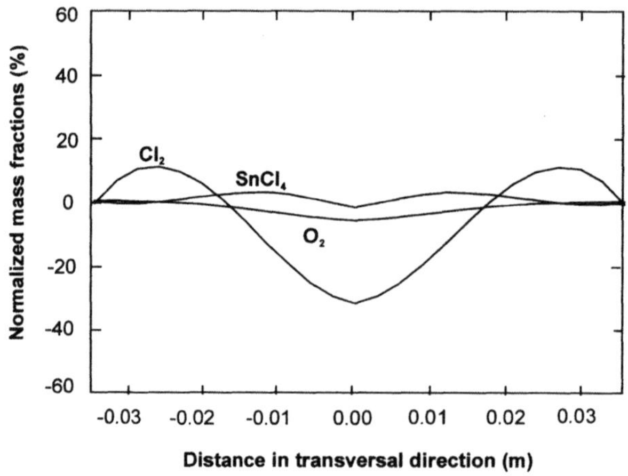

Fig. 3. Normalized species mass fractions across the center line of the wafer

4.1 Influence of the Mixture Composition

At first, the effect of the inlet gas mixture composition on the layer thickness and on its uniformity is investigated.

The simulations are performed at a constant carrier gas flow rate, while the precursor flow rate is changed by adjusting only the bubbler temperature. The results in Fig. 4 indicate that any increase of the $SnCl_4$ mass fraction causes a significant increase of the layer thickness while the thickness uniformity becomes worse.

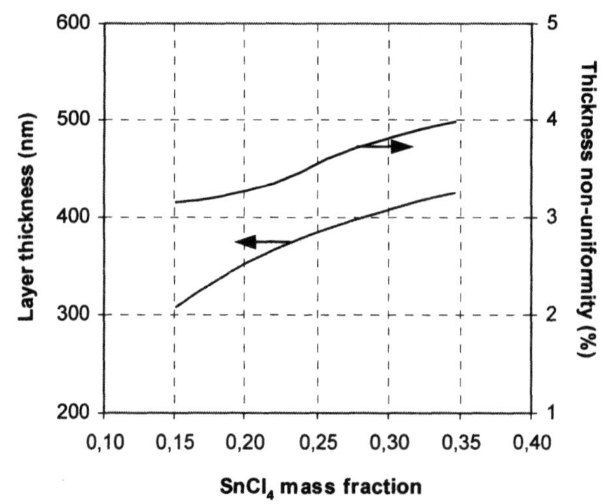

Fig. 4. Layer thickness and thickness non-uniformity as a function of $SnCl_4$ mass fraction

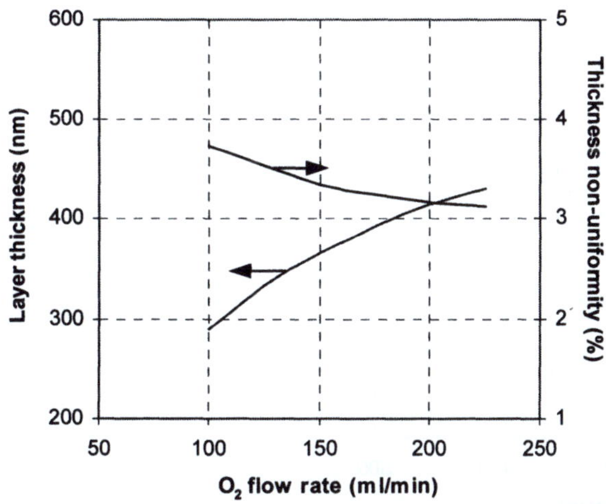

Fig. 5. Layer thickness and thickness non-uniformity as a function of O_2 flow rate

The effect of the O_2 concentration, in terms of inlet flow rate, on the characteristics of the tin oxide layer is shown in Fig. 5. It is evident that the increase of the O_2 flow rate produces a substantial increase of the layer thickness accompanied by a slight improvement of the thickness uniformity. However, the acceptable non-uniformity (i.e. less than 2% across the wafer) is not obtained by any change of the gas mixture composition.

4.2 Influence of the Diluent Gas Flow Rate

In this section, the effect of the diluent gas flow rate (F_d) on the reactor performance is investigated at three different substrate temperatures, where the kinetic coefficients have already been evaluated (see Table 2).

As shown in Fig. 6, the decrease of the diluent gas flow rate, through the central inlet hole, is followed by an increase of the layer thickness and this trend is obtained for the three temperatures examined.

However, the results in Fig. 7 indicate that the effect of the decrease of the diluent gas flow rate on the thickness non-uniformity depends on the substrate temperature. In particular, at the low temperature of 703K, the thickness uniformity is improved with the decrease of the flow rate, F_d, in the range of 200–400ml/min. On the other hand, for the higher temperatures (743 and 763K), the non-uniformity curve meets a minimum at F_d = 250ml/min and then increases as F_d is further decreased.

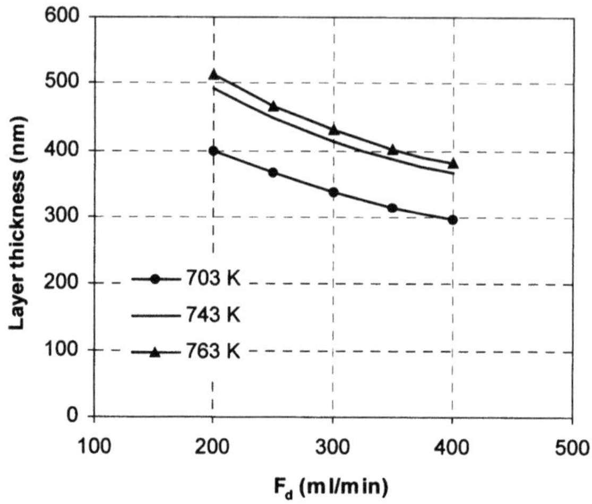

Fig. 6. Layer thickness as a function of the diluent gas flow rate (F_d) at different values of substrate temperature

In order to explain these results, the thickness distribution across the wafer surface is further investigated. Fig. 8a and 8b compare the thickness profile for different values of F_d, at 703K, in the transversal and the axial direction of the wafer, respectively. The same results are presented in Fig. 9a and 9b, for the substrate temperature of 763K. The results at 743K are omitted, as they are very similar to those obtained at 763K.

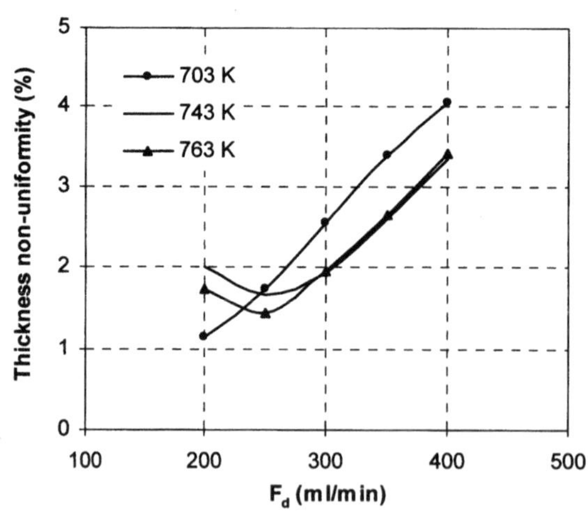

Fig. 7. Thickness non-uniformity as a function of diluent gas flow rate (F_d) at different values of substrate temperature

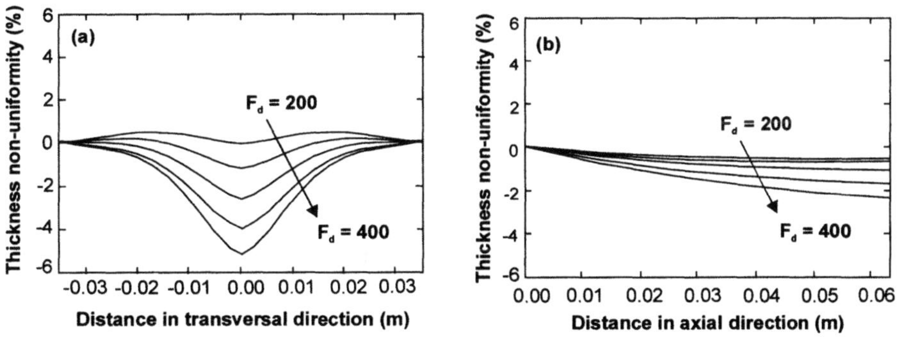

Fig. 8. Layer thickness profile at different values of F_d (interval of 50ml/min) at (a) transversal direction and (b) axial direction. Substrate temperature is 703K

In the transversal direction, the layer thickness is found to be smaller in the wafer center compared to its periphery for almost all the simulation results, as shown in Fig. 8a and 9a.

Although, it is not possible to separate exactly the influence of mass transfer and kinetics on tin oxide deposition, the simulation results in Fig. 8b and 9b clearly suggest that the growth rate in the region around the leading edge of the wafer is more sensitive to the reaction kinetics, while at the trailing edge is more sensitive to the transport properties. This can be explained by the fact that near the leading edge the mass transfer boundary layer is thin, hence the growth rate is governed by the reaction rate. However, at the trailing edge the thickness of the mass transfer boundary layer is significant, hence, the ability to rapidly get the reactants to the wafer surface becomes the controlling factor.

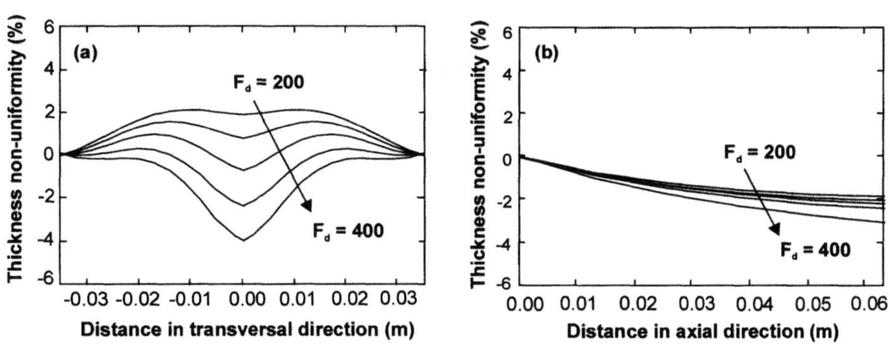

Fig. 9. Layer thickness profile at different values of F_d (interval of 50ml/min) at (a) transversal direction and (b) axial direction. Substrate temperature is 763K

At the lower substrate temperature of 703K (Fig. 8a and 8b), the decrease of the diluent gas flow rate is followed by a significant improvement of the thickness uniformity in both directions.

On the other hand, at the higher temperature of 763K (Fig. 9a and 9b), the substantial improvement of the thickness uniformity in the transversal direction with the decrease of the diluent gas flow rate (Fig. 9a) is responsible for the general trend shown in Fig. 7. However, the thickness profile is changed at F_d=250ml/min and the non-uniformity is increased in the opposite direction at 200ml/min. This observation explains actually the result presented in Fig. 7.

5 Summary

In this paper an integrated method has been developed for the efficient design of CVD reactors. The proposed methodology combines detailed CFD calculations and kinetic coefficients obtained from experimental measurements. This design approach applied to tin oxide deposition by $SnCl_4$ oxidation under atmospheric pressure conditions. From an industrial point of view, the proposed methodology may be useful for solving the non-uniformity problem across the wafer.

The coupled CFD/kinetic simulation model was used for studying the effect of important operating parameters on the reactor performance. The results indicate that the layer thickness and its uniformity across the wafer depend strongly on the gas mixture composition, the substrate temperature and the total inlet flow rate. Specifically, it comes out that by adjusting the gas flow rates through the five inlet holes one could enhance the thickness uniformity of the tin oxide layer.

REFERENCES

1. Pierson HO (1992) Handbook of Chemical Vapor Deposition (CVD): principles, technology and applications. Notes Publications, New Jersey, U.S.A.
2. Komiyama H, Shimogaki Y, Egashira Y (1999) Chemical reaction engineering in the design of CVD Reactors. Chem Eng Sci 54(13-14): 1941-1957
3. Jensen KF (1987) Micro-reaction engineering applications of reaction engineering to processing of electronic and photonic materials. Chem Eng Sci. 42(5): 923-958
4. Kleijn CR, Kuijlaars KJ, Okkerse M, van Santen H, van der Akker HEA (1999) Some recent developments in chemical vapor deposition process and equipment modeling, J Physique IV 9(8): 117-132
5. Xenidou TC, Koukou MK, Boudouvis AG, Markatos NC (2001) Computational analysis of horizontal cold wall CVD reactors at low pressure: application to tungsten deposition from pyrolysis of $W(CO)_6$, J Phys IV 11(3): 183-188
6. Davazoglou D (1997) Optical properties of SnO_2 thin films grown by atmospheric pressure chemical vapor deposition oxidizing $SnCl_4$, Thin Solid Films 302: 204-213

7. Kim KS, Yoon SY, Lee WJ, Kim KH (2001) Surface morphologies and electrical properties of antimony-doped tin oxide films deposited by plasma-enhanced chemical vapor deposition, Surface and Coatings Technology 138: 229-236
8. Koutsogianni A, Tsamakis D (2001) Tin oxide APCVD thin films grown by $SnCl_4$ oxidation on glass and Si substrates in a cold wall reactor, J Phys IV 11(3): 377-383
9. Ghoshtagore RN (1978) Mechanism of CVD thin film SnO_2 formation, J Electrochem Soc: Solid-State Science and Technology 125(1): 110-117
10. Xenidou TC, Diamantis AG, Boudouvis AG, Tsamakis DM, Markatos NC (2003), Tin oxide deposition in a cold-wall CVD reactor: computations and experiments, In Allendorf M, Maury F, Teyssandier F (eds) Chemical Vapor Deposition - XVI and EUROCVD-14, Paris, France, PV 2003-08, pp 202-209
11. Patankar SV (1980) Numerical heat transfer and fluid flow, McGraw-Hill, New York
12. Hirschfelder JO, Curtiss CF, Bird RB (1967), Molecular theory of gases and liquids, John Wiley & Sons, Inc., New York
13. Reid RC, Prausnitz JM, Poling BE (1987), The properties of gases and liquids, 2nd edn, McGraw-Hill, New York
14. Galea ER, Markatos NC (1991) The mathematical modeling and computer simulation of fire development in aircraft, Int J Heat Mass Transfer 34(1): 181-197

Electromagnetically Forced Convection during Solidification of a Binary Metal Alloy

P. NIKRITYUK*, K. ECKERT, R. GRUNDMANN
Institute for Aerospace Engineering
Dresden University of Technology, 01062 Dresden, Germany

*Corresponding author: Tel.: (+351) 463 38092, Fax: (+351) 463 38087
Email: nikrityu@tfd.mw.tu-dresden.de

Abstract The main aim of this work is to study numerically the influence of an external rotating magnetic field on the convection process within liquid binary metal alloy undergoing solidification. Based on the continuum model of two-phase flow the transient transport of momentum, energy and species in presence of a Lorentz force during solidification of Pb25wt%Sn alloy is modeled. The geometry under study is a cylindrical mold with adiabatic walls and cooled bottom. The calculations showed that a swirling flow of the liquid phase (generated by a rotating magnetic field of 1 mT) develops a mushy zone front with convex shape where the maximum is located on the axis of rotation.
Key Words: solidification, stirring, magnetic field.

Nomenclature

$\vec{B}$	magnetic induction, $[T]$
c_p	specific heat at constant pressure, $[J/kg\ K]$
C	concentration, $[wt\%]$
D	mass diffusion coefficient, $[m^2/s]$
F	Lorentz force, $[N/m^3]$
f	mass fraction
h	enthalpy, $[J/kg]$
H	height of cylindrical cavity, $[m]$
R	radius of cylindrical cavity, $[m]$
r, z	cylindrical coordinate axis, $[m]$
$\vec{u}$	velocity, $[m/s]$

L	latent heat, $[J/kg]$
p	pressure, $[Pa]$
Ta	Taylor number
T	temperature, $[K]$
T_e	eutectic temperature, $[K]$
t	time, $[s]$

Greek symbols

λ	thermal conductivity, $[W/m\,K]$
σ	electric conductivity, $[A/V\,m]$
μ	dynamic viscosity, $[N\,s/m^2]$
ρ	density, $[kg/m^3]$
ν	frequency of the electromagnetic field, $[Hz]$
ε	volume fraction of liquid

Subscript

α	constituent α (Sn)
e	eutectic
l	liquid phase
s	solid phase
m	mixture

1. Introduction

Rotating magnetic fields are widely used in metallurgical applications for stirring and mixing liquid metals and electrical conducting liquids in general [1]. Particularly, the stirring, i.e. the generation of swirling flow by a rotating magnetic field, is used during solidification processes to homogenize the liquid phase by modifying the internal convection, namely thermosolutal and shrinkage-driven flow. The homogenization of the liquid zone promotes the isotropy of the solid phase, which leads in turn to isotropic mechanical properties of the solidified material. Certainly the existence of favorable combinations of magnetic induction and cooling rate values for optimal properties of a given alloy can be assumed. To study this question it is necessary to model not only the interaction of liquid alloy phase with magnetic field but also to simulate the solid and the mushy-zone domains.

Mathematical models and numerical procedures to study solidification processes have been improved over the past ten years [2]-[8]. Basically there are two concepts of mathematical models for transport phenomena during solidification. The first concept has been introduced by Beckermann [3],[4] and is called Representative Volume Element (REV). This model uses volume-averaged conservation equations for each phase. The second concept is based on the mixture theory [5]-[8] which describes a two-phase flow as a mixture of solid and liquid phases. In this model microstructure evolution as well as the macroscale transport in solidification of alloys is governed by the progress of the two-phase "mushy-zone". This model is able to predict both, the macrosegregation [9] and the channel formations in the mushy zone [10] in a unidirectional solidification. In this work the second concept is used.

Experimental and theoretical works analyzing the influence of external magnetic fields on the convection of liquid metal without [11]-[15] and with solidification [16]-[20] of binary metal alloy have demonstrated that the motion of the melt depends on the particular type of the magnetic field applied to the molten metal alloy. The influence of a time-harmonic (a.c.) magnetic field on the Pb19%wtSn alloy solidification was studied numerically in a 2D geometry in [19,20]. Particularly in [19] it was shown that turbulence, induced by the electromagnetic stirring, has a significant effect on the solidification.

In this work the convection and the heat transfer phenomena occurring during the solidification of the Pb25%wtSn alloy under influence of a rotating magnetic field are studied.

2. Numerical Formulation

2. 2.1. Assumptions

The continuum model [5] has been adopted for the binary alloy solidification under the following assumptions:

1. All of the properties of the mixture can be obtained from the properties of its components in each phase.
2. All transport properties of each phase, such as thermal and electrical conductivity or viscosity, are constants.
3. The density is constant in each phase and the densities of the two phases are similar.
4. The mushy region is modeled by means of using the mixture viscosity (mushy fluid model [2]).
5. The phases are in local thermodynamic equilibrium. The phase diagram is applied.
6. The flow of the liquid phase is assumed to be laminar.
7. The solid and liquid phase velocities are the same. $\vec{u} = \vec{u}_s$.

2.2. Governing Equations

Based on the assumptions made above, the set of equations has the following form:

Mass Conservation Equation:

$$\frac{\partial \rho_m}{\partial t} + \nabla \cdot (\rho_m \vec{u}) = 0 \cdot \tag{1}$$

Momentum Conservation Equation:

$$\frac{\partial (\rho_m \vec{u})}{\partial t} + \nabla \cdot (\rho_m \vec{u} \cdot \vec{u}) = -\nabla p + \nabla \cdot (\mu_m \nabla \vec{u}) + \vec{F}_L \cdot \tag{2}$$

The projection of the momentum conservation equation in azimuthal direction has the form:

$$\frac{\partial (\rho_m u_\theta)}{\partial t} + \nabla \cdot (\rho_m \vec{u} \cdot u_\theta) = \nabla \cdot (\mu_m \nabla u_\theta) - \frac{\mu_m u_\theta}{r^2} - \frac{\rho u_r u_\theta}{r} + F_L, \tag{3}$$

where the last term in the equations (2) and (3) involves the Lorentz force, which has only the azimuthal component. The mixture viscosity is calculated using the approximation formula taken from [20]:

$$\mu_m = \begin{cases} \varepsilon \mu_l + (1-\varepsilon)\mu_s, \ \varepsilon \le 0.5 \\ k \cdot (1-\varepsilon) + b, \ 0.5 < \varepsilon < 0.6 \ , \\ \mu_l \exp(4.5 \cdot (1-\varepsilon)), \ \ \varepsilon \ge 0.6 \end{cases} \tag{4}$$

where the constants k and b were obtained from a linear interpolation between the points $\varepsilon = 0.5$ and $\varepsilon = 0.6$. Molecular viscosity of the solid phase was set to $100 \ \frac{N\,s}{m^2}$.

Energy Conservation Equation:

$$\frac{\partial (\rho_m h_m)}{\partial t} + \nabla \cdot (\rho_m h_m \cdot \vec{u}) = \nabla \cdot \left(\frac{\lambda_m}{c_{ps}} \nabla h_m \right)$$
$$+ \nabla \cdot \left(\frac{\lambda_m}{c_{ps}} \nabla (h_s - h_m) \right) - \nabla \cdot \rho_m (\vec{u} - \vec{u}_s)(h_l - h_m) \tag{5}$$

Here h_s and h_l refer to the enthalpy of the solid and liquid phase, respectively. The liquid and solid enthalpies are related to the equilibrium temperature through the following thermodynamic relations:

$$h_l = c_{pl} T + (c_{ps} - c_{pl}) T_e + L \ \text{ and } \ h_s = c_{ps} T \, . \tag{6}$$

Species Mass Conservation Equation:

$$\frac{\partial (\rho_m f^\alpha)}{\partial t} + \nabla \cdot (\rho_m f^\alpha \cdot \vec{u}) = \nabla \cdot (\rho_m D_m \nabla f^\alpha) + \nabla \cdot (\rho_m D_m \nabla (f_l^\alpha - f^\alpha))$$
$$- \nabla \cdot \rho_m (\vec{u} - \vec{u}_s)(f_l^\alpha - f^\alpha) \tag{7}$$

Mixture quantities are defined in the following manner:

$$\rho_m = (1-\varepsilon)\rho_s + \varepsilon\rho_l ,$$

$$f_s = (1-\varepsilon)\frac{\rho_s}{\rho_m} , \qquad f_l = \varepsilon \frac{\rho_l}{\rho_m} \tag{8}$$

$$h_m = f_s h_s + f_l h_l , \quad f^\alpha = f_s f_s^\alpha + f_l f_l^\alpha \tag{9}$$

2.3. Lorentz Force

In this study, a rotating magnetic field, which is assumed to be horizontally dominant, with a constant angular frequency $\omega = 2 \cdot \pi \cdot \nu$ is applied to a cylindrical mold with a binary metal alloy. This field induces a time-averaged Lorentz force leading to a rotation of the liquid. In case the rotation velocity of the magnetic field considerably exceeds the induced velocity, i.e. $\frac{U}{\omega \cdot R} << 1$, where U is a characteristic velocity given by $U = B_0 R \sqrt{\sigma\omega / \rho}$, the time-averaged Lorentz force becomes a function of the position and is independent of the velocity of the metal [1]. Under the condition of the infinite cylinder an analytical solution in cylindrical coordinates can be written for the time-averaged Lorentz force [1]:

$$F_{L\infty} = \frac{1}{2}\sigma \cdot \omega \cdot B_0^2 \cdot r . \tag{10}$$

For a finite cylinder there is an analytical solution for the time-averaged azimuthal Lorentz force [14], which has the form:

$$F_L = F_{L\infty} \cdot \left(1 - \frac{2R}{r} \cdot \sum_{k=1}^{\infty} \frac{J_2(\lambda_k) \cdot \lambda_k \cdot J_1\left(\lambda_k \frac{r}{R}\right) \cdot ch\left(\lambda_k \frac{z}{R}\right)}{\left(\lambda_k^2 - 1\right) \cdot J_1^2(\lambda_K) \cdot ch\left(\lambda_k \frac{H}{2R}\right)} \right) , \tag{11}$$

where J_1, J_2 are Bessel functions, λ_k are the roots of $J_1'(x) = 0$. The time-averaged azimuthal Lorentz force calculated by (11) for a cylinder with a height of 63 mm and a radius of 25 mm is shown in **Figure 1**.

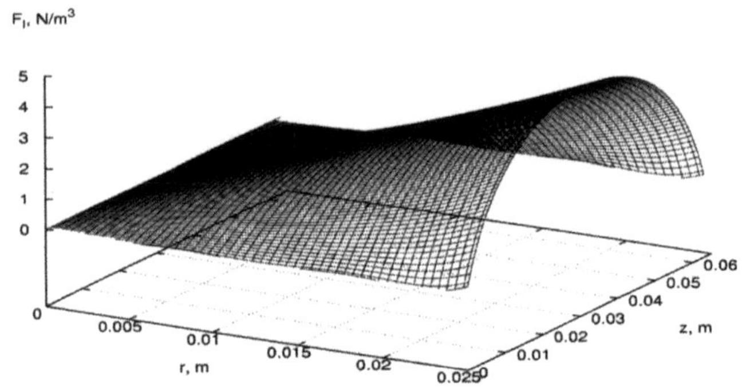

Fig. 1. The time averaged azimuthal Lorentz force induced by rotating magnetic field with $B_0 = 1mT$ for a finite cylinder with a radius of 25 mm and a height of 63 mm.

2.4. Closure of Energy and Species Conservation Equations

The closure of the system of conservation equations requires supplementary relationships between f_s, T and f^{α}. With the assumption of local equilibrium, the expressions required may be obtained from the equilibrium phase diagram [21]. The volume fraction of the liquid ε in the mushy zone is approximated by means of a conduction-dominated solidification rule [22]:

$$\varepsilon = \frac{T - T_s}{T_l - T_s}, \tag{12}$$

where T_s and T_l are the temperatures on the solidus and liquidus line in the phase diagram, respectively. The functional dependencies of T_s and T_l from the mixture concentration have been calculated by means of the least square method:

$$\mathrm{T_l} = \begin{cases} 600.5 - 3.38386 \cdot \mathrm{C} + 0.0479617 \cdot \mathrm{C}^2 - \\ - 0.000500933 \cdot \mathrm{C}^3, \text{if } \mathrm{C} \leq 61.9\,\mathrm{wt\%} \\ \\ 374.001 + 1.34899 \cdot \mathrm{C} - 0.000390725 \cdot \mathrm{C}^2, \\ \text{if } 61.9\,\mathrm{wt\%} < \mathrm{C} \leq 100\,\mathrm{wt\%} \end{cases} \tag{13}$$

$$T_s = \begin{cases} 600.52 - 8.84703 \cdot \mathrm{C} + 0.718111 \cdot \mathrm{C}^2 + \\ + 0.0363395 \cdot \mathrm{C}^3, \text{if } \mathrm{C} \le 18.3 \text{ wt\%} \\ \\ 456 \text{ K}, \quad \text{if } 18.3 \text{ wt\%} < \mathrm{C} \le 100 \text{ wt\%} \end{cases} \tag{14}$$

Where $C = f^{\alpha} \cdot 100$ must be inserted.

In order to define the volume fraction of liquid, the set of equations (6),(8),(9),(12)-(14) is solved iteratively. To avoid oscillations, arising from the volume fraction calculation on the liquid/mush- and solid/mush-boundaries, the following relaxation is used:

$$\varepsilon = (1 - \alpha_{\varepsilon}) \cdot \varepsilon^* + \alpha_{\varepsilon} \cdot \varepsilon \, , \tag{15}$$

where α_{ε} is the relaxation factor which was set equal to 0.5. ε^* is the liquid volume fraction of the previous iteration.

2.5. Calculated Domain and Boundary Conditions.

The equations of conservation of mass, momentum, energy and solute are solved in a two-dimensional domain with boundary conditions shown in **Figure 2**. Following **Figure 2,** AB is the top of the cylinder with a height 63 mm and a radius of 25 mm, BC is the cylindrical wall, CD represents the cooled bottom and AD is the symmetry axis of the cylinder. The time dependent temperature at the bottom is prescribed in the cooling curve and shown in **Figure 3**.

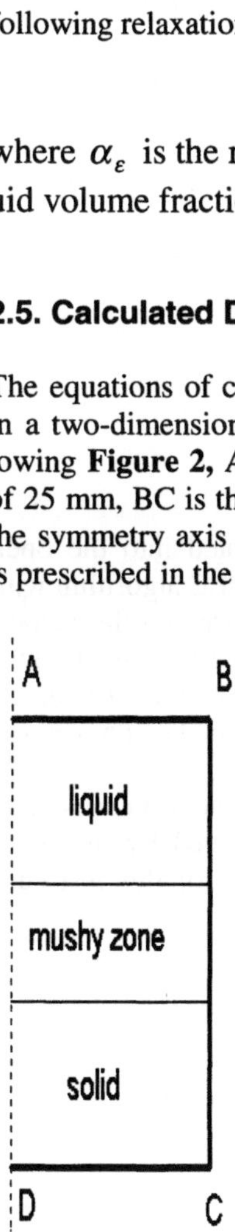

	AB	BC	CD	AD
u_r	$\frac{\partial u_r}{\partial z} = 0$	0	0	0
u_θ	$\frac{\partial u_\theta}{\partial z} = 0$	0	0	0
u_z	0	0	0	$\frac{\partial u_z}{\partial r} = 0$
p'	$\frac{\partial p'}{\partial z} = 0$	$\frac{\partial p'}{\partial r} = 0$	$\frac{\partial p'}{\partial z} = 0$	$\frac{\partial p'}{\partial r} = 0$
h_m	$\frac{\partial T}{\partial z} = 0$	$\frac{\partial T}{\partial r} = 0$	$F(time)$	$\frac{\partial T}{\partial r} = 0$
f^α	$\frac{\partial f^\alpha}{\partial z} = 0$	$\frac{\partial f^\alpha}{\partial r} = 0$	$\frac{\partial f^\alpha}{\partial z} = 0$	$\frac{\partial f^\alpha}{\partial r} = 0$

Fig. 2. Scheme of calculated domain and boundary conditions.

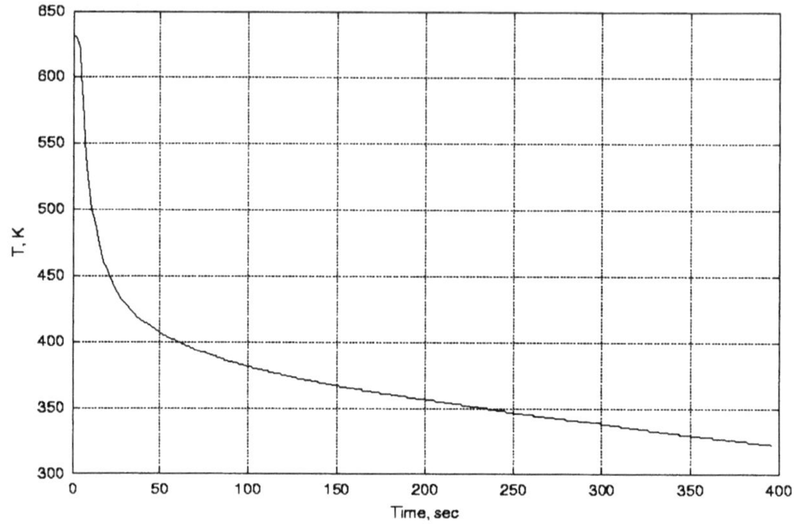

Fig. 3. The time dependence of the bottom temperature.

3. Code Validation

The solidification model described above has been implemented into the open source code of a 2D Navier-Stokes solver [23], where the SIMPLE algorithm with collocated-variables arrangement is used to calculate the pressure and the velocities. The set of equations has been discretized by a finite-volume-finite-difference based method. The system of linear equations has been solved by the SIP solver. The time derivatives are discretized by a three-time-level scheme. The convection terms are discretized by a first order upwind scheme (UDS).

To validate the code a verification test case was performed. The test case corresponds to the laminar steady flow in an infinite cylinder induced by the time-averaged Lorentz force (10). Also an analytical solution exists for this test case [1]:

$$u_\theta = \frac{U^2 \cdot r \cdot \rho}{16 \cdot \mu_l \cdot R^2} \cdot \left(R^2 - r^2\right) \tag{16}$$

The comparison of the calculated azimuthal velocity with the analytically predicted is shown in **Figure 4**. It can be seen that the velocities are identical.

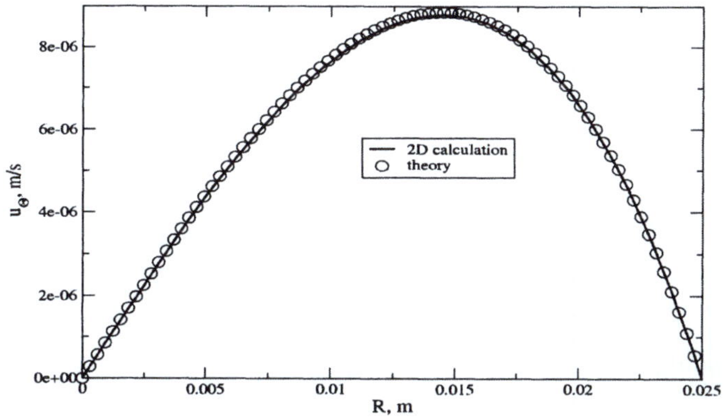

Fig. 4. The radial distribution of azimuthal velocity of a liquid metal induced by RMF with $B_0 = 0.01mT$, $\nu = 50Hz$ for $z = H/2$. The theoretical curve corresponds to the equation (16).

4. Computational Results

The effect of magnetic stirring on Pb25wt%Sn alloy solidification were studied for several magnetic field strengths B_0. In this paper the simulations for B_0 of 0.1 mT and 1 mT are presented corresponding to the Taylor numbers $2.3 \cdot 10^3$ and $2.3 \cdot 10^5$ respectively. The frequency ν was set to 50 Hz and was kept constant. The fixed time step was set to 0.1 sec. The maximal number of outer iterations per each time step was equal to 300. The minimal residual was set to 10^{-4}. The material properties used in simulations were calculated from mixture rule according to the 25 wt% of Sn composition. The transport properties of pure Sn and Pb were taken from [17].

The initialization temperature was set to 630 K. The calculated domain was discretized with a uniform mesh of 130 control volumes (CV) in the vertical z-direction and 50 CV in the radial r-direction.

Figures 5 and **6** show the velocity vectors **(a)** and the isotherms **(b)** in the time 7 sec for a magnetic induction of 0.1 mT and 1 mT, respectively. It can be seen from the isotherms that during 7 sec the temperature did not reach the melting temperature of Pb25wt%Sn. The velocity vectors show that due to rotation of the liquid so called Ekman layers begin to form on the bottom and top of the cylinder. This phenomenon is very well described in the literature [1] and [24].

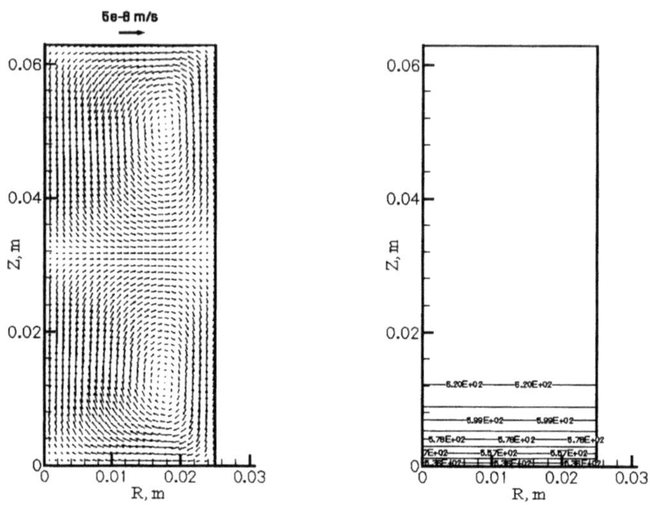

(a) (b)

Fig. 5. Predicted solidification behavior at t=7 sec for magnetic induction $B_0 = 0.1mT$ and frequency $\nu = 50Hz$. a) velocity vectors. b) isotherms

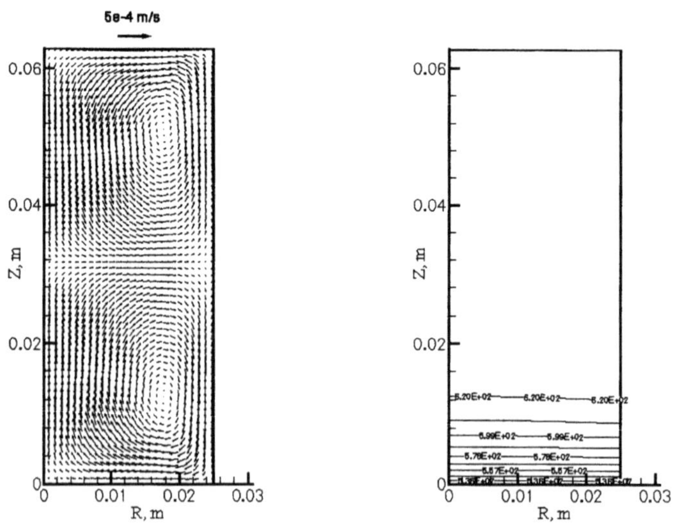

(a) (b)

Fig. 6. Predicted solidification behavior at t=7 sec for magnetic induction $B_0 = 1mT$ and frequency $\nu = 50Hz$. a) velocity vectors. b) isotherms

In **Figures 7** and **8** the velocity vectors, the liquid isovolumes, the isotherms and the azimuthal velocity contours, calculated for magnetic induction of 0.1 mT and 1 mT respectively, are depicted for t=60 sec.

Figure 7 shows that velocities induced by virtue of a magnetic induction of 0.1 mT are too small ($Re << 1$) to affect the mushy zone evolution.

From **Figure 8 (a)** and **(d)** it can be seen that the flow is going to be turbulent with a Reynolds number of the secondary flow about 650. Imposed by the isotherms **(c)** the mushy zone front **(b)** has a convex shape which is the result of a Ekman layer formation in the mushy zone.

The isovolumes of liquid presented in **Figure 8** are in qualitative agreement with results presented in [20] and [25].

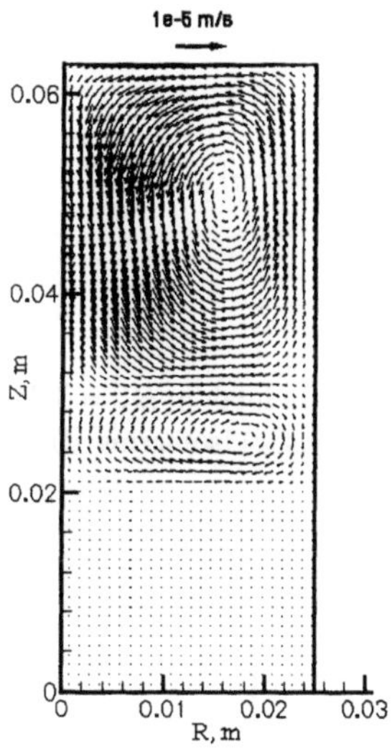

(a)

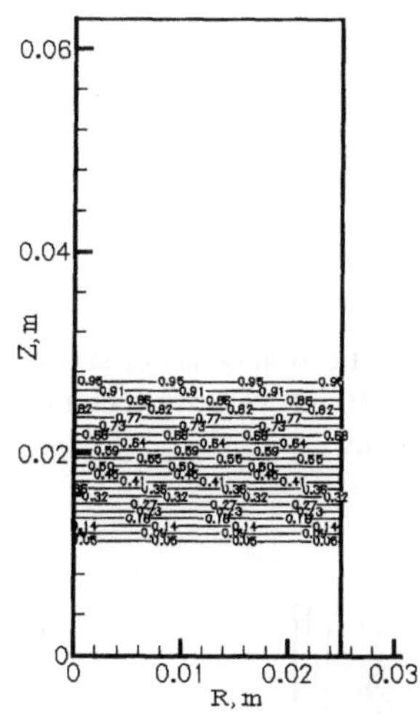

(b)

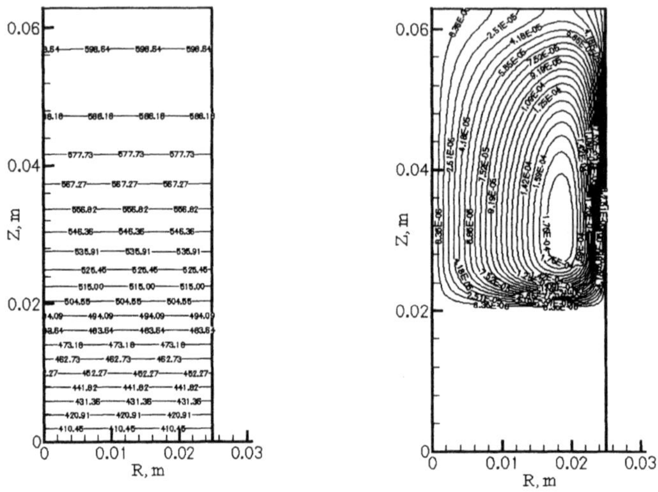

(c) (d)

Fig. 7. Predicted solidification behavior at t=60 sec for magnetic induction $B_0 = 0.1\,mT$ and frequency $\nu = 50Hz$. a) velocity vectors b) liquid isovolumes c) isotherms d) counters of azimuthal velocity

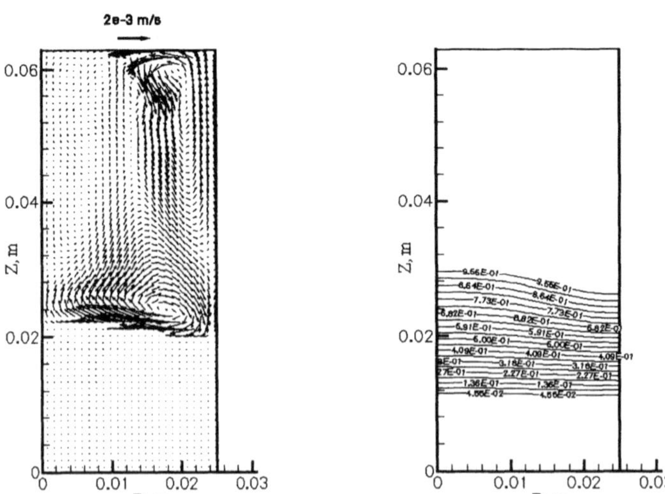

(a) (b)

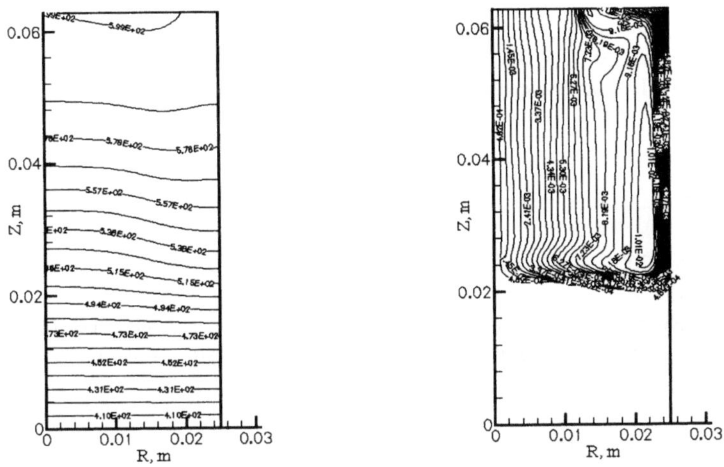

(c) (d)

Fig. 8. Predicted solidification behavior at t=60 sec for magnetic induction $B_0 = 1\,mT$ and frequency $\nu = 50Hz$. a) velocity vectors b) liquid isovolumes c) isotherms d) counters of azimuthal velocity

It should be noted that by modeling of the first 60 sec of solidification no macrosegregations were predicted for both magnetic inductions. Probably it is explained by the high cooling rate that corresponds to the velocity of solidification about 0.2 mm/s or by shrinkage flow effects [26]. A detailed study of PbSn alloy solidification under the influence of rotating magnetic field is in progress [27].

5. Summary

Numerical simulations were performed for the solidification of a Pb25wt%Sn under influence of a rotating magnetic fields with magnetic induction of 0.1 mT and 1 mT. A rotating magnetic field with $B_0 = 1\,\text{mT}$ corresponding to $Ta = 2.3 \cdot 10^5$ produces a turbulent flow in the liquid phase. Furthermore, the liquid-mushy zone and solid-mushy zone interfaces are of convex shape, which is attributed to the formation of Ekman layer in the mush zone.

Acknowledgments

The authors are grateful to Dr. M. Peric for the source code of the 2D Navier-Stokes solver. Financial support by the Deutsche Forschungsgemeinschaft (SFB-609, B2) is gratefully acknowledged.

References

[1] Davidson P.A., 2001. An Introduction to Magnetohydrodynamics. Cambrige University Press. p.431.

[2] Voller V.R., Brent A.D., and Prakash C. 1989. "The modeling of heat, mass and solute transport in solidification systems". Int. J. Heat Mass Transfer. Vol. 32, No. 9, pp. 1719-1731.

[3] Beckermann C., Wang C.Y., 1995. "Multiphase/-scale modeling of alloy solidification". Annual review of Heat Transfer, Vol. 6, pp. 115-198.

[4] Schneider M.C., and Beckermann C. 1995. "A numerical study of the combined effects of microsegregation, mushy zone permeability and flow, caused by volume contraction and thermosolutal convection, on macrosegregation and eutectic formation in binary alloy solidification." Int. J. Heat Mass Transfer. Vol. 38, No. 18, pp. 3455-3473.

[5] Ni J. and Incropera F.P. 1995. "Extension of the continuum model for transport phenomena occurring during metal alloy solidification - I. The conservation equations". Int. J. Heat Mass Transfer. Vol. 38, No. 7, pp. 1271-1284.

[6] Swaminathan C.R., and Voller V.R. 1997. "Towards a general numerical scheme for solidification systems". Int. J. Heat Mass Transfer. Vol. 40, No. 12, pp. 2859-2868.

[7] Swaminathan C.R., and Voller V.R. 1993. "On the enthalpy method". International Journal of Heat and Fluid Flow,3, pp. 233-234.

[8] Bennon W.D., and F.P. Incropera F.P. 1988. "Numerical analysis of binary solid-liquid phase change using a continuum model". Numerical Heat Transfer, 3, pp. 277-296.

[9] Yoo Hoseon and Viskanta R 1992. "Effect of anisotropic permeability on the transport process during solidification of a binary mixture". Int. J. Heat Mass Transfer, Vol. 35, No. 10, pp. 2335-2346

[10] Neilson D.G., and Incropera F.P. 1993. "Three-dimensional considerations of unidirectional solidification in a binary liquid". Numeraical Heat Transfer, Part A, Vol. 23, pp. 1-20.

[11] Moffatt H.K. 1965. "On fluid flow induced by a rotating magnetic field". J. Fluid Mech. 22, pp. 521.

[12] Davidson P.A., Hunt J.C.R. 1986. "Swirling recirculating flow in a liquid-metal column generated by a rotating magnetic field". J. Fluid Mech., Vol. 185, pp. 67-106.

[13] Gelfgat Yu. M., Gorbunov L.A., Kolevzon V. 1993. "Liquid metal flow in a finite-lenght cylinder with a rotating magnetic field". Exp. Fluids. 15, pp.411-416.

[14] Kaiser Th., Benz K.W. 1998. "Taylor vortex instabilities induced by a rotating magnetic field: A numerical approach". Physics of Fluids. Vol. 10, No. 5, pp. 1104-1110.

[15] Moessner R., Gerbeth G. 1999. "Buoyant Melt Flows Under the Influence of Steady and Rotation Magnetic Fields". Journal of Crystal Growth. 197, pp. 341-354.

[16] Presscot P.J. 1992. "Convection Transport Phenomena During Solidification of Binary Metal Alloys and the Effects of Magnetic Fields", Ph.D. Thesis, Purdue University, West Lafayatte, IN, USA.

[17] Prescot P.J., Incropera F.P. 1993. "Magnetically Damped Convection During Solidification of a Binary Metal Alloy". Journal of Heat Transfer, Vol. 115, pp. 302-310.

[18] Prescott P.J., Incropera F.P., Gaskell D.R. 1996. "Influence of electromagnetic stirring on the solidification of a binary metal alloy". Exp. Heat Transfer, Vol. 9, pp. 105-131.

[19] Prescott P.J., Incropera F.P. 1995. "The Effect of Turbulence on Solidification of a Binary Metal Alloy With Electromagnetic Stirring". Journal of Heat Transfer, Vol. 117, pp. 716-724.

[20] Roplekar J.K., and Dantzig J.A. 2001. "A study of solidification with a rotating magnetic field". Int. J. Cast. Metals Research. Vol. 14, No. 2, pp. 79-98.

[21] Karakaya I, Thompson W.T. 1988. Binary Alloy Phase Diagrams. ASM International, Ohio, USA.

[22] Singh A.K., Basu B. 2000. "On Convection in Mushy Phase and Its Effect on Macrosegregation". Metallurgical and Material Transactions A. Volume 31A, pp. 1687-1692.

[23] Ferziger J.H., Peric M. 2002. Computational Methods for Fluid Dynamics. Springer-Verlag. 3rd Edition.

[24] Greenspan. 1968. The Theory of Rotating Fluids. Cambrige University Press. Cambrige.

[25] Neilson D.G., and Incropera F.P. 1993. "Effect of rotation on fluid motin and channel formation during unidirectional solidifcation of a binary alloy". Int. J. Heat Mass Transfer. Vol. 36. No. 2, pp. 489-505.

[26] Chen J.H., Tsai H.L. 1993. "Inverse segregation for a unidirectional solidification of aluminum-cooper alloys". Int. J. Heat Mass Transfer, Vol. 36, pp. 3069-3075.

[27] Nikrityuk P.A., Willers B., Eckert S., Eckert K., Grundmann R. "The influence of rotating magnetic fields on the solidification of PbSn-alloys: experiment vs. theory". In preparation.

Sound Generated by an Unsteady Flow Field, Using a Hybrid Method

Mihai MIHAESCU, László FUCHS
Lund Institute of Technology, Sweden

Abstract The compressible Navier-Stokes equations describe both flow field as well as sound generation. The classical acoustic theory uses the hypothesis that the part of the flow field, which is the source of the acoustic field, has distinct scales so that the acoustic waves do not interfere with the flow. Thus, the velocity vector, the pressure and the density are split into semi-compressible components and an inviscid, irrotational acoustical components. The paper includes the details of the coupling between the turbulent flow (based on Large Eddy Simulation) and the acoustical part (based on Lighthill's acoustic analogy). The approach is applied to the acoustics of a non-isothermal jet.

1. Introduction

The first major step in the development of the acoustics was done by Sir James Lighthill [1, 2]. He formulated and published in 1952 his "acoustic analogy" which represents one of the first theories on aerodynamic noise generation for describing the radiation of the sound field provided by a turbulent flow.

From the fundamental equations of motion namely continuity and momentum for a compressible inviscid fluid, one can obtain the Lighthill's wave equation for the acoustic density (ρ'). Lighthill's equation describes the evolution of the acoustical waves (of pressure perturbation) in the presence of steady or time-dependent source term. In non-dimensional form it can be written as:

$$\frac{\partial^2 \rho'}{\partial t^2} - \frac{1}{M^2}\frac{\partial^2 \rho'}{\partial x_i \partial x_i} = \frac{\partial^2 T_{ij}}{\partial x_i \partial x_j} \tag{1.1}$$

The Lighthill's stress tensor T_{ij} is given by:

$$T_{ij} = \rho u_i u_j + \delta_{ij}\left(p - \frac{1}{M^2}\rho\right) \tag{1.2}$$

where M is the local Mach number and δ_{ij} is Kronecker delta function.

The total acoustic source term T_{total} (the forcing term from right-hand side of Lighthill's wave equation) is decomposed into the flow non-uniformities (such as turbulence) related acoustic source term T_{turb} and entropy acoustic source term T_{entr} [3].

$$T_{total} = T_{turb} + T_{entr} \Leftrightarrow \frac{\partial^2 T_{ij}}{\partial x_i \partial x_j} = \frac{\partial^2}{\partial x_i \partial x_j}\left(\rho u_i u_j\right) + \frac{\partial^2}{\partial x_i \partial x_j}\left[\delta_{ij}\left(p - \frac{1}{M^2}\rho\right)\right] \tag{1.3}$$

With this decomposition one can notice directly the contribution of the velocity fluctuations and compressibility effects to the acoustical source. The acoustical (density or pressure) perturbation can be integrated (in time) on a separate grid independently of the flow calculations. This approach is the basis for the one-way coupling (i.e. the flow field affects the acoustics but not vice versa).

The noise generation is given by the total acoustic source term which can be determined using data provided by solving the semi-compressible (density is temperature dependent only) equations. The simplest method is using a Reynolds Averaged Navier-Stokes approach (RANS) with appropriate turbulence model [3]. The drawback of the RANS approach is limited to predicting the mean of the Reynolds stresses (which is much smoother than the corresponding non-averaged tensor, $u_i u_j$). A more appropriate approach is using Large Eddy Simulation (LES) [5] or Direct Numerical Simulation (DNS) [6-8]. When using LES or DNS, one can use different computational domains and temporal- and spatial- resolution for handling the flow and the acoustics separately. This in turn yields enhanced computational efficiency. In addition, in the LES/DNS framework one may also allow for a two-way coupling between the two systems, which in turn is an efficient method for solving the compressible Navier-Stokes equations at low Mach numbers.

In the present paper, we study the acoustics of a hot circular-jet near the ground. The turbulent flow is handled by LES. The shear layer of the turbulent jet and the mixing of the hot exhaust gas with the cold surrounding air form the acoustical source in Lighthill equation. The one-way coupling approach between the flow solver and the acoustical part is used.

2. Numerical Methods

2.1. The Flow Solver

The flow is considered to be semi-compressible. The governing equations are the equation of continuity, momentum and energy, this later written in terms of temperature (see Eqs. (2.1), (2.2), (2.3)).

$$\frac{\partial \rho}{\partial t} + \frac{\partial}{\partial x_i}\left(\rho u_i\right) = 0 \tag{2.1}$$

$$\frac{\partial}{\partial t}\left(\rho u_i\right) + \frac{\partial}{\partial x_i}\left(\rho u_i u_j\right) = -\frac{\partial p}{\partial x_i} + \frac{1}{\mathrm{Re}}\frac{\partial^2 u_i}{\partial x_j \partial x_j} \tag{2.2}$$

$$\frac{\partial T}{\partial t} + u_i \frac{\partial T}{\partial x_i} = \frac{1}{\mathrm{Pr}\cdot\mathrm{Re}}\frac{\partial^2 T}{\partial x_i \partial x_i} \tag{2.3}$$

A polynomial dependence between density and temperature is used in order to compute the density field.

Turbulence is modelled based on the LES approach. The subgrid scale (SGS) terms are not modelled explicitly. Instead, the dissipation of the numerical scheme amounts for the SGS dissipation [9-10]. This approach is reasonable provided that the spatial resolution is adequate (i.e. at least some portion of the inertial subrange is resolved). The spatial dicretization, on locally refined, Cartesian staggered grids, use third/fourth order finite differences, for the convective- and all the other- terms. The integration in time is done through an implicit finite difference scheme.

The numerical efficiency is enhanced by the use of a Multi-Grid method.

The flow solver provides the necessary data to compute the turbulent related acoustic source term T_{turb} and the entropy acoustic source term T_{entr}. The source terms are discretized using second-order accurate central-difference expressions in space. Since the acoustics are computed on a different (larger but coarser) grid than the LES solver, the acoustical sources are interpolated to the acoustical grid.

2.2. The Acoustical Solver

For Lighthill's wave equation, a finite difference scheme can be derived from a Taylor-series expansion in the following manner:

$$\rho'^{n+1}_{i,j,k} = \rho'^{n}_{i,j,k} + \frac{\partial \rho'^{n}_{i,j,k}}{\partial t}(\Delta t) + \frac{\partial^2 \rho'^{n}_{i,j,k}}{\partial t^2}\frac{(\Delta t)^2}{2!} + \frac{\partial^3 \rho'^{n}_{i,j,k}}{\partial t^3}\frac{(\Delta t)^3}{3!} + \dots \tag{2.4}$$

$$\rho'^{n-1}_{i,j,k} = \rho'^{n}_{i,j,k} - \frac{\partial \rho'^{n}_{i,j,k}}{\partial t}(\Delta t) + \frac{\partial^2 \rho'^{n}_{i,j,k}}{\partial t^2}\frac{(\Delta t)^2}{2!} - \frac{\partial^3 \rho'^{n}_{i,j,k}}{\partial t^3}\frac{(\Delta t)^3}{3!} + \dots \tag{2.5}$$

where Δt represents the time step. Adding these two equations one can obtain an explicit scheme for the computation of acoustic density fluctuation:

$$\rho'^{n+1}_{i,j,k} = 2\rho'^{n}_{i,j,k} - \rho'^{n-1}_{i,j,k} + (\Delta t)^2 \frac{\partial^2 \rho'^{n}_{i,j,k}}{\partial t^2} + \dots \tag{2.6}$$

Eq. (1.1) gives us the second time derivative of the acoustic density fluctuation:

$$\frac{\partial^2 \rho'}{\partial t^2} = \frac{1}{M^2}\frac{\partial^2 \rho'}{\partial x_i \partial x_i} + \frac{\partial^2 T_{ij}}{\partial x_i \partial x_j} \tag{2.7}$$

which replaced in Eq. (2.6) leads to an explicit expression (of Lax-Wendroff type) for the acoustic fluctuation at the new time level:

$$\rho'^{n+1}_{i,j,k} = 2\rho'^{n}_{i,j,k} - \rho'^{n-1}_{i,j,k} + \frac{(\Delta t)^2}{M^2}\frac{\partial^2 \rho'^{n}_{i,j,k}}{\partial x_i \partial x_i} + (\Delta t)^2 \frac{\partial^2 T_{ij}}{\partial x_i \partial x_j} \tag{2.8}$$

Data (the total acoustic source term) is transferred from the flow solver grid to the acoustical grid at each time step. The acoustical field is solved only on its own grid, the acoustic density perturbation being computed independently of the flow calculations. In this way we are able to consider a larger domain and we are able to predict the sound in far field. The decoupling of the flow from the acoustics, eliminates the need for un-necessarily, very small time steps for the (computationally heavy) LES calculations.

3. Boundary Conditions

In order to get correct results from our computations it is necessary to define proper boundary conditions (BC).

3.1. Flow Field Boundary Conditions

The semi-compressible solver requires boundary conditions on all boundary points. For the case of the near ground hot jet, the fluid velocity and temperature at the nozzle are specified. No-slip BC for the velocity is set on solid walls and flux conserving, zero gradient conditions are set at the outlet. Constant temperature condition was taken at the walls.

3.2. Acoustic Field Boundary Conditions

Two types of boundary conditions have been considered for the acoustic field: wall boundary condition (totally reflected wave) and absorbing boundary conditions (non-reflective condition for the acoustic waves). Both of them are sketched in Fig. 3.1.

Wall boundary condition requires a zero projection of the acoustic fluctuation onto the wall normal:

$$\rho' \cdot \vec{n} = 0 \Rightarrow \frac{\partial \rho'}{\partial n} = 0 \tag{3.1}$$

where $\frac{\partial \rho'}{\partial n}$ is the space derivative of the acoustic density fluctuation in the normal direction to the wall n. This type of boundary condition will reflect totally the acoustic waves from the wall into the interior of computational domain.

The absorbing boundary conditions on the other hand do not reflect the acoustic waves. The information reaching the boundary is let out without causing reflection into the computational domain. If we consider the wave equation of planar waves, the sign of the characteristic speed c at the boundary provides us the information regarding direction of propagation. The out going characteristic at the wall satisfies:

$$\frac{\partial \rho'}{\partial t} + c\frac{\partial \rho'}{\partial n} = 0 \tag{3.2}$$

This information is used to set up the non-reflective condition for the acoustic waves. Setting the characteristic propagation speed positive, it will extrapolate the information from interior of the domain outside of it.

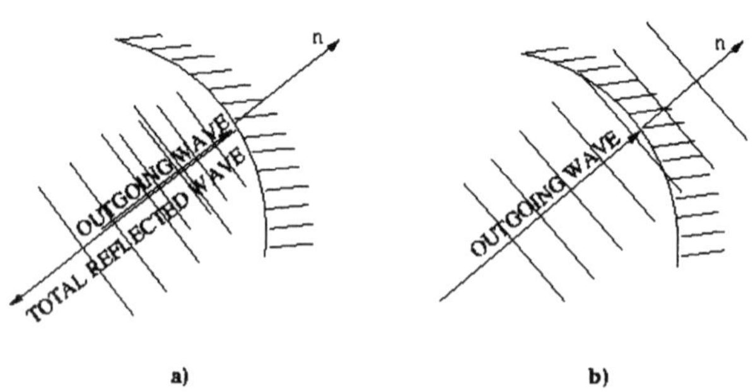

Fig. 3.1. Boundary conditions for acoustic waves; a) total reflected wave; b) non-reflected wave

Using upwind finite difference schemes in both time and space one can obtain from Eq. (3.2) the expression for the acoustic fluctuation at time step *n+1*, on the boundary:

$$\rho'^{n+1}_{i,j,k} = \rho'^{n}_{i,j,k} - c(\Delta t)\frac{\rho'^{n-1}_{i,j,k} - \rho'^{n-1}_{i-1,j,k}}{\Delta x} \tag{3.3}$$

where Δx is the spatial variable in the direction normal to the boundary. This condition gives a complete absorption of the waves that are propagating normally towards a specified boundary.

4. Method Validation

The first step of method validation has been to compute the resonance frequencies for a simple rectangular geometry, using wall boundary conditions (totally reflected wave) on the sides.

In this case, with no acoustical sources, we simply solve the 3D homogeneous wave equation. The initial perturbation function in the acoustical field is given by:

$$\rho'(z,0) = \sin\left[\frac{\pi(z - z_{max}/2)}{z_1 - z_0}\right], if, z_0 \le z \le z_1$$
$$\rho'(z,0) = 0, otherwise \tag{3.4}$$

The perturbation is initialised on half of the domain and the main direction of propagation is considered *z* direction (see Fig. 4.1.).

The theoretical frequency for the propagating planar wave in a rectangular box is given by:

$$f = \frac{c}{2L} \tag{3.5}$$

where *c* is the wave propagation speed and *L* is the length of box assuming closed ends. The theoretical primary frequency in the present case is: *f=2.5 Hz*.

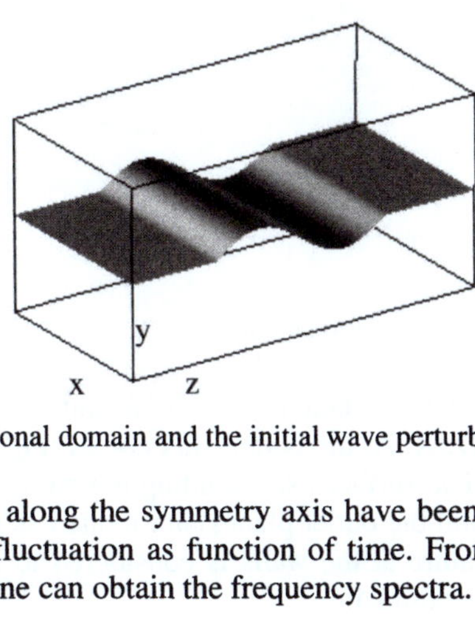

Fig. 4.1. The computational domain and the initial wave perturbation; [42 x 42 x 82] cells

Monitoring points along the symmetry axis have been used to study the evolution of the acoustic fluctuation as function of time. From the time history of the acoustic fluctuation one can obtain the frequency spectra.

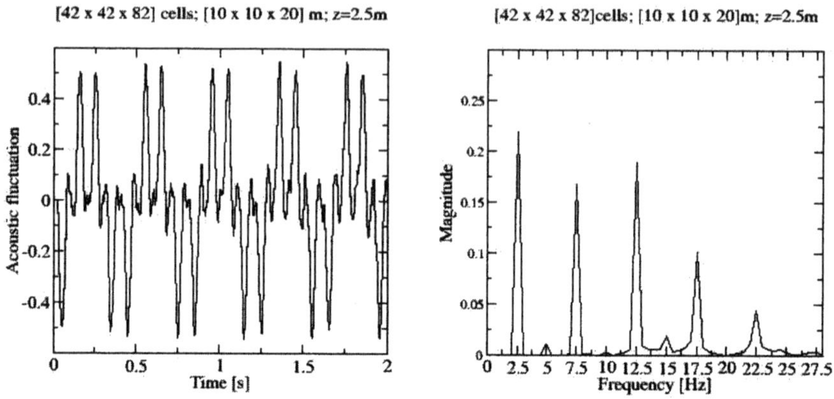

Fig. 4.2. Time evolution of the acoustic fluctuation (left) and its frequency spectrum at the point $z=L/8$ (right)

As showed in Fig. 4.2, the theoretical frequency computed for the rectangular box, is well captured in these computations.

5. Results and Discussions

Simulations were performed for a hot turbulent axisymmetric air jet (*Re=10000*) discharging from a circular pipe into a wide channel with cold air. The geometry of the computational domain is presented in Fig. 5.1.

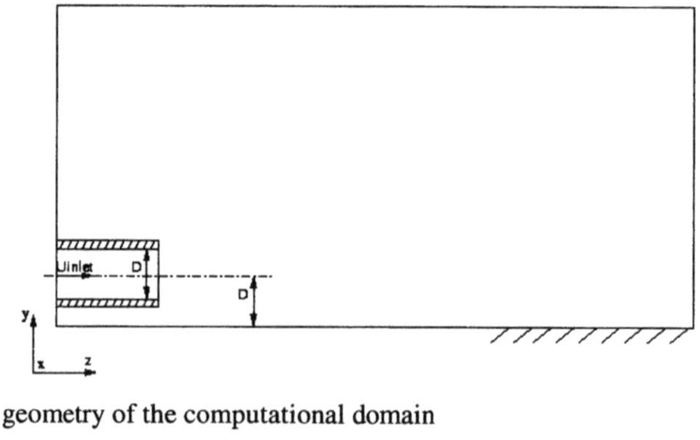

Fig. 5.1. The geometry of the computational domain

The temperature of the jet was taken to be *500K* and the initial temperature in the flow field was set at *273K*. At the outlet, flux conserving zero gradient boundary condition is applied. The surfaces surrounding the computational domain (exception making the inlet and the outlet planes) are treated as non-slip walls for the velocities and the temperature at the walls is considered to be constant, equal to the initial temperature in the domain.

For the calculation of the acoustic field wall boundary condition (total reflected wave) has been used on the bottom surface ("ground level") and absorbing boundary conditions (non-reflective condition for the acoustic waves) on the other boundaries.

The LES of the jet flow provides the required data about the acoustical sources. Figures 5.2 and 5.3 show the distribution of the instantaneous turbulent acoustic source term and the entropy acoustic source term, respectively (Eq. (1.3)).

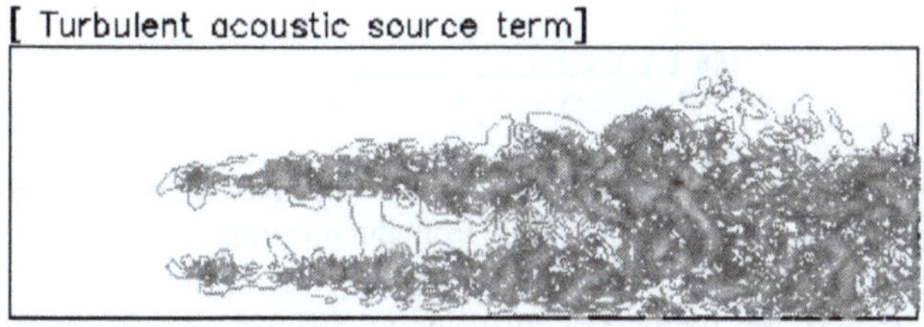

Fig. 5.2. Contour plots of the turbulent acoustic source term; (jet mid-plane)

These source terms are plotted in mid-plane (y-z plane) of the channel containing the symmetry axis of the jet. Dominant sound sources are found as expected in the jet shear layer and in the region near the end of the jet potential core. The location of these maxima near the end of the potential core region correlates very well with the location of maximum sound pressure level as shown in the following.

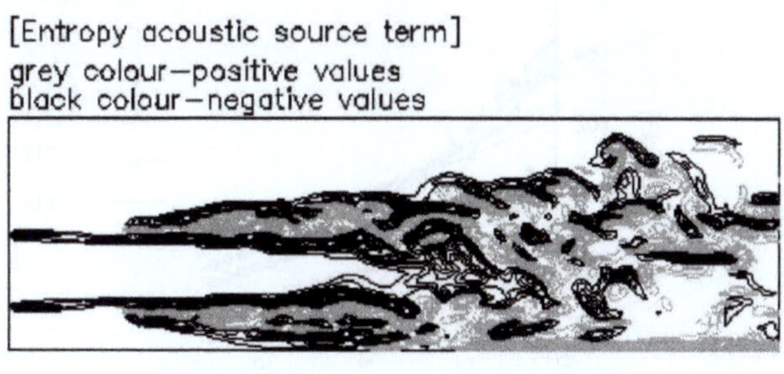

Fig. 5.3. Contour plots of the entropy acoustic source term; (jet mid-plane)

The entropy term shows negative and positive values in the regions with high temperature gradients (the edges of the potential core).

The decibel scale for pressure amplitude is the Sound Pressure Level (SPL). The SPL is measured in decibels with zero decibel being the threshold of human hearing (the pressure amplitude for the threshold of hearing is $P_{ref} = 2 \times 10^{-5}$ N/m^2). The decibel SPL value for a sound with acoustic pressure fluctuation P', is given by the relation:

$$SPL = 20 \cdot \log_{10}\left(\frac{P'}{P_{ref}}\right) \tag{5.1}$$

The SPL values and distribution yield information related to the sound production, its magnitude and spreading.

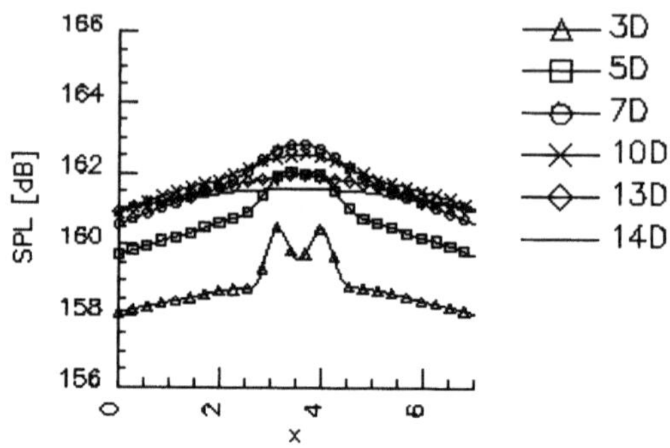

Fig. 5.4. Sound Pressure Level Profiles along radial lines parallel with the ground level, at different axial locations (*3, 5, 7, 10, 13, 14 D* from the nozzle)

Figure 5.4 depicts the mean SPL profiles at different axial distances from the nozzle along radial lines parallel with the bottom surface of the geometry-"ground level" (x-z plane). Near the inlet one finds low values for the sound pressure level. Largest radial values are found in the shear layer and axially downstream of the end of the potential core (roughly *8D* distance from the nozzle). However, the amplitude of sound decreases as the jet expands and the level of production decreases.

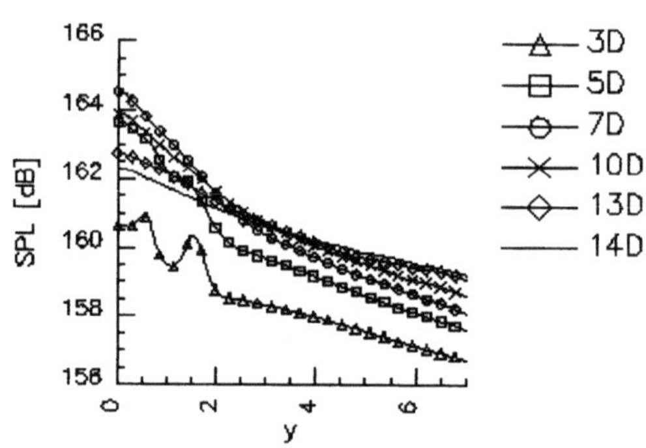

Fig. 5.5. Sound Pressure Level Profiles along radial lines normal to the ground level, at different axial locations (*3, 5, 7, 10, 13, 14 D* from the nozzle)

In order to see the influence of the bottom surface (treated as wall) over the sound pressure level, the SPL has been calculated at different downstream locations on radial lines normal to the "ground level" (mid-plane), as it can be seen in Fig. 5.5.

The largest values are found near the ground, where high values for the acoustic source terms are observed (Figs. 5.2 and 5.3), due to large gradients in velocity and density. As the jet expands and the wall shear layer weakens, the acoustical source weakens too and SPL decreases. As we are travelling downstream after the end of the potential core the value of the sound pressure level near the ground starts to decrease.

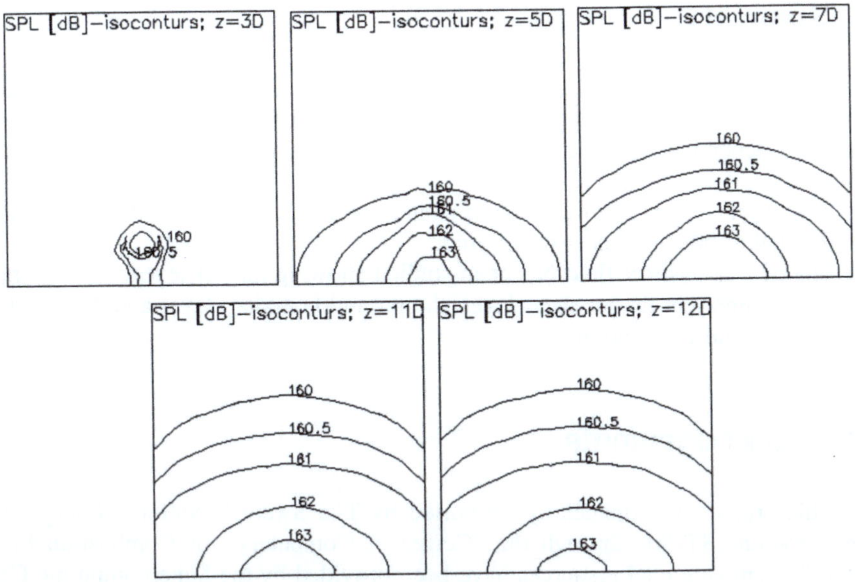

Fig. 5.6. Cross sections (x-y plane) with isocontours (*160, 160.5, 161, 162 and 163*) of mean Sound Pressure Level [dB], at different downstream locations

Figure 5.6 shows isocontours of mean SPL in decibels. The isocontours are presented in cross sections at different downstream locations, namely *3, 5, 7, 11, 12* diameters from the inlet plane.

These plots show very well the extent of each "SPL isocontour bubble" in space. It is evident that they surround the potential core of the jet.

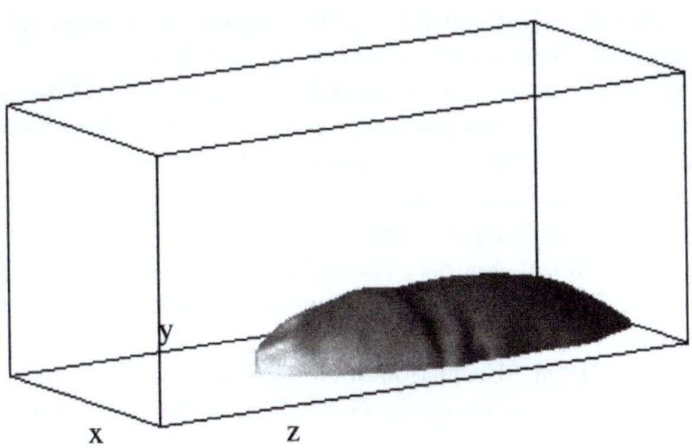

Fig. 5.7. Sound Pressure Level, isosurface *SPL=162 dB*

A 3-D picture which shows the *162 dB* isosurface for sound pressure level is presented in Fig. 5.7.

6. Summary

A method for computing sound wave production and propagation in low Mach numbers has been developed. The method is based on Lighthill's acoustic analogy, the time-dependent acoustic source terms being computed from Large Eddy Simulation data.

The approach has been applied to simulating the acoustical field due to a non-isothermal turbulent jet near the ground.

The near acoustical field has been studied showing largest levels of noise production, namely in the jet shear layer, the ground boundary layer and downstream of the potential core region.

Acknowledgements

This project was financially supported by The Swedish National Energy Administration (STEM) through the "Center of Competence of Combustion Processes". Computational resources have been provided by the Supercomputing Center at Lund University (LUNARC). All these contributions are highly appreciated.

References

[1] Lighthill, M.J., 1952, "On Sound Generated Aerodynamically: I. General Theory", Proceedings of the Royal Society of London, Series A, Vol. 211, pp. 564-587.

[2] Lighthill, M.J., 1954, "On Sound Generated Aerodynamically: II. Turbulence as a Source of Sound", Proceedings of the Royal Society of London, Series A, Vol. 222, pp. 1-32.

[3] Heron, N., Bailly, C., and Candel, S., 2001, "Supersonic Coaxial Jets Noise Prediction", C. R. Acad. Sci. Paris, t. 329, serie II b, pp. 497-502.

[4] Tanahashi, M., et al., 2002, "On the Sound Generation and its Controls in Turbulent Combustion Field", The 3rd Symposium on Smart Control of Turbulence, Tokyo, 2002.

[5] Bogey, C., and Bailly, C., 2002, "Downstream Subsonic Jet Noise: Link with Vortical Structures Intruding into the Jet Core", C. R. Mecanique, Vol. 330 (3), pp. 527-533.

[6] Lilley, G.M., 1996, "The Radiated Noise from Isotropic Turbulence with Applications to the Theory of Jet Noise", Journal of Sound and Vibration, Vol. 190 (3), pp. 463-476.

[7] Colonius, T., and Freund, J. B., 2000, "Application of Lighthill's Equation to Mach 1.92 Turbulent Jet", AIAA Journal, Vol. 38 (2), pp. 368-370.

[8] Colonius, T., et al., 1998, "Evaluation of Noise Radiation Mechanisms in a Turbulent Jet", Center for Turbulence Research, Proceedings of the Summer Program, pp. 159-167.

[9] Olsson, M., and Fuchs, L., 1996, "Large Eddy Simulation of the Proximal Region of a Spatially Developing Circular Jet", Phys. Fluids, Vol. 8, pp. 2125-2137.

[10] Olsson, M., and Fuchs, L., 1998, "Large Eddy Simulation of a Forced Semi-Confined Circular Impinging Jet", Phys. Fluids, Vol. 10, pp. 476-486.

Influence of Cross-Sectional Configuration on Kármán Vortex Excitation

Mizuyasu KOIDE
Life Engineering Research Center, Niigata Sangyo University
4730 Karuigawa, Kashiwazaki, Niigata, 945-1393
Japan

Tsutomu TAKAHASHI, Masataka SHIRAKASHI
Department of Mechanical Engineering, Nagaoka University of Technology
1603-1 Kamitomioka, Nagaoka, Niigata, 940-2188
Japan

Abstract Three cylindrical bodies with different cross sectional configurations, i.e. a circular, semi-circular and triangular cylinder, are used as the test cylinders, in order to investigate the influence of movement of separation point on the Kármán vortex excitation. The cylinders were supported elastically by plate springs. The synchronization of Kármán vortex shedding occurs on all three cylinders over almost equal ranges of oscillation amplitude and frequency given by the mechanical oscillator. However, the Kármán vortex excitation behavior differs drastically among the three cylinders in spite that the cylinders are supported elastically with virtually equal structure parameters.

Nomenclature

A	[m]	Oscillation amplitude of the cylinder, $=\sqrt{2}Z_{rms}$
C_{L0}	[-]	Lift coefficient of the stationary cylinder
C_{LR}	[-]	Lift coefficient of the oscillating cylinder due to the Kármán vortex excitation
d	[m]	Characteristic length, i.e. the cylinder height, see Table 1
F_L	[N]	Lift force
f_c	[Hz]	Oscillation frequency of the cylinder
f_v	[Hz]	Kármán vortex shedding frequency
f_{v0}	[Hz]	Kármán vortex shedding frequency for the stationary cylinder
m_e	[kg]	Effective mass
Re	[-]	Reynolds number, $= Ud/\nu$

U	[m/s]	Free stream velocity
U_0	[m/s]	Free stream velocity at which f_{v0} is equal to f_c
U_{Amax}	[m/s]	Free stream velocity at which A due to Kármán vortex excitation is the maximum
u	[m/s]	Velocity fluctuation in the near wake of the cylinder
Z	[m]	Cylinder displacement
δ	[-]	Logarithmic damping factor
ϕ_{uZ}	[rad]	Phase difference between u and Z
ρ	[kg/m^3]	Density of air
σ_ϕ	[rad]	Standard deviation of ϕ_{uZ}

1. Introduction

The Kármán vortex excitation is an oscillation caused by the hydrodynamic force which originates from periodic Kármán vortex formation and shedding, and it has basically the nature of resonance of linear system. That is, the cylinder is oscillated by the alternating force due to Kármán vortex shedding and the amplitude is large when the vortex shedding frequency f_v is equal to the natural frequency of the structure f_n. The cylinder motion affects the flow around it and then the magnitude and frequency of exciting force in turn.

Therefore, the Kármán vortex excitation is regarded as a non-linear self-excitation caused by the feedback loop which is constituted by two interactions, i.e. the effect of cylinder motion on the fluid force due to the periodic vortex shedding and the response of the structure to the resulting fluid force. Hence, its mechanism can be discussed by separating the process into the two interactions.

Funakawa [1] showed that the separation point on an oscillating circular cylinder moves forward and backward, synchronizing with the cross-flow oscillation. Based on this observation he suggested a self-excitation mechanism for the Kármán vortex excitation.

In this work, influence of the cross-sectional configuration of cylindrical body on the Kármán vortex excitation was investigated experimentally by using a circular cylinder, a semi-circular cylinder and a triangular cylinder, in order to investigate the role of separation point movement in the mechanism of the Kármán vortex excitation.

2. Experimental apparatus and measurements

Table 1 shows cross-sectional configurations of cylinders used in this investigation. They were set in a uniform flow from left to right (in the x-direction) in a wind tunnel. When the cylinders are set in a uniform flow, the separation point movement is free on the circular cylinder surface, restricted to the upstream circular arc of the semi-circular cylinder and fixed at the downstream vertices of the triangular cylinder [6].

Measurements were carried out in a blow-down type wind tunnel with a measuring section of 320(H) x 320(W) x 1000(L) mm. The turbulence level at the center of the measuring section is less than 0.6 % at $U = 2.0$ m/s. A ring type vortex anemometer [5] was applied to measure U.

Figure 1 shows the experimental apparatus arrangement for the measuring section. The test cylinder was set horizontally and perpendicularly to the free stream. The height d of each cylinder was 26 mm and the blocking ratio was about 8 %. The cylinder was passing through slots on the side walls and supported at both ends outside the measuring section, as shown in Fig. 1. End plates were attached to the cylinder to remove influence of flow through the slots [7].

The elastically supported system is composed of two cantilever plate springs. Since the length of cantilever plate is much longer than the oscillation amplitude A, the cylinder motion is regarded as a translational oscillation in the vertical direction (z-direction), i.e. a cross-flow oscillation. Two laser displacement meters were used to measure cylinder displacement Z at both ends of the cylinder, as shown in Fig. 1. The oscillation amplitude A was calculated from root-mean square (rms) value of Z, i.e. $A = \sqrt{2}Z_{rms}$.

The natural frequency f_n, effective mass m_e and logarithmic damping factor δ of the system were determined by free damping oscillation in otherwise quiescent air. The values of f_n, m_e and δ were set virtually equal for all the three cylinders.

Table 1. Cross-sectional configuration and effective mass m_e of the elastically supported cylinders ($d = 26\times10^{-3}$ m)

	Circular cylinder	Semi-circular cylinder	Triangular cylinder
Effective Mass [kg]	53.7×10^{-3}	50.1×10^{-3}	50.7×10^{-3}
Cross-sectional configuration	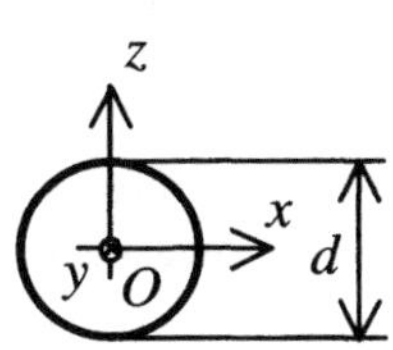	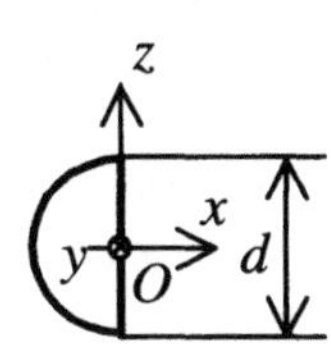	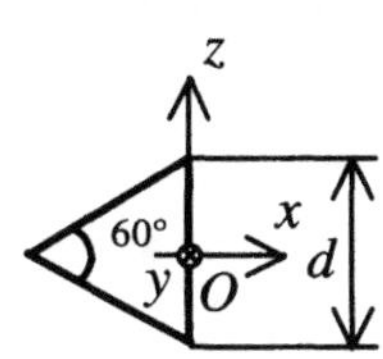

Flow direction: from left to right

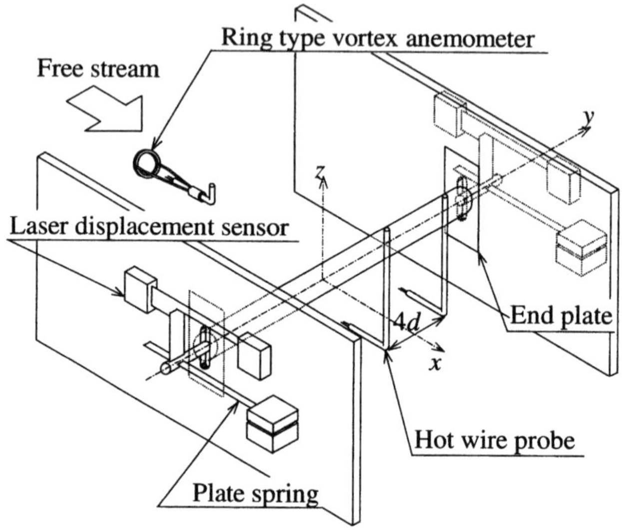

Fig. 1. Arrangement of the experimental apparatus (elastically-supported system) and the coordinate system

The vortex shedding frequency f_v was obtained by applying FFT analysis to the streamwise fluctuating velocity u detected by a hot wire probe at a location in the near wake of the cylinder ($x = 2d$, $z = d$). The spectrum of u was averaged over 20 data to reduce the effect of turbulence. The vortex shedding frequency f_v was determined as the frequency at which the spectrum of u had the maximum peak. Furthermore, the phase difference between u and Z was calculated and its average value ϕ_{uZ} was obtained form 20 data. The standard deviation of ϕ_{uZ}, σ_ϕ, was also obtained. The cross correlation coefficient R_{uu} was obtained from two u signals detected by two hot wire probes set with $4d$ spanwise separation. The cross correlation coefficient R_{uu} increases due to Kármán vortex excitation since it shows an enhancement of spanwise coherency of the Kármán vortex [6]. In this paper, the increase of R_{uu} is adopted as one of criteria to decide whether Kármán vortex excitation occurs or not.

The lift force F_L exerting on the stationary cylinder was measured by load transducers at both ends of the cylinder outside the measuring section.

3. Results and discussion

3.1. Behavior of the cylinder oscillation and vortex shedding in Kármán vortex excitation

Figures 2, 3 and 4 show behavior of the cylinder oscillation and vortex shedding against the free stream velocity U for the circular cylinder, the semi-circular cylinder and the triangular cylinder, respectively. In one run of the wind tunnel experiment, U was first increased stepwise from the lowest value of around 1.5 m/s to the velocity beyond the Kármán vortex excitation range, and then decreased again to the lowest velocity. The closed symbols are for increasing U and the open symbols for decreasing U respectively.

The vortex shedding frequency f_v and the non-dimensional oscillation amplitude A/d are plotted in Figs. 2(a), 3(a) and 4(a). The Kármán vortex shedding frequency from the stationary cylinder f_{v0} is also plotted to compare with f_v. In these figures, U_0 is the velocity at which f_{v0} coincides with the natural frequency of the system f_n, and U_{Amax} indicates the velocity at which A/d is maximum. Figs. 2(b), 3(b) and 4(b) show the phase difference ϕ_{uZ} between velocity fluctuation u and the cylinder displacement Z and its standard deviation σ_ϕ. The shaded region is the Kármán vortex excitation region when U is increased.

3.1.1. Circular cylinder

Figure 2 shows the results of the circular cylinder. As shown in Fig. 2(a), the oscillation amplitude A/d is large and the vortex shedding frequency f_v coincides with f_n over a certain velocity range including U_0, showing that the Kármán vortex excitation occurs there. In the case of circular cylinder, the velocity U_0 is very close to the lower velocity edge of the Kármán vortex excitation range. In contrast, the synchronization of Kármán vortex shedding was observed at both sides of U_0 when the circular cylinder was oscillated with a certain amplitude and frequency using a mechanical oscillator [6].

The amplitude of the circular cylinder due to the Kármán vortex excitation has maximum value of $A/d = 0.4$ at U_{Amax} which is considerably larger than U_0. In a linear oscillation model, the oscillation amplitude reaches its maximum when the frequency of excitation force coincides with the natural frequency, which means that A/d would be maximum at U_0 in the present experiment. From discrepancy between U_0 and U_{Amax}, it is inferred that the alternating lift force F_L exerting on the oscillating cylinder is larger than that on the stationary cylinder and enhanced with the oscillation amplitude due to influence of the cylinder oscillation on the Kármán vortex shedding.

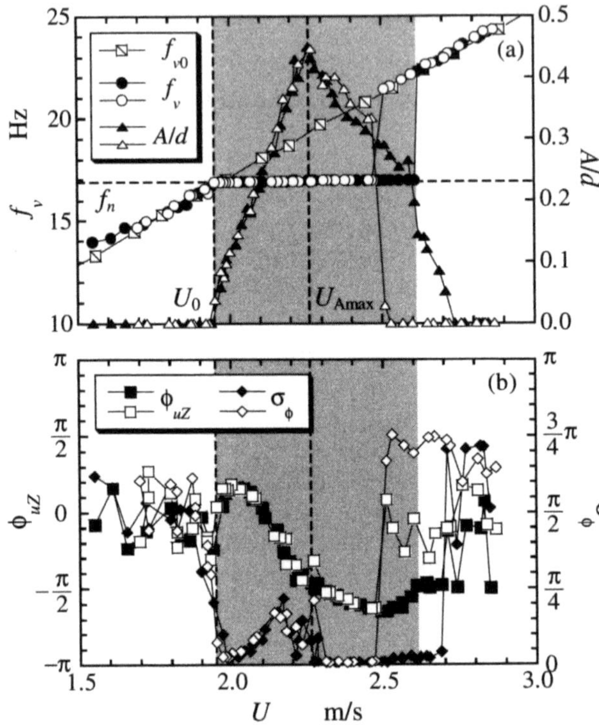

Fig. 2. Vortex shedding frequency f_v, A/d, ϕ_{uZ} and σ_ϕ versus U for the circular cylinder. $f_n = 16.93$ Hz, $\delta = 0.00958$. (Closed symbols: U increased, open symbols: U decreased)

A remarkable hysteresis is observed near the higher velocity edge of the excitation range, where both A/d and f_v at a same value of U differs depending on whether U is increased or decreased. When the circular cylinder was oscillated using a mechanical oscillator, such hysteresis was not observed. The phase difference ϕ_{uZ} changes gradually in the excitation range around U_{Amax}, and the total amount of the change is approximately π, as seen in Fig. 2(b). This shows that the timing of the vortex shedding becomes anti-phase against the cylinder oscillation. The standard deviation σ_ϕ represents the extent of disturbance in ϕ_{uZ}, and it is an index of coherency between the cylinder oscillation and the vortex shedding. In Fig. 2(b), ϕ_{uZ} varies continuously in the range $U_0 < U < U_{Amax}$, and σ_ϕ considerably increases at the same time. This result shows that the coherency of vortex shedding with the cylinder oscillation becomes lower with U in this range. In contrast, ϕ_{uZ} is nearly constant and σ_ϕ is almost suppressed in the range of $U > U_{Amax}$ in the excitation range, showing that the vortex shedding completely synchronizes with the cylinder oscillation.

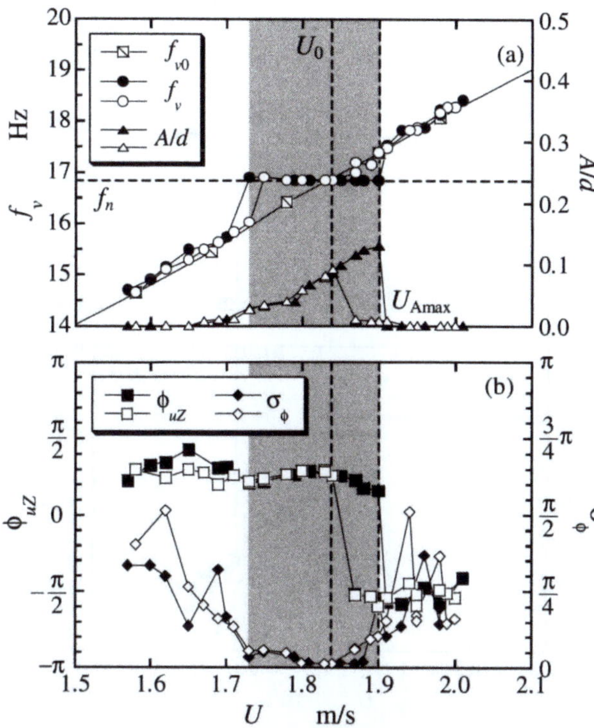

Fig. 3. Vortex shedding frequency f_v, A/d, ϕ_{uZ} and σ_ϕ versus U for the semi-circular cylinder. f_n = 16.86 Hz, δ = 0.0116. (Closed symbols: U increased, open symbols: U decreased)

3.1.2. Semi-circular cylinder

The Kármán vortex excitation occurs over a considerable velocity range in the case of the semi-circular cylinder as seen in Fig. 3. However, the maximum amplitude for the semi-circular cylinder is 1/3 of that for the circular cylinder, in spite of the fact that the height d, the effective mass m_e, the logarithmic damping factor δ and velocity U are almost the same in the two cases. When U increases, the nondimensional amplitude A/d continues increasing in the excitation range, and the oscillation suddenly ceases soon after A/d reaches its maximum. The excitation range of the semi-circular cylinder lies on both sides of U_0, while that of the circular cylinder exists only on $U \geq U_0$.

A definite hysteresis is observed for the semi-circular cylinder in f_v and $A/d \sim U$ curves at higher velocity edge of the excitation range, and the repeatability of the hysteresis behavior is much higher for the semi-circular cylinder than for the circular cylinder.

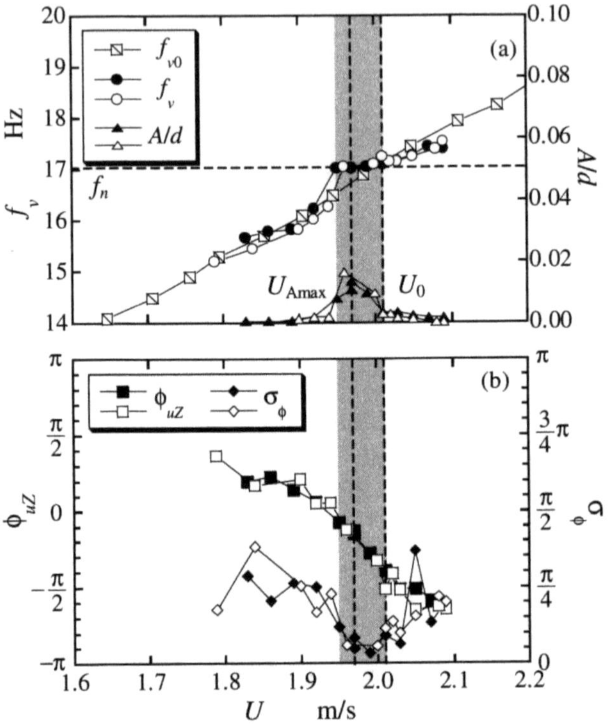

Fig. 4. Vortex shedding frequency f_v, A/d, ϕ_{uZ} and σ_ϕ versus U for the triangular cylinder. $f_n = 17.03$ Hz, $\delta = 0.0141$. (Closed symbols: U increased, open symbols: U decreased)

In the excitation range, ϕ_{uZ} is virtually constant and σ_ϕ is very small, showing that the vortex shedding is well synchronizes with the cylinder oscillation. A sudden change of ϕ_{uZ} is observed simultaneously with the cessation of the oscillation at the higher velocity edge. However, it cannot be concluded whether this change of ϕ_{uZ} is caused by the same phenomenon that caused the change of ϕ_{uZ} in the excitation range of the circular cylinder.

3.1.3. Triangular cylinder

Although experimental conditions for the triangular cylinder are the same as those of the circular cylinder and the semi-circular cylinder, the maximum value of A/d is 1/30 of the circular cylinder as shown in Fig. 4(a). The vortex excitation is observed over a narrow velocity range. It seems that the vortex excitation range lies on $U \leq U_0$ in contrast to the cases of the circular and the semi-circular cylinders. The hysteresis is not discerned clearly, maybe because A/d is small and the vortex excitation range is narrow.

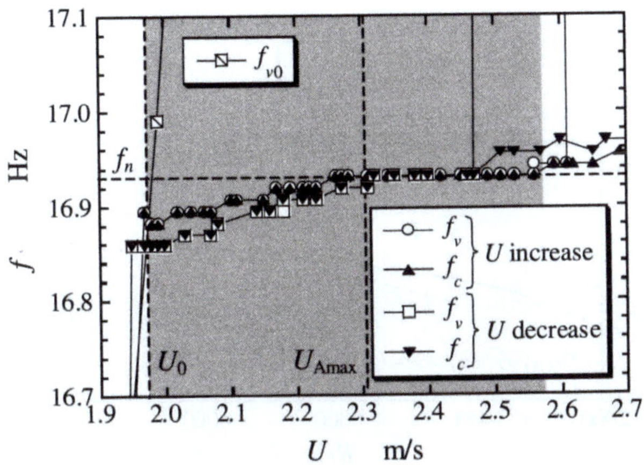

Fig. 5. Lift coefficient on stationary cylinder $(C_{L0})_{rms}$ versus Reynolds number *Re*. ··········, Vickery; ———, Bishop and Hassan; — · —, Keefe; for circular cylinder [2]. Shaded region shows the range of Kármán vortex excitation experiments for circular cylinder

Figure 4(b) shows that ϕ_{uZ} steeply decreases with U in the vortex excitation range. However, the standard deviation σ_ϕ in Fig. 4(b) is small over the same range, which indicates the occurrence of definite synchronization of Kármán vortex shedding in spite of the large change in ϕ_{uZ}.

3.2. Lift force acting on the cylinder

3.2.1. Lift force acting on the stationary cylinder

The alternating lift force F_L, which acts on the fixed cylinders, was measured over a free stream velocity U range including the vortex excitation range presented in the Section 3.1. Since F_L was composed of the contributions of Kármán vortex shedding and other unsteady turbulence, the root mean square value of F_L, $(F_L)_{rms}$, was used to estimate the magnitude of F_L. Hence, the fluctuating lift coefficient of a stationary cylinder $(C_{L0})_{rms}$ was calculated by the following equation.

$$(C_{L0})_{rms} = \frac{(F_L)_{rms}}{\frac{1}{2}\rho U^2 dl} \tag{1}$$

where ρ is air density, d is the height of the cylinder and l is the effective length of the cylinder defined as the distance between the two end plates (see Fig. 1).

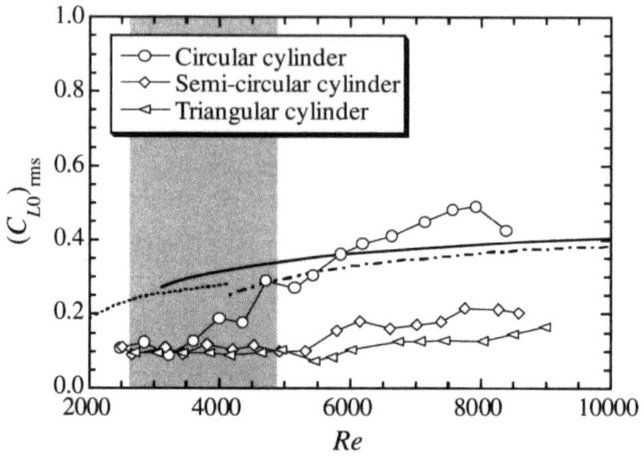

Fig. 6. The details of vortex shedding frequency f_v and oscillation frequency f_c of circular cylinder in Kármán vortex excitation region

The lift coefficients $(C_{L0})_{rms}$ for the three cylinders were plotted against the Reynolds number Re in Fig. 5. The shaded region shows the range where the oscillation measurement for the circular cylinder was carried out. $(C_{L0})_{rms}$ for circular cylinder measured by Vickery, Bishop and Hassan, Keefe [2] was also plotted in Fig. 5 for comparison. The lift coefficient $(C_{L0})_{rms}$ of circular cylinder obtained in this work agree well with those by others when $Re > 5000$, but considerably lower when $2500 < Re < 4500$.

The lift coefficient $(C_{L0})_{rms}$ of the circular cylinder increases gradually with Re in the range of the oscillation experiment, say from 0.1 at $Re = 2500$ to 0.3 at $Re = 5000$. While $(C_{L0})_{rms}$ for the semi-circular cylinder and the triangular cylinder are equal to that of the circular cylinder when $Re = 2500$, they remain constant in the same Re range. Hence, $(C_{L0})_{rms}$ of the circular cylinder is larger than those of the other two cylinders by the factor of 3 at $Re = 5000$.

3.2.2. Lift force acting on the oscillating cylinder excited by Kármán vortex excitation

In order to estimate the alternating fluid force exerted on the cylinder under the Kármán vortex excitation, it is assumed that the cylinder supported elastically by two plate springs is a linear system with constant mass, spring and damping. Based on this assumption, the following equation determines the cylinder motion Z.

$$m_e\ddot{Z} + c\dot{Z} + kZ = F_L \tag{2}$$

where m_e is the effective mass of the cylinder, c is the damping factor and k is the spring constant. F_L is the lift force caused by the periodic Kármán vortex shedding, and is assumed to be sinusoidal. That is,

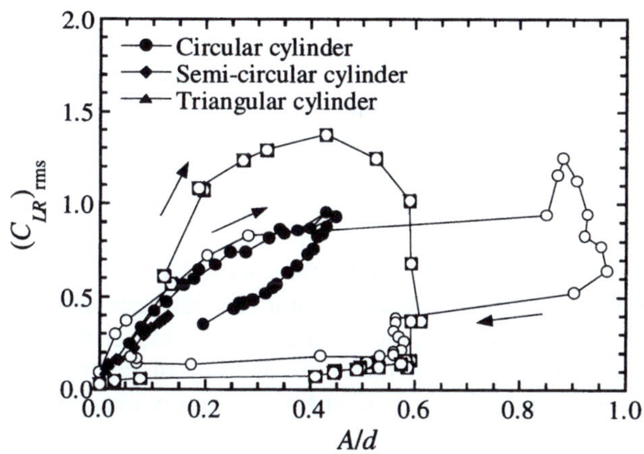

Fig. 7. Lift coefficient C_{LR} for oscillating cylinder versus non-dimensional oscillation amplitude A/d under Kármán vortex excitation, when U increases. —O—, Khalak and Williamson [4], —□—, K. Hayashi et al. [3], for circular cylinder

$$F_L = F_0 \sin(2\pi f_v t) \tag{3}$$

From the solution of Eqs. (2) and (3) together with the definition of the lift coefficient $(C_{LR})_{rms}$ as Eq. (1), we obtain

$$(C_{LR})_{rms} = \frac{8\pi^2 m_e f_n^2 A\sqrt{(1-\eta^2)^2 + (2\zeta\eta)^2}}{\sqrt{2}\rho U^2 dl} \tag{4}$$

where $\eta = f_v / f_n$ and ζ is the damping ratio (= $\delta/(2\pi)$). From Eq. (4) it is seen that a slight difference between f_v and f_n in the vortex excitation range affects $(C_{LR})_{rms}$ drastically when the damping ratio ζ is very small like the experiments in this work. Therefore, the results for the circular cylinder in Fig. 2(a) were examined more precisely. Fig. 6 shows the oscillation frequency f_c and the vortex shedding frequency f_v in the vortex excitation range in detail. In this figure, it is seen that f_c coincides with f_v but they do not coincide with f_n when U is smaller than U_{Amax}. The slight difference between f_v and f_c reflects the large value of σ_ϕ in the range $U_0 < U < U_{Amax}$ shown in Fig. 2(b). When $U > U_{Amax}$ in the vortex excitation range, f_c and f_v are strictly equal to f_n.

The lift coefficient for oscillating cylinder $(C_{LR})_{rms}$ was calculated using Eq. (4), based on the results that are described in the Section 3.1. In Fig. 7, $(C_{LR})_{rms}$ of the circular cylinder, the semi-circular cylinder and the triangular cylinder are plotted against A/d for the cases of increasing U. In the figure, the arrows indicate the order of measurement. The alternating lift coefficients of oscillating circular cylinder by Khalak and Williamson [4] and Hayashi et al. [3] are added for comparison. The former was measured directly in water and the latter measured in air.

The lift coefficient $(C_{LR})_{rms}$ of the circular cylinder and the semi-circular cylinder are equal to each other and increase with A/d when A/d <0.13. However, the maximum value of $(C_{LR})_{rms}$ of the semi-circular cylinder is around 0.4 at A/d = 0.13 while $(C_{LR})_{rms}$ of the circular cylinder continues to increase till it attains the maximum value around unity, at $A/d = 0.45$. The lift coefficient $(C_{LR})_{rms}$ of the triangular cylinder at very small amplitude is nearly equal to that of the other two cylinders, and no definite increase is observed since the range of A/d is limited. Although the experimental conditions for the three cylinders are almost the same, the maximum value of $(C_{LR})_{rms}$ is quite different among them, e.g. maximum value of $(C_{LR})_{rms}$ for the circular cylinder is about 10 times that of the triangular cylinder.

The difference between $(C_{LR})_{rms}$ of the three cylinders with different cross sectional configuration is attributed to the separation point movement of these cylinders [6]. For the circular cylinder, the separation point moves widely on the arc of surface synchronizing with the cylinder oscillation. In contrast, the separation point on the triangular cylinder is fixed at the rear vertex irrespective of the cylinder oscillation.

The lift coefficient $(C_{LR})_{rms}$ of the circular cylinder based on the linear analysis in this work coincides with the direct measurements by Khalak et al. and Hayashi et al. in the range of $A/d < 0.13$. When $0.13 < A/d < 0.45$, $(C_{LR})_{rms}$ of the higher stem (i.e. $U < U_{Amax}$) in this work agrees well with that by Khalak et al., while that by Hayashi et al. is considerably higher. When A/d is still higher, or begins to decrease passing its maximum, results of the three works are largely different from each other. This may be because of the fact that the assumption of linear model does not hold when A/d is large, or because of the difference between experimental conditions and ways of measurements.

3.3. Occurrence region of the Kármán vortex excitation

The synchronization region of the Kármán vortex shedding was investigated for all the three cylinders by giving them a controlled oscillation using a mechanical oscillator by the present authors [6]. Since the behavior of Kármán vortex excitation is disturbed by irregular modulation and its magnitude is drastically different among the three cylinders, the criterion of Kármán vortex excitation should be defined more precisely. According to the earlier study, occurrence of the Kármán vortex excitation is defined by a criterion composed of the following four conditions: i) an increase of the oscillation amplitude, ii) $f_c = f_v$, iii) an increase of the cross-correlation coefficient R_{uZ} between the velocity u and the cylinder displacement Z, and iv) an increase of the cross-correlation coefficient R_{uu}. When all of these four conditions are satisfied, the Kármán vortex excitation, and hence the synchronization of vortex shedding with the cylinder oscillation, i.e. "Lock-in", is decided to occur. When none of the four conditions are satisfied, it is taken to be no vortex excitation and "No Lock-in". The remaining cases are defined as "Transition".

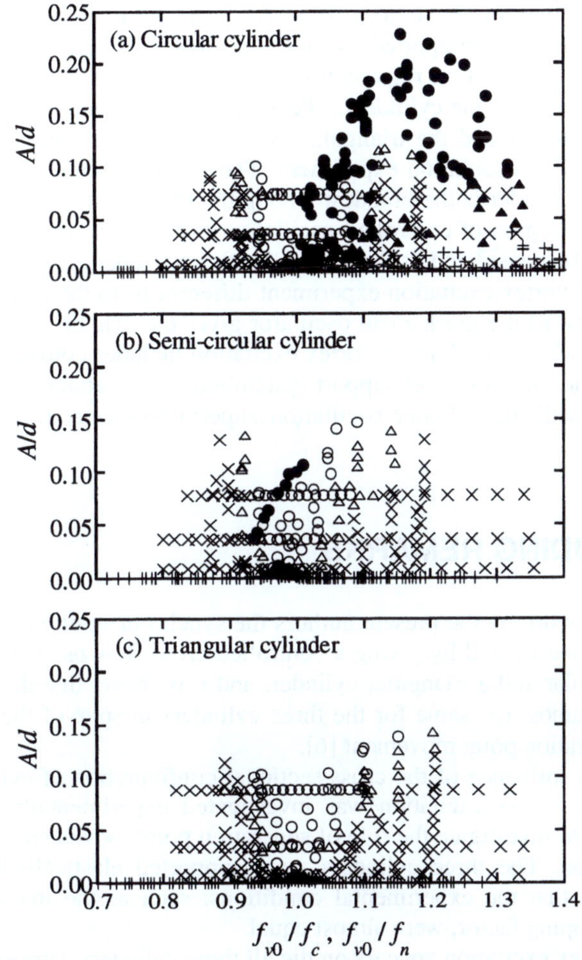

Fig. 8. Kármán vortex excitation region on a plane of frequency ratio and non-dimensional amplitude. ●: Kármán vortex excitation, +: No excitation, ▲: Transition; Controlled oscillation, ○: Lock-in, ×: No Lock-in, △: Transition. [6]

The occurrence regions of the Kármán vortex excitation on the plane of frequency ratio f_{v0} / f_n and the non-dimensional amplitude A/d are plotted in Figs. 8(a), 8(b) and 8(c) for each of the cylinders, where the synchronization data obtained by the controlled oscillation experiment are also added [6]. In this controlled oscillation experiments, *Re* was fixed at about 3500, i.e. *U* was constant at 2 m/s, to exclude the influence of *Re*. Since the measurement of Kármán vortex excitation was carried out for increasing or decreasing *U*, *Re* is not constant in these figures. Some of the data were obtained by using systems with different damping factor.

As seen by comparing Figs. 8(a), (b) and (c), the Kármán vortex excitation regions for the elastically supported cylinder are drastically different among them, although the synchronization regions for controlled oscillation experiment are almost the same for all of the cylinders. The Kármán vortex excitation region for the semi-circular cylinder and the triangular cylinder lies within the synchronization region of controlled oscillation experiment. The Kármán vortex excitation region for the circular cylinder is the broadest among the three cylinders and exceeds beyond the lock-in region of controlled oscillation experiment when $f_{v0} / f_n > 1$. The followings can be considered as the causes for the discrepancy: i) *Re* was not kept constant for the vortex excitation experiment different from the controlled oscillation experiments, ii) the mechanical oscillator gives the cylinder a precise sinusoidal oscillation, while, the Kármán vortex excitation includes considerable modulation of amplitude, iii) cantilever support system induces an attack angle fluctuation synchronizing with the cylinder oscillation superimposed on the z-displacement oscillation.

4. CONCLUDING REMARKS

In the previous work by the present authors the synchronization of Kármán vortex shedding was investigated by giving a controlled cross-flow oscillation to a circular, a semi-circular and a triangular cylinder, and it is shown that the synchronization region is almost the same for the three cylinders in spite of the different behaviors of separation point movement [6].

In this work, influence of the cross-sectional configuration of cylindrical body on the Kármán vortex excitation was investigated experimentally by using the same cylinders to investigate the role of separation point movement in the Kármán vortex excitation. The three cylinders were supported elastically by cantilever plate spring so that the experimental conditions, such as the mass, natural frequency and damping factor, were almost equal.

Kármán vortex excitation appears on the all three cylinders. However, the oscillation behavior was drastically different among them.

The alternating lift coefficient is equal for the three cylinders when they are at rest. However, the lift coefficient $(C_{LR})_{rms}$ obtained from measured oscillation amplitude A/d are largely different each other. Although $(C_{LR})_{rms}$ increases with A/d for all the three cylinders, the maximum value is largely different among them, that is, around $(C_{LR})_{rms}$ is around 1.0 at $A/d = 0.45$ for the circular cylinder, 0.4 at $A/d = 0.15$ for the semi-circular cylinder and 0.1 at $A/d = 0.05$, for the triangular cylinder. This tendency of enhancement of lift coefficient by the oscillation corresponds with the separation point movement on these cylinders shown by the controlled oscillation experiment in the previous work.

In conclusion, the mechanism of the Kármán vortex excitation can be explained as follows. A cylinder is oscillated by the alternating lift force due to the periodic Kármán vortex shedding, and the amplitude reaches its maximum when the flow velocity is around U_0. The resulting cylinder oscillation causes the synchronization of the vortex shedding, which makes the resonance range of flow velocity broader than otherwise. At the same time, the cylinder oscillation influences the movement of separation point and enhances the lift, depending on the cross sectional configuration. This enhancement effect is largest for the circular cylinder and negligibly small for the triangular cylinder.

References

[1] Funakawa M (1970) Excitation mechanism of elastically supported circular cylinder in the flow (in Japanese). Bulletin of the Japan Society of Mechanical Engineers, 36, 285:303-312

[2] Gartshore I S (1984) Some Effects of Upstream Turbulence on the Unsteady Lift Forces Imposed on Prismatic Two Dimensional Bodies. Journal of Fluids Engineering, 106:418-424

[3] Hayashi K, Tanaka K, Fujima K, Shigemura T (1997) Fluid Force Acting on a Vortex Excited Circular Cylinder Vibrating in Steady Flow (in Japanese). National Defense Academy Science and Engineering research report, 34, 2:11-23

[4] Khalak A, Williamson C H K (1999) Motions, forces and mode transitions in vortex-induced vibrations at low mass-damping. Journal of Fluids and Structures, 13:813-853

[5] Koide M, Takahashi T, Shirakashi M (2001) Development of a ring-type vortex anemometer for low-velocity wind tunnel experiments (in Japanese). Bulletin of Japan Society of Mechanical Engineers Series B, 67, 657:1105-1111

[6] Koide M, Tomida S, Takahashi T, L Baranyi, Shirakashi M (2002) Influence of Cross-sectional Configuration on the Synchronization of Kármán Vortex Shedding with the Cylinder Oscillation. Japan Society of Mechanical Engineers International Journal Series B, 45, 2:249-258

[7] Shirakashi M, Ishida Y, Wakiya S (1985) Higher velocity resonance of circular cylinder in cross flow. Transaction of the ASME Journal of Fluids Engineering, 107:392-396

TURBULENCE MODELLING AND NUMERICAL METHODS

A New Wall-Function Strategy Applied to Mixed Convection in Vertical Annular Passages

T. J. CRAFT, A. V. GERASIMOV*, H. IACOVIDES and B. E. LAUNDER
Mechanical, Aerospace and Manufacturing Engineering Department, UMIST, Manchester, UK

*Corresponding author: George Begg Bldg, UMIST, PO Box 88, Manchester, M60 1QD, United Kingdom
Tel.: +44 (0) 161 200 4547, Email: A. Gerasimov@postgrad.umist.ac.uk

Abstract The paper summarizes a new analytical formulation of the near-wall region in turbulent flow that has been found to account markedly better for the influence of buoyancy on the flow structure than conventional logarithmic laws. The paper summarizes the key features of this approach and prior applications of the scheme. It then considers the case of vertically downward flow through an annular passage, a configuration often considered in relation to safety studies in nuclear reactors. Four different test-cases are examined for values of the buoyancy parameter, Bo, ranging from 0.22 to 2.89. All show reasonably close agreement with experimental data and with computations using the far more elaborate and costly low-Reynolds-number form of the k-ε model.
Key Words: buoyant turbulent flows, near-wall region, wall functions

1. Introduction

In using CFD to simulate complex industrial flows, it is usually not practicable to adopt the exceedingly fine grids needed to resolve the viscous sub-layer. Instead, it is assumed that across this sub-layer and, indeed, extending some way out into the fully turbulent region, the turbulence is in a state of simple shear where the production and dissipation rates of turbulence energy are equal. Unfortunately, this idealized condition of local equilibrium seldom exists outside the university laboratory or flow-simulation computer program. The presence of buoyant forces caused by substantial heating or cooling at the wall is just one of many possible causes of serious departures from local equilibrium in this wall adjacent layer.

The Turbulence Mechanics Group at UMIST has committed considerable effort over the past three years to developing more accurate approaches for creating wall functions. The most recent summary of these approaches is provided in [1].

Briefly, two schemes have been created, one based on a one-dimensional numerical solution of the equations for the dependent variables across the sub-layer [2] and the other on a simpler analytic formulation [3]. While the former is the more general, the latter approach has a number of attractions, not least the fact that the separate contributions of different physical constituents are readily apparent in the resultant analytical form. It is this latter scheme with which the present paper is solely concerned. It has been successfully applied in [4] to buoyantly-aided up-flow in vertical pipes and has been shown to capture with great fidelity the inhibition of heat transfer coefficients that may arise in this case, leading to a severe overheating of the tube wall. An example of such an application appears in Fig. 1 which shows the variation of Nusselt number along the pipe following the application of strong heating at x/d = 50. The horizontal line represents the level of Nu that would be expected according to the Dittus-Boelter correlation for purely forced convection pipe flow. The experimental data however show a marked drop below this line due to the partial laminarization of the flow, a trend that is well mimicked by both the 'low-Reynolds-number' solution and the wall function adopted herein. A particularly pleasing feature of the result is that the predicted Nusselt number depends only weakly on the thickness of the near-wall control volume over which the wall function is applied. This is in marked contrast with the conventional log law.

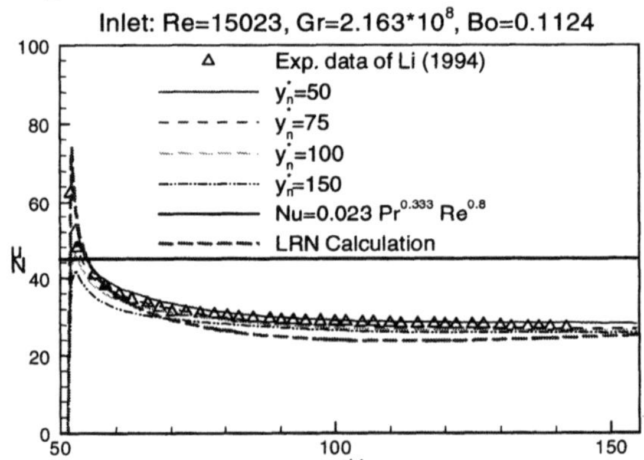

Fig. 1. Nusselt number for buoyancy-aided flow of air in a vertical pipe. Lines labelled yn*=50 to 150 are predictions with the present scheme, using different sized near-wall cells.

The present paper summarizes the main planks of the analytical approach and shows new applications of the strategy to the problem of *downward* directed flow in a vertical annulus in which the core tube is strongly heated, thus causing a substantial retardation and even a limited reversal in direction of the near-wall fluid. Computational results generated by the new analytical wall functions are compared with those resulting from a conventional 'log-law' approach and from a so-called 'low-Reynolds-number' treatment of the flow. While this last is, in principle, the most reliable of the approaches, as indicated above, its use in computing an industrial flow is usually not acceptable since it would typically require two orders of magnitude more computing time than a wall-function approach [2]. The flow problem here considered, besides providing an opportunity for testing the analytical wall-function (AWF) approach in a quite different physical situation from any hitherto examined, is also important in its own right since the nuclear industry sees it as a generic test-case for judging how well a particular turbulence model is likely to perform in predicting the flow in a nuclear reactor following the failure of the coolant circulation pumps.

Section 2 below presents the main features of the scheme adopted here for approximating the near-wall region while numerical practices and the turbulence modelling over the region of flow outside of that covered by the wall function are given in Section 3. Applications of the resultant solver for the case of downward flow in a heated annulus are presented and discussed in Section 4.

2. The Analytical wall function

2.1. Conventional Wall Functions

The usual approach to the CFD treatment of industrial near-wall flows is to place the wall-adjacent node sufficiently far from the wall for it to be located in fully turbulent fluid. It is assumed that there the production of turbulent kinetic energy by shear balances the energy dissipation rate, ε, and that the mean flow is in simple shear. These assumptions enable boundary conditions for the dependent variables to be obtained in terms of the wall shear stress, τ_w, (and, in the case of temperature, T, the heat flux through the wall and the wall temperature, T_w) the distance from the wall and the fluid's molecular properties. This leads to the following widely employed form [5]:

$$U^* = (1/\kappa)ln(E^*y^*); \quad T^* = \sigma_T(U^* + \boldsymbol{P}^*) \qquad \text{(1a, b)}$$

$$\varepsilon = (\tau_w/\rho)^{3/2}/(\kappa y) \qquad \text{(2)}$$

where $U^* = Uk^{½}/(\tau_w/\rho)$; $T^* = \rho C_p k^{½}(T_w - T)/q_w$ and $y^* = yk^{½}/\nu$, the modified wall-law variables proposed by Spalding (see [6]). The coefficients κ, E^* and σ_T are constants while $\boldsymbol{P}^*$ is an empirical function of the molecular Prandtl number.

While the turbulent kinetic energy, k, could be consistently obtained from its local equilibrium value

$$k = (\tau_w/\rho)^{½}/c_\mu^{½}, \qquad \text{(3)}$$

it is usually determined by solving its budget equation using Eq.(2), approximating the k- production rate by taking the effective fluid shear stress equal to τ_w/ρ and using Eq.(1a) to obtain the velocity gradient. This inconsistent elaboration avoids the absurd result that the turbulent heat transfer rate falls to zero at a stagnation point (where, of course the wall shear stress is zero) but as may be readily anticipated, the approach contains so many plainly wrong assumptions that it delivers quite unreliable levels of wall stress and heat transfer coefficients in complex flows.

2.2. The Analytical Wall Function

The variation of effective turbulent viscosity with distance from the wall is known to be more generally valid than the variation of mean velocity. For that reason it is the viscosity that provides the starting point for the present approach. However, to pursue this analysis, the assumed distribution of turbulent viscosity must be kept simple. We adopt a viscous sublayer of thickness y_v where the turbulent viscosity is zero while, for larger values of the wall-normal distance, y, a linear increase of viscosity across the near wall sublayer is assumed, Fig. 2a:

$$\mu_t = \rho c_\mu \sqrt{k} l = \mu_v c_\mu c_l \left(y^* - y_v^*\right) \quad for \quad y^* \geq y_v^* \tag{4}$$

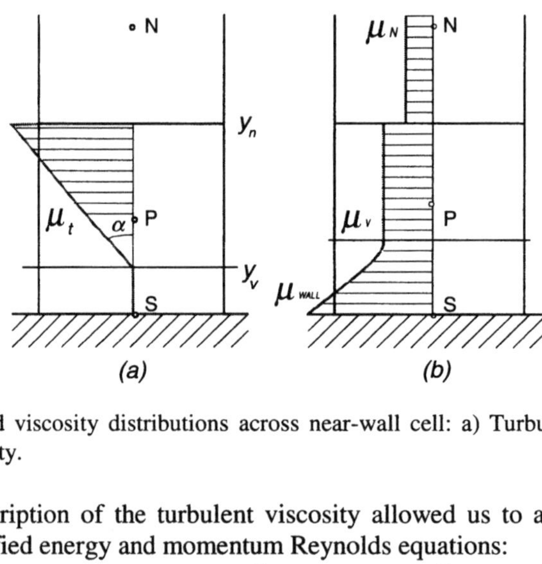

Fig. 2. Prescribed viscosity distributions across near-wall cell: a) Turbulent viscosity; b) Molecular viscosity.

Such a prescription of the turbulent viscosity allowed us to analytically integrate the simplified energy and momentum Reynolds equations:

$$\rho U \frac{\partial T}{\partial x} + \rho V \frac{\partial T}{\partial y} = \frac{\partial}{\partial y}\left[\left(\frac{\mu}{Pr} + \frac{\mu_t}{\sigma_t}\right)\frac{\partial T}{\partial y}\right] \tag{5}$$

$$\rho U \frac{\partial U}{\partial x} + \frac{\partial P}{\partial x} - g\left(\rho - \rho_{ref}\right) = \frac{\partial}{\partial y}\left[\left(\mu + \mu_t\right)\frac{\partial U}{\partial y}\right] \tag{6}$$

In flows with appreciable wall heating the variation of the molecular viscosity across the viscous sublayer needs to be taken into account (Fig. 2b). The precise function used to approximate this variation of μ was found to have a strong effect on the solution stability. The most stable approach tested was to assume a parabolic variation across the sublayer:

$$\mu = \frac{\mu_{wall}}{1 + b_\mu y^* \left(y^* - 2 y_v^* \right)} \tag{7}$$

where

$$b_\mu = \frac{1}{y_v^{*2}} \left(1 - \frac{\mu_{wall}}{\mu_v} \right) \tag{8}$$

The resulting expressions for the velocity and temperature variations are given in [7].

Another area where it was felt appropriate to improve on standard practice was in obtaining k at the near-wall node. As outlined above, the solution of the k equation over the near-wall cell requires one to compute the average production and dissipation rates across the cell. Whilst the former can be evaluated using the prescribed turbulent viscosity profile and stress $\partial U/\partial y$ obtained from the analytical solution of the mean momentum equation, the latter is normally based on a two-part profile across the wall adjacent cell (Fig. 3a).

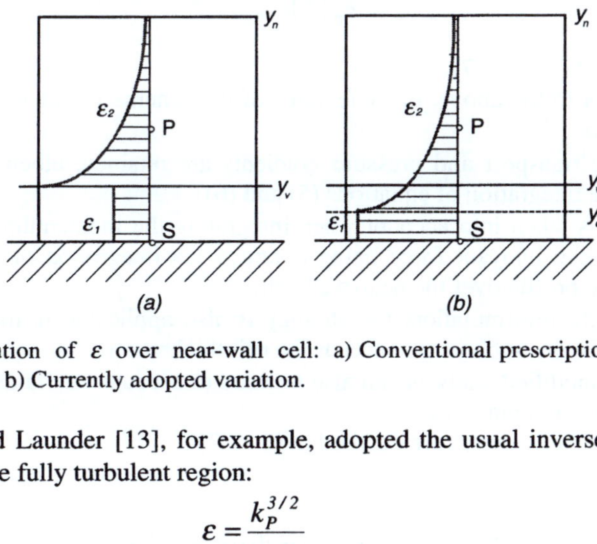

Fig. 3. Distribution of ε over near-wall cell: a) Conventional prescription (Chieng and Launder, [13]); b) Currently adopted variation.

Chieng and Launder [13], for example, adopted the usual inverse dependence of ε only in the fully turbulent region:

$$\varepsilon = \frac{k_P^{3/2}}{c_l y} \tag{9}$$

whilst in the vicinity of the wall the exact wall-limiting result (Jones and Launder, [14]) was approximated:

$$\varepsilon_v = \nu \left(\frac{\partial k^{1/2}}{\partial y} \right)^2 \approx \frac{2 \nu k}{y^2} = \frac{2 \nu k_P}{y_v^{\,2}} \tag{10}$$

However, while k varies as y^2 very close to the wall it levels out at the edge of the viscous sublayer. Consequently, the last form in Eq. (10) gave sublayer dissipation levels lower than those in the adjacent fully-turbulent zone, a result which contradicts DNS data. Therefore, in the present work, the inverse variation of ε with the wall distance is continued closer to the wall (Fig. 3b) to a position y_d which has been chosen to ensure continuity of ε at the matching point. This leads to $y_d^*=5.1$ and results in an average dissipation rate:

$$\bar{\varepsilon} = \frac{1}{y_n}\left[\frac{2k_P^{3/2}}{y_d^*} + \frac{k_P^{3/2}}{c_l} ln\left(\frac{y_n}{y_d}\right)\right] \tag{11}$$

In the non-equilibrium flows being considered, the non-dimensional viscous sublayer thickness, y_v^*, is far from a universal constant, and another important element of the present model is that it attempts to account for this feature. Analysis of existing data suggested that the ratio of the shear stress at the wall to that at the edge of the sublayer, λ, was a good indicator of sublayer thickening or thinning, and our initial plan was simply to make y_v^* a function of the parameter λ. However, this practice caused stability problems, and after examining a number of alternatives, our adopted approach is to simulate the effect of variable viscous sublayer thickness by adjusting the mean level of dissipation rate obtained from (11) via a scaling function F_ε, which is dependent on the parameter λ:

$$\bar{\varepsilon}_{new} = F_\varepsilon(\lambda)\bar{\varepsilon}_{old} \tag{12}$$

Details are contained in [7].

In the addition to the above, other features of the scheme pertinent to the current application are:

- Convective transport and pressure gradients are retained, albeit in simplified form, in the integration of equations (5) and (6).
- Buoyancy is taken into account when integrating the momentum equation (6), with the mean buoyant force being obtained via integration of the analytical temperature profile over the near-wall cell.
- With a slight reformulation, the strategy is also applicable in those situations when the viscous sublayer y_v exceeds the cell thickness y_n.
- A slightly modified analysis can also be used to account for fluids with Prandtl numbers greater than unity.

These and other features are detailed in [3, 4, 7]

3. Other elements of the numerical and physical model

3.1. Turbulence Model in Main-Flow Region

In the main flow domain the Launder-Sharma k-ε model [11] has been employed, in which the transport equations solved are

$$\frac{Dk}{Dt}=\frac{\partial}{\partial x_j}\left[(\nu+\nu_t)\frac{\partial k}{\partial x_j}\right]+P_k+P_g-\varepsilon-2\nu\left(\frac{\partial\sqrt{k}}{\partial x_j}\right)^2 \tag{13}$$

$$\frac{D\varepsilon}{Dt}=\frac{\partial}{\partial x_j}\left[\left(\nu+\frac{\nu_t}{\sigma_\varepsilon}\right)\frac{\partial\varepsilon}{\partial x_j}\right]+c_{\varepsilon 1}\frac{\varepsilon}{k}\left(P_k+P_g\right)-$$

$$c_{\varepsilon 2}f_2\frac{\varepsilon^2}{k}+2\nu\nu_t\left(\frac{\partial^2 U_i}{\partial x_j\partial x_k}\right)^2 \tag{14}$$

where

$$P_k=-\overline{u_i u_j}\frac{\partial U_i}{\partial x_j};\quad P_g=-\beta g_i\overline{u_i\theta}$$

are the expressions for the production of the turbulent kinetic energy due to shear and buoyancy and

$$\nu_t=c_\mu f_\mu\frac{k^2}{\varepsilon}$$

is the expression for the turbulent viscosity.

The stresses and turbulent heat fluxes are given by

$$\overline{u_i u_j}=\frac{2}{3}k\delta_{ij}-\nu_t\left(\frac{\partial U_i}{\partial x_j}+\frac{\partial U_j}{\partial x_i}\right);\quad \overline{u_i\theta}=-\frac{\nu_t}{\sigma_t}\frac{\partial T}{\partial x_i}$$

and the model coefficients are summarized in Table 1.

Table 1. Constants and functions of the Launder-Sharma k-ε model.

$c_\mu=0.09$	$c_{\varepsilon 1}=1.44$	$c_{\varepsilon 2}=1.92$	$\sigma_t=0.9$	$\sigma_\varepsilon=1.3$
$f_2=1-0.3\,exp\left(-Re_t^2\right)$			$f_2=1-0.3\,exp\left(-Re_t^2\right)$ $Re_t=\frac{k^2}{\nu\varepsilon}$	

Although linear EVM's are known to be unsuitable in some complex flows, this particular scheme, when applied as a low-Reynolds-number model, does actually return predictions in good agreement with experimental data for the flows studied here, and hence provides a good modelling framework within which to test the wall function treatment.

3.2. Numerical Treatment and Boundary Conditions

The computations have been carried out using the TEAM computer code [8] which is a finite-volume solver employing a staggered storage arrangement on a rectangular plane or axisymmetric Cartesian grid. The SIMPLE pressure correction scheme is employed, whilst convection is approximated via the QUICK scheme [9] in equations for mean quantities, but with the first order upwind scheme in the turbulence equations.

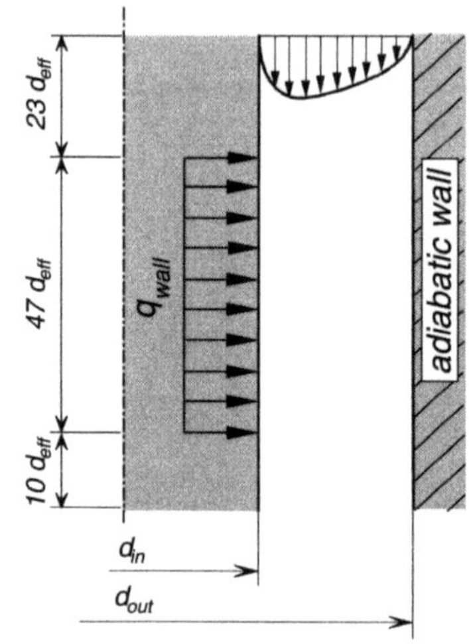

Fig. 4. Buoyancy-opposed flow in an annulus.

The domain employed has inner and outer radii of 38 mm and 70 mm and extends for a streamwise length of 80 effective diameters, $d_{eff} = d_{out} - d_{in}$. A grid of 252 (streamwise) by 60 (radial) nodes is employed for a low-Reynolds-number model, whilst with wall-function approaches the number of radial cells is reduced to 12. Fully developed inlet conditions are applied and adiabatic wall conditions are used for the first 23 diameters of development, to match the experimental arrangement. A prescribed wall heat flux is then applied for a distance of 47 diameters, followed by a recovery distance of 10 diameters where adiabatic conditions are again applied (see Fig.4).

4. Presentation and discussion of results

We focus principally on the variation of Nusselt number along the annulus. The labelling on the figures has the following meaning: AWF – the analytical wall functions that are here proposed; StWF – Standard wall functions, as described in Section 2.1; LRN are fine-grid results adopting the k-ε model of Launder & Sharma [11] containing terms designed to make the model applicable in the viscous sublayer of the flow. For the last model the computations were also run for the case where the gravitational acceleration was set to zero in order to see the contribution of (principally) the buoyant term of the vertical momentum equation to the predicted Nusselt number. The experimental data are those generated by Jackson *et al* [10].

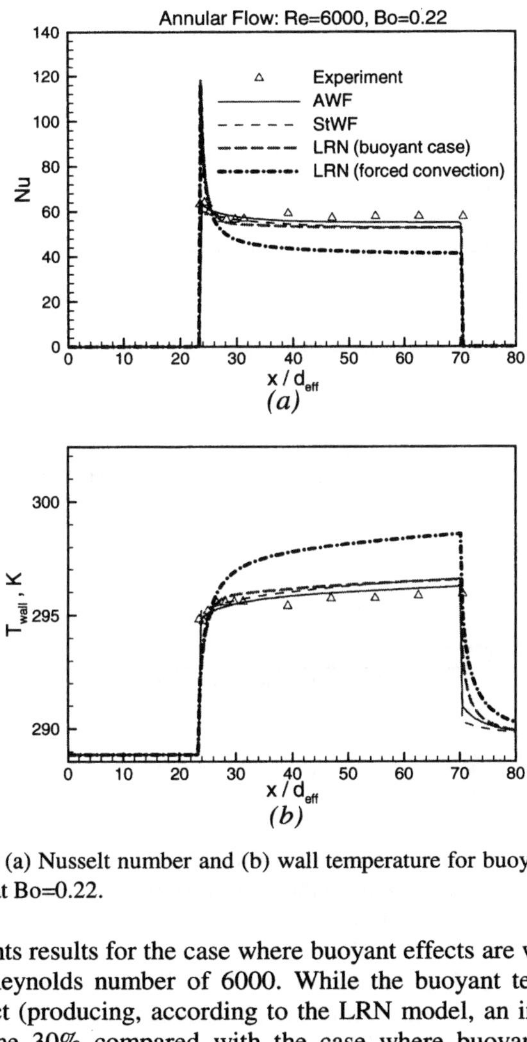

Fig. 5. Predicted (a) Nusselt number and (b) wall temperature for buoyancy-opposed water flow in annulus at Bo=0.22.

Fig. 5 presents results for the case where buoyant effects are weak, Bo = 0.22 at a mean pipe Reynolds number of 6000. While the buoyant terms are evidently having an effect (producing, according to the LRN model, an increase of Nusselt number of some 30% compared with the case where buoyant forces are suppressed) the different wall function results give closely the same distribution of Nu which accords well with the low Reynolds number computations and the experimental data.

A strikingly different picture emerges when the buoyancy parameter, Bo, is increased to 0.78, Fig. 6. At this level of heating the LRN model indicates that buoyant effects raise the level of Nusselt number above the level in purely forced convection by some 70% and the resultant level of Nu predicted by this scheme is in close agreement with the measured values.

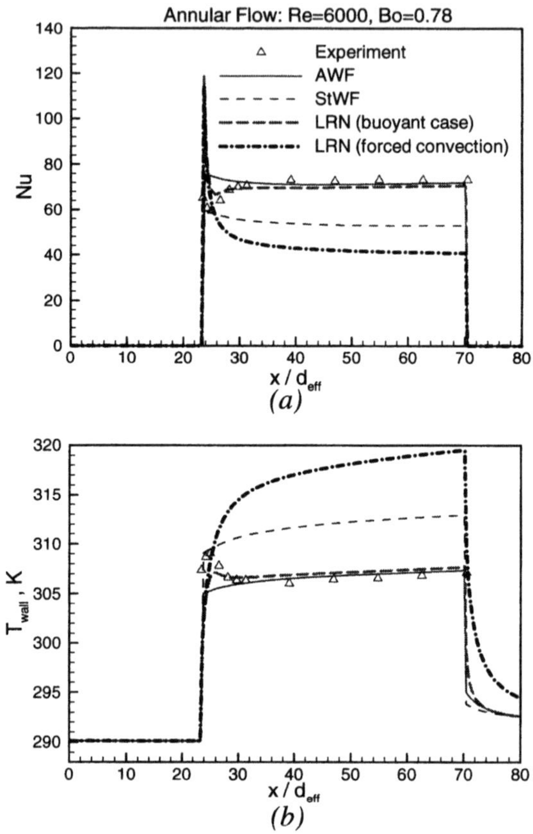

Fig. 6. Predicted (a) Nusselt number and (b) wall temperature for buoyancy-opposed water flow in annulus at Bo=0.78.

Computations employing 'standard' wall functions, however, capture only about 40% of the increase relative to the non-buoyant case. The computations employing the analytical wall functions (AWF) do very much better.

If the Reynolds number of the flow is reduced further while keeping Bo roughly the same, Fig. 7, the AWF approach again reproduces the experimental data with good accuracy over most of the development length following the application of heating whereas now the 'standard' approach actually returns levels below the forced convection LRN scheme result. The primary cause of the above behaviour seems to be that there is no mechanism in this approach for the sublayer-thinning that arises when the peak of energy generation is located beyond the viscous sublayer (see the discussion in Johnson & Launder [12]) and which is present in the AWF through Eq. (12).

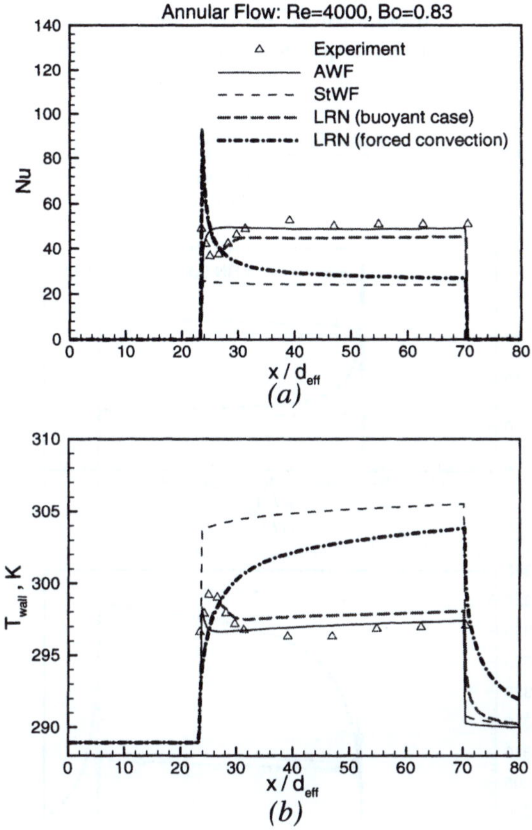

Fig. 7. Predicted (a) Nusselt number and (b) wall temperature for buoyancy-opposed water flow in annulus at Bo=0.83 and Re=4000.

It is worth pausing to note that, although overall the AWF achieves a satisfactory prediction of the measured behaviour, in both Figs. 6 and 7 this scheme does not reproduce the measured minimum in the Nusselt number that occurs about four effective diameters after the start of heating. The physical cause of this minimum may be inferred from the two low-Reynolds-number computations. It is noted that for the first two or three diameters of development both the buoyant and non-buoyant computations are the same, exhibiting a rapid decline of Nu with downstream distance as the thermal boundary layer progressively increases in thickness. Then, as the thermal boundary layer for the computation including buoyant effects grows out into the fully turbulent region, buoyant forces have their indirect effect on the established turbulence structure (since the flow has developed isothermally for the first 23 diameters) producing an increase in Nusselt number.

Since the LRN scheme does partly capture the behaviour exhibited by the measurements, it may be inferred that the imperfections of the AWF approach immediately following the application of heating are associated with the assumption that there is, from the outset, a quasi-developed temperature profile extending out into the turbulent region.

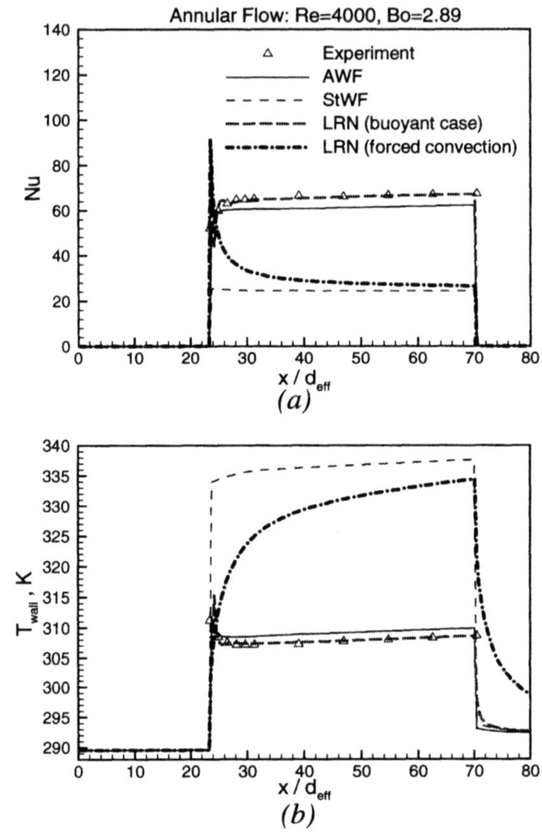

Fig. 8. Predicted (a) Nusselt number and (b) wall temperature for buoyancy-opposed water flow in annulus at Bo=2.89.

The final comparison is for the situation where Bo is increased to 2.89, representing very strong buoyant effects (although even at this level local flow reversal due to heating does not occur). In this case, according to the LRN results, the action of buoyancy has raised the Nusselt number by 150% relative to the non-buoyant situation. Even under these extreme conditions the AWF scheme still gives a satisfactory account of the variation of heat-transfer coefficient returning levels within 10% of the measurements over most of the flow. The 'standard' wall-function approach, in contrast, leads to predicted levels of Nu less than 40% of the measured levels!

5. Concluding remarks

The present enquiry has shown that the analytical wall functions, developed by the authors initially to give an economical yet reasonably accurate scheme for predicting buoyancy-aided flow through pipes where the flow partially laminarizes [3,4], also performs very satisfactorily in predicting the augmentation in Nusselt number that occurs when buoyant effects oppose the flow's motion. In contrast, the standard wall-function approach gives a quite inadequate account of the effects of buoyancy-opposed motion. This aberrant behaviour is consistent with the fact that the standard scheme also entirely misses the damping of Nu that arises in the buoyancy-aided case.

It is relevant to mention that the authors have also recently completed an examination of the development of a hot wall jet injected into a cold opposing stream [15], the object being to predict the distance of penetration of the jet and its spreading rate. In this case the overall behaviour is especially sensitive to the modelling away from the wall where an eddy-viscosity treatment does not adequately capture streamline curvature effects. There is nevertheless a marked improvement in the computed behaviour from using the AWF approach compared with a standard wall function. Indeed, if the AWF scheme is interfaced with our best 2nd moment closure for the main flow region (the TCL model [16]), agreement with the data base (from a large-eddy simulation [17]) is even closer (see Fig. 9).

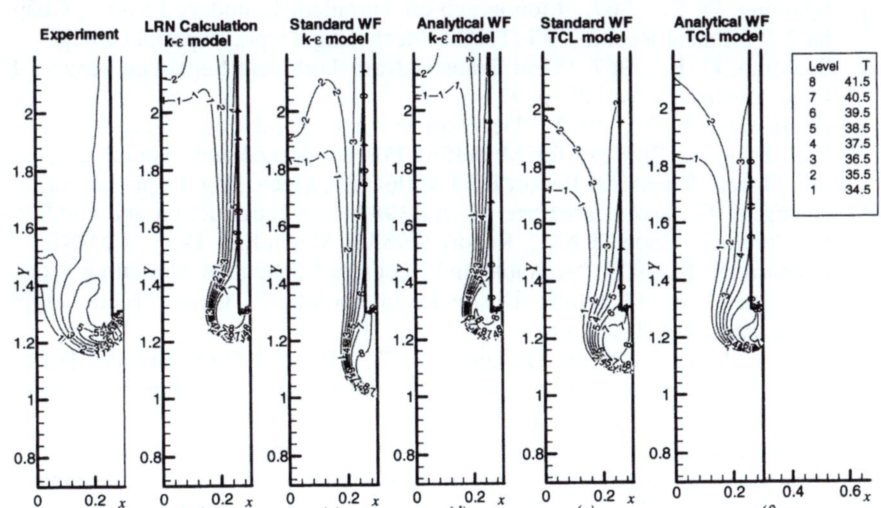

Fig. 9. Predicted vertical velocity contours for opposed downward wall jet flow using k-ε and second -moment-closure (TCL) model with different wall functions.

Acknowledgments

Financial support for this work has been coordinated by British Energy plc through contract no. BWD40064589. AVG also acknowledges the support of a UK ORS scholarship.

Authors' names are listed alphabetically.

References

[1] Craft, T. J., Gant, S. E, Gerasimov, A. V., Iacovides, H., and Launder, B. E., 2002, "Wall Function Strategies for Use in Turbulent Flow CFD", Invited Keynote Paper, 12th Int. Heat Transfer Conference, Grenoble, France.

[2] Craft, T. J., Gant, S. E., Iacovides, H., and Launder, B. E., 2001, "Development and Application of a New Wall Function for Complex Turbulent Flows", ECCOMAS CFD Conference, Swansea, Wales, UK.

[3] Craft, T. J., Gerasimov, A. V., Iacovides, H., and Launder, B. E., 2002, "Progress in the Generalization of Wall Function Treatments", Int. J. Heat and Fluid Flow, Vol. 23, No. 2, pp. 148-160.

[4] Gerasimov, A. V., 2001, "Development of a New Analytical Wall Function Approach for Modelling Mixed Convection Flow Phenomena", Technical Report BWD40029343, Mech. Eng. Dept., UMIST

[5] Spalding, D. B., 1967, "Monograph on Turbulent Boundary Layers", Chapter 2, Technical Report TWF/TN/33, Mech. Eng. Dept., Imperial College

[6] Spalding, D. B., 1967, "Heat Transfer from Turbulent Separated Flows", J. Fluid Mechanics, vol. 27, p. 97

[7] Gerasimov, A. V., 2002, "Development and Validation of an Analytical Wall Function Strategy for Modelling Forced, Mixed and Natural Convection Flows", Technical Report BWD40064589, Mech. Eng. Dept., UMIST

[8] Huang, P. G., and Leschziner, M. A., 1983, "An Introduction and Guide to the Computer Code TEAM", Report TF/83/9, Mech. Eng. Dept., UMIST

[9] Leonard, B. P., 1979, "A Stable and Accurate Convective Modelling Procedure Based on Quadratic Upstream Interpolation", Comp. Meth. Appl. Mech. Eng., vol. 19, p. 59

[10] Jackson, J. D., He, S., Xu, Z., and Wu, T., 2002, "CFD Quality and Trust – Generic Studies of Thermal Convection", Technical Report HTH/GNSR/5029, University of Manchester

[11] Launder, B. E., and Sharma B. I., 1974, "Application of the Energy-Dissipation Model of Turbulence to the Calculation of Flow near a Spinning Disk", Letters in Heat and Mass Transfer, vol. 1, pp. 131-138

[12] Johnson, R. W., and Launder, B. E., 1980, "Discussion of 'On the Calculation of Turbulent Heat Transport Downstream from an Abrupt Pipe Expansion'", Numerical Heat Transfer, vol. 5, pp. 493-486

[13] Chieng, C. C., and Launder, B. E., 1982, "On the Calculation of Turbulent Heat Transport Downstream from an Abrupt Pipe Expansion", Numerical Heat Transfer, vol. 3, pp. 189-207

[14] Jones, W. P., and Launder, B. E., 1972, "The Prediction of Laminarization with a Two-Equation Model of Turbulence", Int. J. Heat Mass Transfer, vol. 15, pp. 301-314

[15] Craft, T. J., Gerasimov, A. V., Iacovides, H., and Launder, B. E., 2004, "The Negatively Buoyant Turbulent Jet: Performance of Alternative Options in RANS modelling", Submitted for publication in Int. J. of Heat and Fluid Flow.

[16] Craft, T. J., Ince, N. Z., and Launder B. E., 1996, "Recent Developments in Second-Moment Closure for Buoyancy-Affected Flows", Dynamics of Atmospheres and Oceans, vol. 23, pp. 99-114.

[17] Addad, Y., Benhamadouche S., and Laurence D., 2004, "The Negatively Buoyant Wall-Jet: LES Database", Submitted for publication in Int. J. of Heat and Fluid Flow.

RANS Simulations for Film-Cooling Analysis and Design

Gorazd MEDIC*, research associate, Paul DURBIN, professor
Flow Physics and Computation Division
Department of Mechanical Engineering Stanford University

*Corresponding author: Bldg. 500, Rm. 500V, Stanford, CA 94305-3030, USA
Tel.: (+1 650) 723 8476 Fax: (+1 650) 725 3525, Email: gmedic@stanford.edu

Abstract In this paper we present the application of Reynolds-Averaged Navier-Stokes (RANS) equations with the eddy-viscosity turbulence models for turbomachinery problems with the particular emphasis on the heat transfer and film cooling of gas turbines. We discuss several RANS-related issues, such as the overproduction of turbulent kinetic energy, as well as the problems related to the simulation of unsteady flows.

Finally, we present a case of optimal shape design coupled with a RANS solver used to setup a turbine cascade experiment.
Key Words: Reynolds-Averaged Navier-Stokes equations, turbomachinery heat transfer, turbulence models

Nomenclature

c	$[m]$	blade chord
h	$[W/(m^2K)]$	heat transfer coefficient
j	[]	cost function (dimensions vary)
m	[]	dimensionless blowing ratio
$\dot{q}$	$[W/m^2]$	heat flux rate
s	$[m]$	curvilinear coordinate
y	[m]	wall distance
H	[m]	channel height
M	[-]	dimensionless Mach number
Pr	[-]	dimensionless Prandtl number
Re	[-]	dimensionless Reynolds number
T	$[K]$	temperature
$\mathcal{T}$	[s]	turbulence time scale
Tu	[-]	dimensionless turbulence intensity
θ	[m]	momentum thickness

Subscripts and Superscripts

in	inflow
ex	exit
w	wall
0	total conditions
is	isentropic conditions
∞	free stream
c	coolant

1. Introduction

Three-dimensional numerical simulation of heat transfer and film cooling is becoming a part of complex procedure of gas turbine blade design and is slowly replacing simpler two-dimensional methods mainly based on boundary layer computations and relying strongly on experimental databases. However, many fundamental and practical developments are needed before the full three-dimensional computational analysis becomes reliable.

In this paper we present three applications of RANS-based numerical simulations for turbomachinery problems. First, we discuss how the overprediction of turbulent kinetic energy effects the heat transfer predictions for an experimentally document test case of a film-cooled gas turbine rotor blade (VKI blade test case, see [9-11]) and how different eddy-viscosity based turbulence models (k-ε, k-ω and v^2-f models) deal with that problem. The second test case represents an unsteady three-dimensional slot jet, where we are currently conducting detailed analysis of unsteady effects (including jet pulsations [12]) using RANS methodology coupled with eddy-viscosity turbulence models.

Finally, the third example is a case of inverse design that was used in a process of designing an experimental rig at Stanford University [13] meant to help further assess how well the different turbulence models are predicting turbulence in a turbine cascade.

2. Turbulence Models

In the analysis of flow through jet engine gas turbine blade cascades we are dealing with transonic compressible flow. The equations used to describe the flow are the Favre averaged, compressible Navier-Stokes equations with turbulence models that provide eddy viscosities. The effects of turbulence are taken into account through a constitutive model:

$$R = -\frac{2}{3}\rho k I + \mu_t \left[\left(\nabla U + \nabla U^T \right) - \frac{2}{3} (\nabla \cdot U) I \right] \tag{1}$$

$$\dot{q}_t = \frac{\mu_t C_P}{\Pr_t} \nabla T \tag{2}$$

for the Reynolds stress and heat flux tensors. To predict the eddy viscosity μ_t we used the following turbulence models: two-layer k-ε model with a bound on turbulence time scale $\mathcal{T}$, v^2-f model and variants of k-ω model. The heat flux was computed with a turbulent Prandtl number $\Pr_t = 0.9$.

2.1. Two-layer formulation of k-ε model

The two-layer formulation of k-ε model consists in patching together the k-l and k-ε models. The standard model equation for turbulent kinetic energy is:

$$\partial_t(\rho k) + \nabla \cdot (\rho U k) = \rho P_k - \rho\varepsilon + \nabla \cdot ((\mu + \mu_t)\nabla k) \tag{3}$$

where

$$\rho P_k = R : \nabla U = -\frac{2}{3}\rho k(\nabla \cdot U) + 2\mu_t |S|^2 - \frac{2}{3}\mu_t (\nabla \cdot U)^2 \tag{4}$$

with $S_{ij} = 1/2(\partial_j U_i + \partial_i U_j)$. It is used in both k-l and k-ε formulation.

In the k-ε model, the equation for turbulent kinetic energy Eq. (3) is supplemented by

$$\partial_t(\rho\varepsilon) + \nabla \cdot (\rho U \varepsilon) = \frac{C_{\varepsilon 1}\rho P_k - C_{\varepsilon 2}\rho\varepsilon}{\mathcal{T}} + \nabla \cdot \left(\left(\mu + \frac{\mu_t}{\sigma_\varepsilon}\right)\nabla\varepsilon\right) \tag{5}$$

with the turbulence time-scale $\mathcal{T} = k/\varepsilon$ and the eddy viscosity $\mu_t = C_\mu \rho k \mathcal{T}$.

The standard model constants are:

$$C_{\varepsilon 1} = 1.44; \quad C_{\varepsilon 2} = 1.92; \quad \sigma_\varepsilon = 1.3; \quad C_\mu = 0.09$$

In the k-l formulation, the dissipation rate is represented by $\varepsilon = k^{3/2}/l_\varepsilon$ and the eddy viscosity by $\mu_t = C_\mu \rho \sqrt{k}\, l_v$. These supplement the k-equation.

The Van Driest form for the length scales will be adopted here

$$l_\varepsilon = C_1 y\left(1 - e^{-R_y/A_\varepsilon}\right) \quad l_v = C_1 y\left(1 - e^{-R_y/A_v}\right) \tag{6}$$

where $R_y = y\rho\sqrt{k}/\mu$ is a wall-distance based Reynolds number. The log-layer solution requires $C_1 = \kappa / C_\mu^{3/4}$, with κ the von Karman constant → κ = 0.41 gives $C_1 = 2.5$. The requirement $k \to \varepsilon y^2/2\nu$ as $y \to 0$ implies $A_\varepsilon = 2C_1 = 5.0$. The only free constant A_v was chosen as $A_v = 62.5$. The two-layer formulation consists simply of using the k-l formulation near a wall, and switching abruptly to a k-ε formulation at a patching point. For solid walls, the boundary condition $k(0) = 0$ is applied.

This standard two-layer k-ε formulation has a problem of producing excessive levels of turbulent kinetic energy in regions of large rate of strain. Various fixes for that problem have been proposed and we will use the concept of a bound on turbulence time-scale $\mathcal{T}$, proposed in [1] and analyzed in more details in [2]:

$$\mathcal{T} = \min\left[\frac{k}{\varepsilon}, \frac{\alpha}{\sqrt{6}C_\mu |S|}\right] \tag{7}$$

where the value of the coefficient α was selected as $\alpha = 0.6$ in [4] for the v^2-f model, and we apply the same value here for the k-ε model.

Other types of limiters could be used. In [2] we analyzed Kato-Launder approach of modifying the production term ρP_k, Eq. (4), as well as the modification of the dissipation rate equation, Eq. (5), proposed by Chen. Those modifications yield similar results to the ones obtained using the bound on turbulence time-scale $\mathcal{T}$, Eq. (7).

2.2. v^2-f model

The eddy viscosity is now defined as $\mu_t = C_\mu \rho \overline{v^2} \mathcal{T}$ and the turbulence time-scale as:

$$\mathcal{T} = \min\left[\max\left[\frac{k}{\varepsilon}, 6\sqrt{\frac{\nu}{\varepsilon}}\right], \frac{\alpha k}{\sqrt{6\overline{v^2}}C_\mu |S|}\right] \tag{8}$$

with $\alpha = 0.6$. The equations for k and ε are now supplemented with the equation for $\overline{v^2}$:

$$\partial_t\left(\rho\overline{v^2}\right) + \nabla\cdot\left(\rho U\overline{v^2}\right) = \rho k f - \rho N\frac{\overline{v^2}}{k}\varepsilon + \nabla\cdot\left((\mu+\mu_t)\nabla\overline{v^2}\right) \tag{9}$$

with f representing the nonlocal effects

$$f - L^2\Delta f = (C_{f1}-1)\frac{2/3-\overline{v^2}/k}{\mathcal{T}} + C_{f2}\frac{P_k}{k} + (N-1)\frac{\overline{v^2}}{k\mathcal{T}} \tag{10}$$

where the turbulence length scale L is

$$L = C_L \max\left[\min\left[\frac{k^{3/2}}{\varepsilon}, \frac{k^{3/2}}{\sqrt{6\overline{v^2}}C_\mu |S|}\right], C_\eta \frac{\nu^{3/4}}{\varepsilon^{1/4}}\right] \tag{11}$$

For solid walls, when $y \to 0$ this yields

$$k(0)=0; \overline{v^2}(0)=0; \varepsilon \to \frac{2\nu k}{y^2}; f \to \frac{4(6-N)\nu^2\overline{v^2}}{\varepsilon y^4} \tag{12}$$

The original v^2-f model with $N = 1$ was later modified in order to avoid the numerical difficulties due to strong nonlinear coupling of turbulence variables through the boundary conditions – the value of $N = 6$ was chosen (for more details see [4,5]).

An additional modification was to eliminate the wall distance from the equation for $C_{\varepsilon 1}$ resulting in the following values for the model constants:

$$C_\mu = 0.22; \quad C_{\varepsilon 1} = 1.4\left(1 + 0.050\sqrt{k/v^2}\right) \quad C_{\varepsilon 2} = 1.9; \quad \sigma_\varepsilon = 1.3;$$

$$C_{f1} = 1.4; \quad C_{f2} = 0.3; \quad C_L = 0.23; \quad C_\eta = 70;$$

2.3. Variants of k-ω model

In [2] we have also studied the standard Wilcox k-ω model [6] and we have proposed adding the bound on turbulence time scale in that model as well. With the eddy viscosity $\mu_t = C_\mu \rho k \mathcal{T}$, we proposed to use:

$$\mathcal{T} = \min\left[\frac{1}{C_\mu \omega}, \frac{\alpha}{\sqrt{6} C_\mu |S|}\right] \tag{13}$$

Another undesirable property of the standard k-ω model is its strong sensitivity with respect to free stream conditions. To circumvent those problems a model that blends together the k-ω model used near the surface and the k-ε model in the outer region was developed in [7]. The overprediction of the eddy viscosity is still a problem for the model defined in this way and in [7] a limiter was introduced in the definition of the eddy viscosity:

$$\mu_t = C_\mu \rho k \mathcal{T}, \qquad \mathcal{T} = \min\left[\frac{1}{C_\mu \omega}, \frac{c_1}{C_\mu |S| F_2}\right] \tag{14}$$

with F_2 a blending function restricting the limiter to the boundary layer. That model is usually referred to as the shear stress transport (SST) k-ω model [7].

The boundary conditions applied at the solid wall are:

$$k(0) = 0; \quad \omega \to \frac{6v}{\beta_1 y^2} \text{ as } y \to 0 \tag{15}$$

2.4. Numerical method

For numerical solution of the flow equations we have employed a standard commercial software package STAR-CD with turbulence models implemented through user defined subroutines.

STAR-CD is an implicit finite volume solver that employs a variant of the well-known SIMPLE method [8], with the turbulence equations decoupled and solved sequentially. It can be used as a general convection-diffusion equation solver via user-defined subroutines. Two-layer k-ε model with a bound on turbulence time scale $\mathcal{T}$ and the v^2-f model were programmed through these user defined subroutines, as well as variants of Wilcox k-ω model.

3. Anomalous Turbulent Energy

Computational analysis for the VKI experiment (see [2,3,9-11]) represents a good illustration of the effect of the overestimation of turbulent kinetic energy on the heat transfer predictions. This test case consists of a film-cooled gas turbine rotor blade. Here, we will present numerical results for the heat transfer coefficient for a case without film-cooling, as well as for a case of suction side cooling.

The blade chord length, c, equals 80 mm, with pitch-to-chord ratio equal to 0.67 and the stagger angle of 38.5 deg. The flow through cascade is transonic with the inflow conditions: M_{in} = 0.25, $Re_{c,in}$ = 8.5 105, $T_{0\infty}$ = 409.5 K and the exit Mach number is $M_{ex,is}$ = 0.92. The inflow angle is 30 deg and the wall temperature is T_w = 298 K. The inflow turbulence intensity is Tu_{in} = 5% and the length scale $C_\mu^{3/4} k^{3/2} / \varepsilon$ was estimated as ≈ 1 cm.

The geometry of the blade and the details of the film cooling holes are discussed in more details in [3].

3.1. Heat Transfer Coefficient

The following concentrates on the results that concern the blade heat transfer when there is no film cooling: the surface isentropic Mach number, M_{is}, and the heat transfer coefficient, $h_t = \dot{q}_w / (T_{0\infty} - T_w)$. These quantities are presented with respect to curvilinear coordinate s (normalized by the blade chord c) with the origin at the stagnation point.

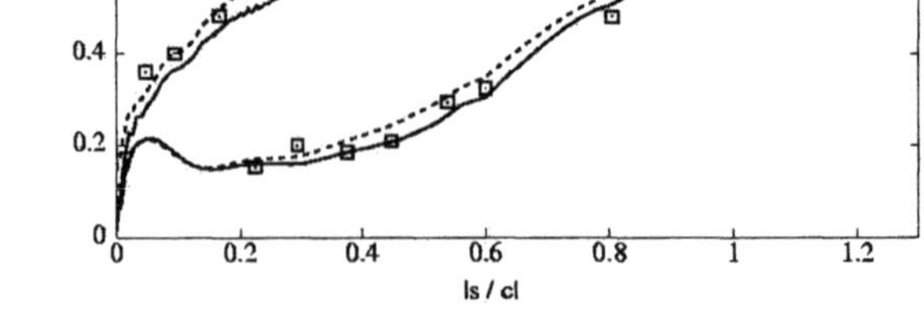

Fig. 1. Isentropic Mach number, Mis, vs. normalized curvilinear coordinate ls/cl

In terms of heat transfer, here we will present only the results obtained with the standard k-ε model, k-ε model with the bound on $\mathcal{T}$ and v^2-f model. Results obtained using k-ω models were reported in [2,3] and they follow similar trends (with respect to the effect of limiters).

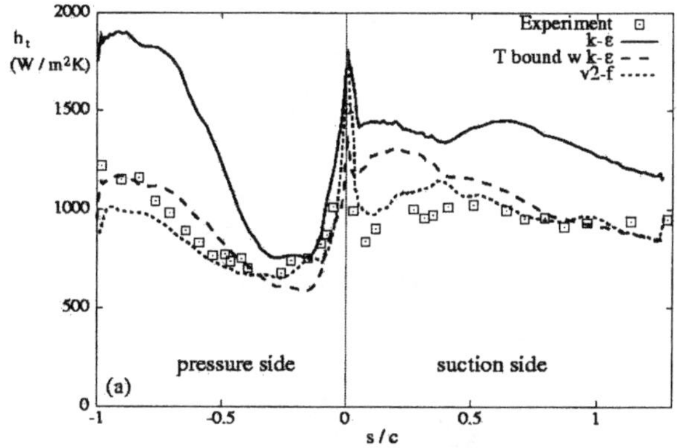

Fig. 2. Heat transfer coefficient ht [W/(m2K)] vs. normalized curvilinear coordinate s/c

3.2. Turbulent kinetic energy

The reason for the difference in heat transfer predictions obtained with different models must be related to the quantity that links turbulence equations to the mean flow and temperature – the eddy viscosity. For the presented models the eddy viscosity is related to the turbulent energy. In order to understand the heat transfer predictions we consider this quantity.

Examining the values of turbulent kinetic energy predicted in the center of the passage, we see that switching from the standard k-ε model to a model that includes a limiter ($\mathcal{T}$ bound with k- ε or v^2-f model) reduces k by a factor on the order of 30 times. This is illustrated in Figs. 3 to 5 by plotting the turbulent intensity $\mathrm{Tu} = \sqrt{2/3k}/|U|$:

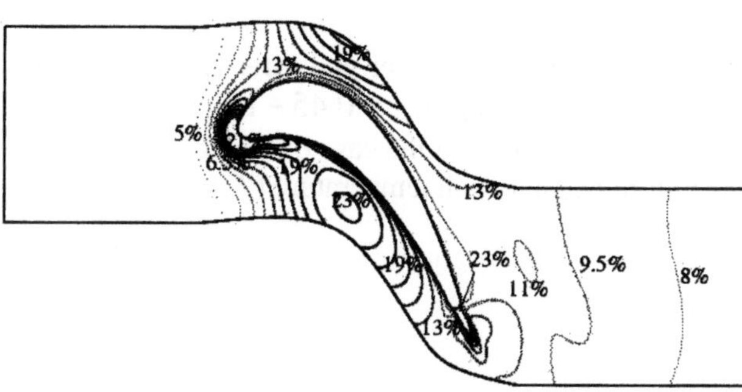

Fig. 3. Turbulence intensity Tu, k-ε model.

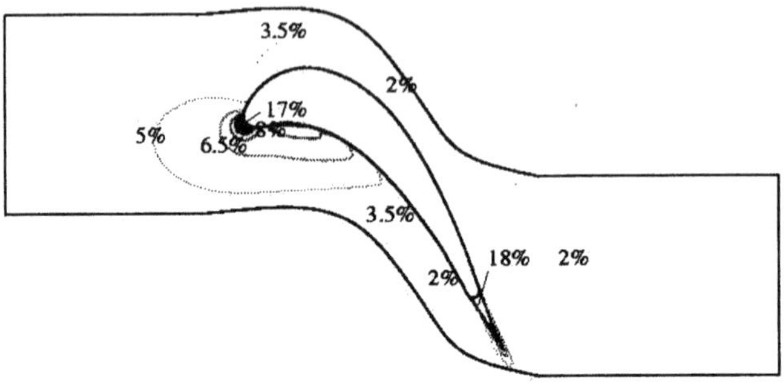

Fig. 4. Turbulence intensity Tu, T bound with k-ε model.

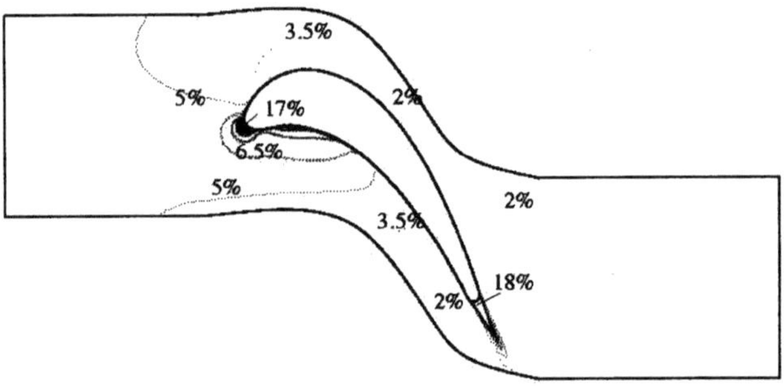

Fig. 5. Turbulence intensity Tu, v2-f model.

In Fig. 2 we see that there is an additional difference between the results obtained with the k-ε model with a bound on $\mathcal{T}$ and v^2-f model. v^2-f model gives results that agree better with the data on the suction side of the blade, while at the same time it fails to recover to the right heat transfer coefficient level at the pressure side close to the trailing edge. These effects are consistent with the results reported in [5] where there was some speculation about the modified v^2-f ($N = 6$) model relaminarizing prematurely.

3.3. Film Cooling

The coolant inflow conditions of Camci and Arts [9-11] were in the following range: blowing ratios $m = \rho_c U_c / \rho_\infty U_\infty = 0.45 - 1.0$ and coolant temperatures $T_c / T_w = 0.6\text{-}1.0$. Here we will focus on the case of high blowing ratio (m = 1.0) with the coolant temperature equal to wall temperature ($T_c = T_w$).

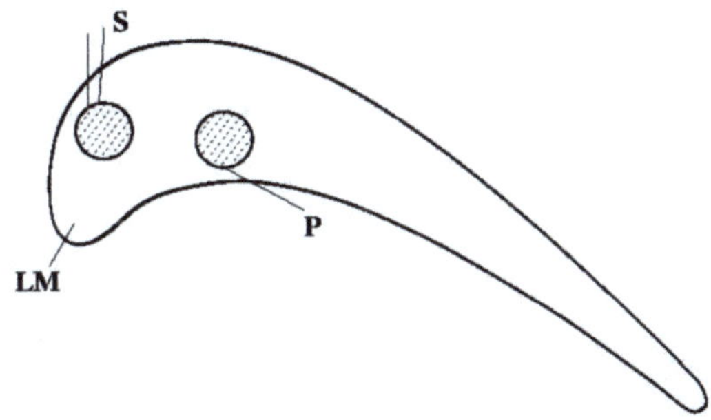

Fig. 6. Geometry of cooling holes.

Mesh generation for this flow problem represents a major challenge. We have opted for multi-block structured mesh with nonmatching (nonconforming) interfaces. This approach gives significant flexibility in generating generic mesh-blocks for different film cooling configuration while retaining sufficient control over the near-wall (near-hole) spacing. A final computational mesh with approximately 750,000 nodes was used in the computations for the suction side cooling with $y^+_1 < 1$ for all solid walls.

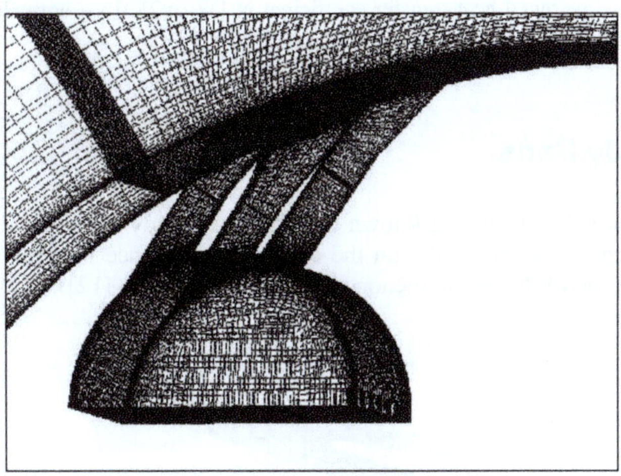

Fig. 7. Detail of the computational mesh, suction side cooling holes (S).

Overproduction of turbulent kinetic energy in the center of the passage seen in the computations with standard k-ε model causes the ht predictions using the standard k-ε model to be completely erroneous. On the other hand, the heat transfer in the region in the vicinity of cooling holes seems to be strongly influenced by the flow pattern within the cooling tubes (see [3]). The differences in the magnitude of the crossflow, as well as the higher velocities in the core of the cooling jets, influence the extent of jet penetration into the mainstream and the lateral spreading of cool fluid. As a result, the v^2-f model seems to be predicting lower values of spanwise averaged heat transfer coefficient, as presented in Fig. 8. These differences originate inside the hole (as reported in [3]), and the uncertainties with regard to the inflow conditions for the coolant plenum might have a significant influence over those predictions.

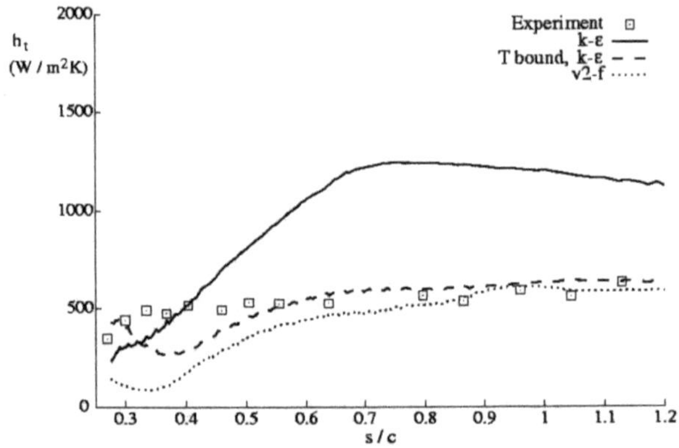

Fig. 8. Spanwise averaged heat transfer coefficient ht [W/(m2K)] vs. normalized curvilinear coordinate s/c, suction side cooling, m = 1.0, Tc = Tw

4. Unsteady Rans

Three dimensional slot jets are known to exhibit unsteady behavior. Here we will present some preliminary results on the effect of turbulence modeling on the numerical resolution of this phenomenon (for more details, see [12]).

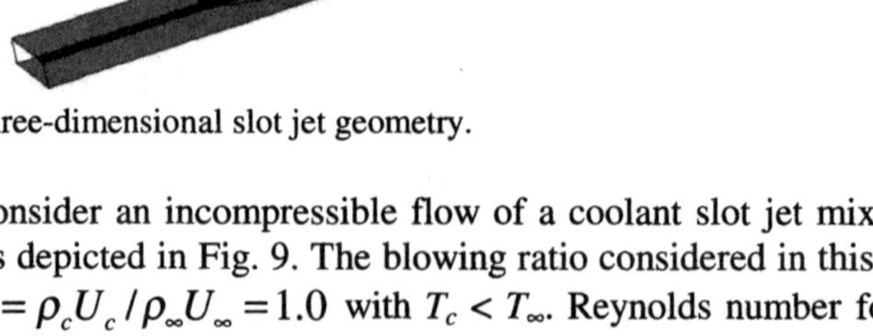

Fig. 9. Three-dimensional slot jet geometry.

We consider an incompressible flow of a coolant slot jet mixing with the free stream as depicted in Fig. 9. The blowing ratio considered in this particular analysis is $m = \rho_c U_c / \rho_\infty U_\infty = 1.0$ with $T_c < T_\infty$. Reynolds number for the coolant jet based on the channel height H was $Re_{c,h} = 20{,}000$, and momentum thickness Reynolds number for the free stream was chosen as $Re_{\infty,\theta} = 5000$. The computational mesh, chosen after a detailed mesh independence study, contains approximately 650,000 nodes with $y^+_1 < 1$ for all solid walls.

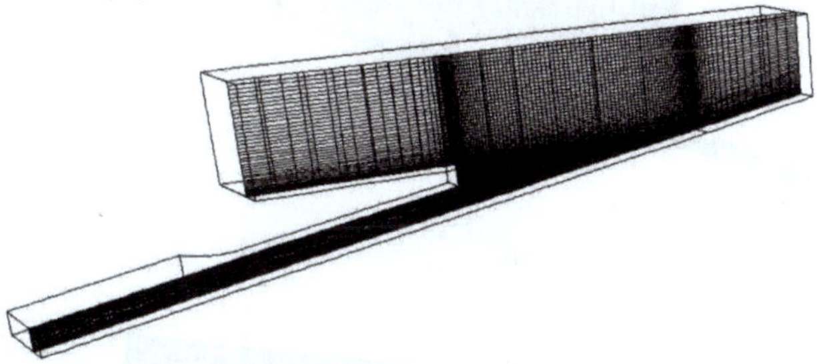

Fig. 10. Cross-section of computational mesh.

An unsteady flow occurs with a vortex shedding Strouhal number of approximately $St = f\,H\,/\,U = 0.2$ (see [12]). This unsteadiness strongly enhances the mixing of the coolant with the free stream as indicated in Fig. 11b.

Obviously, standard k-ε model fails to produce unsteady results – there is an overestimation of turbulent kinetic energy in the region where we expect to see the unsteady vortex shedding. Initially, we used Menter's SST k-ω model, but we obtained virtually the same results with the two-layer k-ε model with a bound on turbulence time scale $\mathcal{T}$ and those results are presented herein.

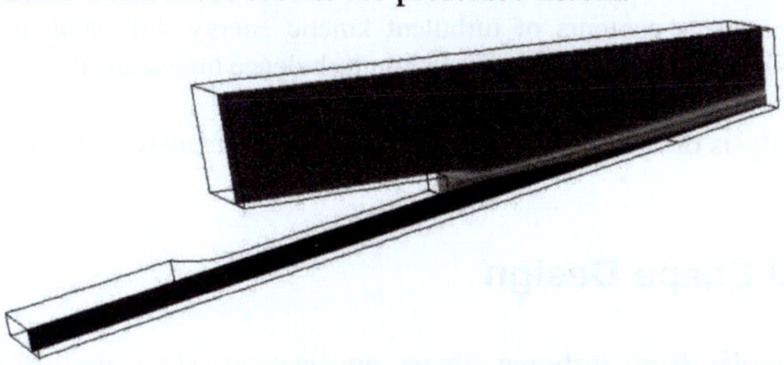

(a)

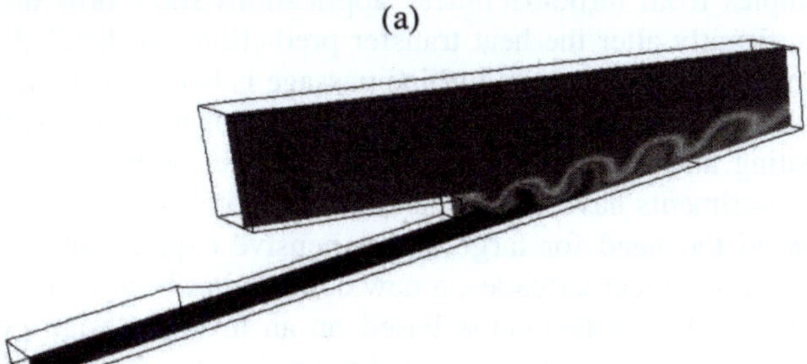

(b)

Fig. 11. Instantaneous contours of temperature: (a) standard two-layer k-ε model; (b) two-layer k-ε model with a bound on turbulence time scale T

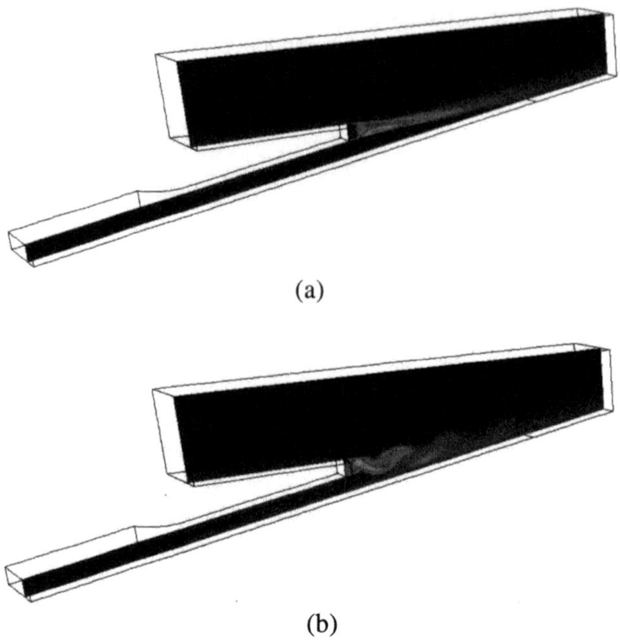

Fig. 12. Instantaneous contours of turbulent kinetic energy: (a) standard two-layer k-ε model; (b) two-layer k-ε model with a bound on turbulence time scale T

Further analysis of these unsteady phenomena is currently under way [12].

5. Optimal Shape Design

Previous examples from turbomachinery applications show how different turbulence can significantly alter the heat transfer predictions. A detailed experimental investigation of turbulence in a gas turbine passage is being currently conducted at Stanford University aiming at further understanding of the underlying physics, as well as validating and improving existing turbulence models. Heat transfer and film cooling experiments have also been conducted on a similar model (see [14]). In order to avoid the need for large and expensive experimental facilities used when analyzing large linear cascades, a new concept has been conceived (see Laskowski et al [13]). A new technique based on an inverse design procedure was used to define a double passage cascade that produces essentially the same flow conditions on the blade and in the major part of the passage as the ones seen in an infinite cascade.

We initially decided to try to match the blade surface pressure in the form of surface isentropic Mach number of the blade in the double passage to that of the blade in an infinite cascade. Thus the problem statement is

$$\min_{\phi}\{j(\phi,U(\phi)) : E(\phi,U(\phi))=0\} \quad (16)$$

where j is the cost function, U is the state variable vector and E is the set of governing equations to be solved. ϕ is the vector of control variables that were chosen here as the spline points used in constructing the double passage walls. The cost function is defined as:

$$j=\sum_{i=1}^{I}\alpha_i\gamma_i(U) \quad (17)$$

where α_i are weight coefficients, I is the number of the cost function components, and γ_i are the cost function components. Initially, as previously explained, I was taken to be 1, α_I was set to 1 and γ_I was defined as:

$$\gamma_1 = \int_{\Gamma_{blade}} \beta(s)\left|M_{is} - M_{is}^{*}\right|ds \quad (18)$$

with $\beta(s)$ a step-function type scaling factor, M_{is} and M_{is}^* the blade surface isentropic Mach number distributions from the double passage simulation and from the target infinite cascade simulation, respectively. In the course of analysis, further components were added to the cost function: blade surface shear stress; shear stress along the double passage walls (more details can be found in [13]).

The method of steepest descent coupled with an approximate line search was used as the optimization method and the gradients were assembled using finite differences. Two-layer k-ε model with a bound on turbulence time scale $\mathcal{T}$ was used in the flow solver.

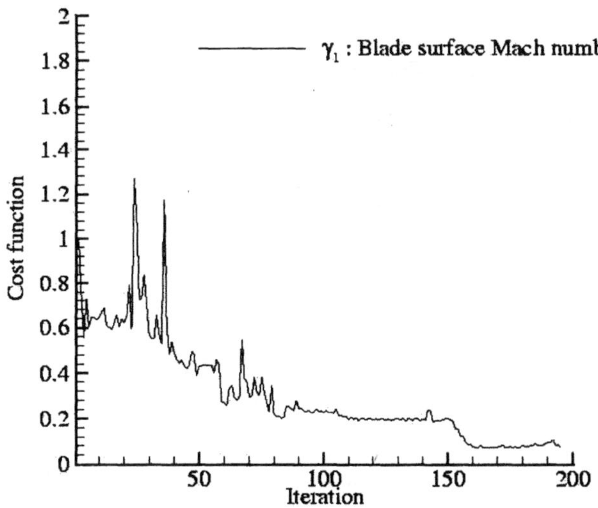

Fig. 13. Convergence of the optimization procedure.

As shown in Figs. 13 to 15, the optimization procedure converges to a local solution for the geometry (in fact, three different designs are presented in [13]) with excellent agreement for the blade surface isentropic Mach number, as well as a reasonably fine agreement throughout the passage.

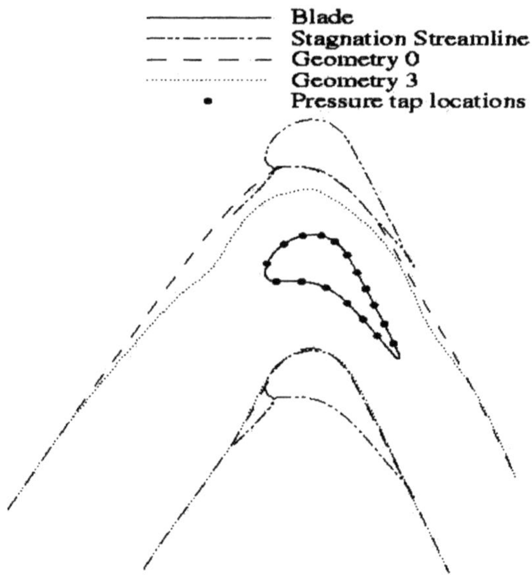

Fig. 14. Initial design for the double passage -Geometry 0; and the final result of the optimization procedure – Geometry 3

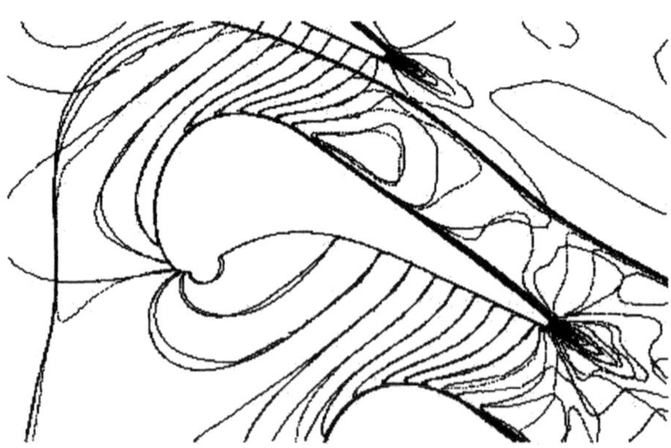

Fig. 15. Mach number contours comparison between the final result – Geometry 3 and the infinite cascade

Once the experimental rig has been built, the blade surface pressure has been measured and compared to the computational results – the measurements presented in [13] show perfect agreement between computational results for both infinite cascade and double passage and the experimental data.

6. Summary

We have shown herein that substantially improved heat transfer and film cooling predictions can be obtained when eddy-viscosity based turbulence models are carefully used. Limiters, such as the bound on turbulence time scale $\mathcal{T}$, that prevent anomalous levels of turbulent kinetic energy predictions in a turbine passage are an example of simple and efficient improvement to widely used turbulence models. As a final illustration we presented an application of these improved eddy-viscosity models as a part of an optimal shape design algorithm used for the design of an experimental rig.

Acknowledgments

This work was sponsored by General Electric Aircraft Engines.

The authors wish to thank Greg Laskowski for providing the figures for the section on optimal shape design.

References

[1] Durbin, P.A., 1996, "On the k-ε Stagnation Point Anomaly", Int. J. Heat Fluid Flow, Vol. 17, pp. 89-90.

[2] Medic, G. and Durbin, P.A., 2002, "Toward Improved Prediction of Heat Transfer on Turbine Blades", ASME J. Turbomachinery, Vol. 124, pp. 187-192.

[3] Medic, G. and Durbin, P.A., 2002, "Toward Improved Film-Cooling Prediction", ASME J. Turbomachinery, Vol. 124, pp. 193-199.

[4] Behnia, M., Parneix, S., Shabany, Y. and Durbin, P.A., 1999, "Numerical Study of Turbulent Heat Transfer in Confined and Unconfined Impinging Jets", Int. J. Heat Fluid Flow, Vol. 20, pp. 1-9.

[5] Lien, F.S. and Kalitzin, G., 2001, "Computations of Transonic Flows with v^2-f Turbulence Models", Int. J. Heat Fluid Flow, Vol. 22, pp. 53-61.

[6] Wilcox, D.C., 1993, "Turbulence Modeling for CFD", DCW Industries, Inc., La Canada, CA.

[7] Menter, F.R., 1994, "Two Equation Eddy-Viscosity Turbulence Models for Engineering Applications", AIAA Journal, Vol. 32(8), pp. 1598-1605.

[8] STAR-CD Version 3.15 – Methodology, 2002, Computational Dynamics Limited.

[9] Camci, C. and Arts T., 1985, "Short Duration Measurements and Numerical Simulation of Heat Transfer Along the Suction Side of a Film-Cooled Gas Turbine Blade", ASME J. Eng. Power, Vol. 107, pp. 991-997.

[10] Camci, C. and Arts T., 1985, "Experimental Heat Transfer Investigation Around the Film-Cooled Leading Edge of a High Pressure Gas Turbine Rotor Blade", ASME J. Eng. Power, Vol. 107, pp. 1016-1021.

[11] Camci, C. and Arts T., 1990, "An Experimental Convective Heat Transfer Investigation Around a Film-Cooled Gas Turbine Blade", ASME J. Turbomachinery, Vol. 112, pp. 497-503.

[12] Medic, G. and Durbin P.A., 2003, "Numerical Simulation of Pulsating Slot Jets", in preparation.

[13] Laskowski, G., Medic, G., Durbin, P.A., Vicharelli, A., Elkins, C.L. and Eaton, J.K., 2003, "CFD-Based Inverse Design of a Double Passage Transonic Turbine Cascade Model", submitted to ASME J. Turbomachinery.

[14] Kodzwa, P.M., Elkins, C.L. and Eaton J.K., 2003, "Measurements of Film Cooling Performance in a Transonic Single Passage Model", 2nd International Conference on Heat Transfer, Fluid Mechanics and Thermodynamics, June 23-26.

Very Large Eddy Simulation for the Prediction of Unsteady Vortex Motion

Albert RUPRECHT*, Thomas HELMRICH, Ivana BUNTIC
Institute of Fluid Mechanics and Hydraulic Machinery
University of Stuttgart

*Corresponding author: Pfaffrnwaldring 10, D-70550 Stuttgart, Germany
Tel: (+49) 711 685 3256, Fax: (+49) 711 685 32 55
E-mail: ruprecht@ihs.uni-stuttgart.de

Abstract A new turbulence model for Very Large Eddy Simulation, based on the extended k-ε model of Chen and Kim is developed and presented in this paper. Introducing an adaptive filtering technique, the model can distinguish between numerically resolved and unresolved parts of flow. It is applied to the simulation of unstable vortex motion in a pipe trifurcation. This flow phenomenon cannot be predicted with classical RANS methods and commonly used turbulence models. Using the VLES method with the new turbulence model, the phenomenon is well predicted and the results agree reasonably well with measurement data.

Nomenclature

f	filter function
$h_{\max}$	local grid size
k	turbulent kinetic energy
L	Kolmogorov length scale
P_k	production term
U	local velocity
α	model constant
Δ	resolved length scale
Δt	time step
ε	dissipation rate
ν	kinematic viscosity
ν_t	turbulent viscosity

1. Introduction

The fundamental problem of Computational Fluid Dynamics (CFD) remains the prediction of the turbulence effects. Over the decade using the Reynolds-averaged Navier-Stokes (RANS) equations has become a standard industrial simulation tool for the design and analysis of engineering systems involving fluid flow. It means that the complete turbulence behaviour is expressed by an appropriate turbulence model which has to take into account all turbulent scales ranging from the largest eddies down to the Kolmogorov scale. Consequently it is extremely difficult to define such a model suitable for prediction of all flow phenomena, especially for prediction of unsteady vortex motions.

For resolving all turbulent scales down to the Kolmogorov scale, a Direct Numerical Simulation (DNS) is nowadays one of the possibilities with the highest accuracy. As the smallest scales strongly decrease with increasing Reynolds number, a very fine grid is essential for DNS. Carrying out 3D simulations for complex geometries is in that case time consuming and requires high computing power. Therefore, DNS cannot be applied to the flows of practical relevance in the foreseeable future.

Large Eddy Simulation (LES) is likely to be a compromise. It is viewed as a mature technique for studying complex flow, although its major limitation is still excessive computational costs. In the "real" LES (from the turbulence research point of view) all anisotropic turbulence structures are resolved in the computation and only the smallest isotropic scales are modelled. Therefore, the used turbulence models can be simpler compared to RANS models, since they only have to describe the influence of the isotropic scales on the resolved anisotropic flow filed. Unfortunately, with increasing Reynolds number the anisotropic scales decrease becoming isotropic and not resolvable. There are many "LES" applications to the engineering relevant flows reported in the literature, although from the turbulence research point of view these simulations are mostly unsteady RANS (URANS), since they only resolve the unsteady mean flow not taking into account or resolving any turbulence structure.

If there is a gap in the turbulence spectrum between the unsteady mean flow and the turbulent flow, a "classical" URANS models (e.g. k-ε) can be applied, as they are developed for modelling the whole range of turbulent scales (see Fig. 1.1). To the contrary these models cannot be employed properly. Clearly "classical" RANS is not suitable for prediction and analysis of many unsteady vortex phenomena.

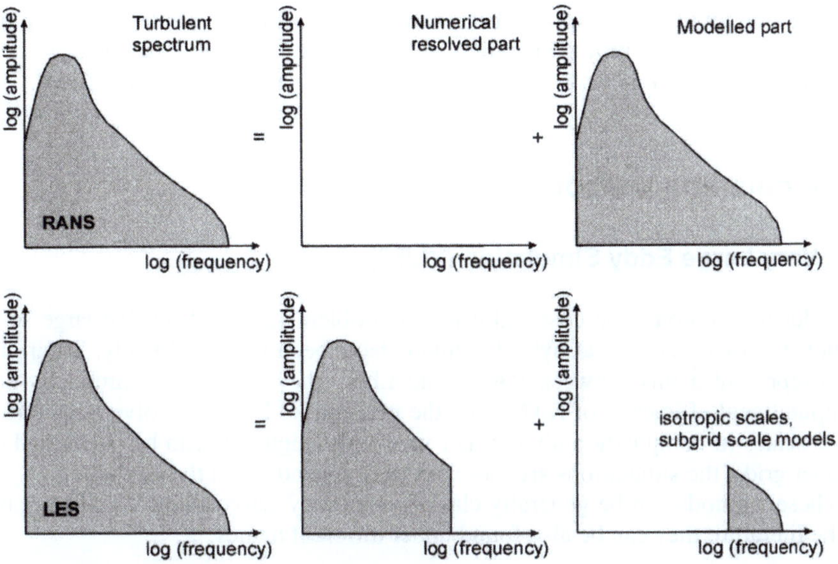

Fig. 1.1. Modelling approaches for RANS and LES

Currently VLES seems to be a promising way for simulations of industrial flow problems. It is similar to the LES, only that a smaller part of the turbulence spectrum is resolved. Applied turbulence model has to be more advanced since it has to express the influence of a larger part of the spectrum. In Table 1.1 the availability of the different approaches for the simulation of the flow around an aircraft is summarised. Other complex flow problems can similarly refer to these data.

Table 1.1. Application of the different approaches to the flow around an aircraft [3]

Approach	Re-number dependency	Empirical	Required grid size	Required time steps	Available
RANS	Weak	High	10^7	$10^{3.5}$	1995
VLES	Weak	High	10^8	10^4	2000
LES	Weak	Small	$10^{11.5}$	$10^{6.7}$	2045
DNS	Strong	Zero	10^{16}	$10^{7.7}$	2080

Development of a VLES turbulence model is presented in this paper. It is based on the extended k-ε model of Chen and Kim [1]. By adding an appropriate filtering technique, which depends on the local grid spacing and the computational time step, the new turbulence model distinguishes between resolved and modelled part of the turbulent spectrum. It has an adaptive characteristic and thus can be applied for the whole range of turbulent modelling approaches from the RANS up to the DNS.

Here presented application of the new adaptive turbulence model is the simulation of the flow in a pipe trifurcation of a water power plant. An unsteady non-periodic vortex motion has been observed and computationally well predicted.

2. Simulation Method

2.1. Very Large Eddy Simulation (VLES)

In order to overcome the cost and capacity problems and still resolve large scale turbulent structures several hybrid methods have been proposed lately. Their aim is to represent a link between RANS and LES. They try to keep simplicity and computational efficiency of RANS and the potential of LES to resolve large turbulence scales in computationally cheaper way. Although they can be performed on coarser grids, the simulations are more strongly dependent on the modelling.

These methods can be generally classified as Very Large Eddy Simulation, but in the literature they can be also found under different names:

- Semi-Deterministic Simulations (SDS)
- Coherent Structure Capturing (CSC)
- Detached Eddy Simulation (DES)
- Hybrid RANS/LES
- Limited Numerical Scales (LNS).

All above mentioned methods slightly differ in filtering techniques and interpretation of the resolved motion, but broadly speaking they all have a tendency to solve complex unsteady turbulent flows at high Reynolds number implying a principle "solve less – model more" (shown in Fig. 2.1 and Table 2.1). It means that the relevant part of the flow (unsteadiness) is resolved and the rest is modelled.

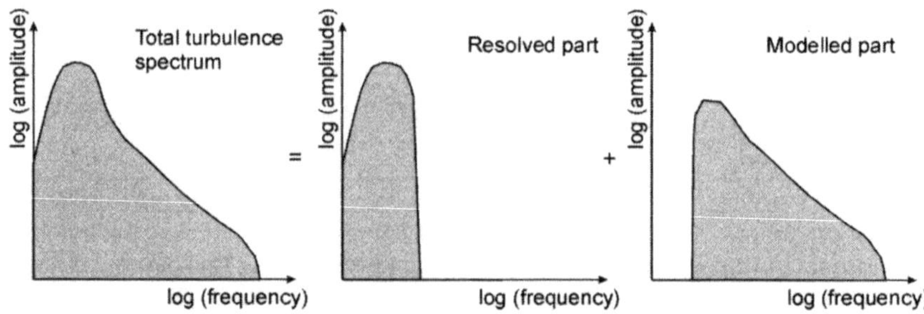

Fig. 2.1 Turbulence treatment in VLES

Table 2.1. Resolution in DNS, LES and VLES [8]

Model	Resolution
Direct numerical simulation (DNS)	All turbulent scales are resolved
Large eddy simulation with near-wall resolution	Grid size and filtering are sufficient to resolve 80% of the energy
Large eddy simulation with near-wall modelling	Grid size and filtering are sufficient to resolve 80% of the energy distant from the wall, but not in the near-wall region
Very large eddy simulation (VLES)	Grid size and filtering are not sufficiently fine to resolve 80% of the energy

2.2. Numerical Method

The calculation are carried out using the program FENFLOSS (Finite Element based Numerical FLOw Simulation System) which has been developed at the Institute of Fluid Mechanics and Hydraulic Machinery, University of Stuttgart, for more than a decade [9,10].

The partial differential equations are solved by Galerkin Finite Element Method. The spatial discretisation of the domain is performed by 8-node hexahedral elements. For the velocity components and the turbulence quantities a tri-linear approximation is applied. The pressure is assumed to be constant within each element. For advection dominated flow a Petrov-Galerkin formulation of 2nd order with skewed upwind orientated weighting function is used. The time discretisation involves a three-level fully implicit finite difference approximation of 2nd order.

For the solution of the momentum and continuity equations a segregated algorithm is used. Each momentum equation is handled independently. They are linearised by successive substitution. The linear systems are solved by the BICGSTAB2 algorithm of van der Vorst [12] with an incomplete LU decomposition (ILU) for preconditioning. The pressure is treated by the modified Uzawa pressure correction scheme [14]. The pressure correction is performed in a local iteration loop without reassembling the system matrices until the continuity error is reduced to a given order (usually 6–10 iterations are needed).

After solving the momentum and continuity equations, the turbulence quantities are calculated and a new turbulence viscosity is gained. The k and ε–equations are also linearised and by successive substitution and the linear systems are solve by the BICGSTAB2 algorithm with ILU preconditioning. The whole procedure is carried out in a global iteration until convergence is obtained. For unsteady simulations the global iteration has to be carried out for each time step.

The code is parallelised and computational domain is decomposed using double overlapping grids. In that case the linear equation solver BICGSTAB2 has a parallel performance and the data exchange between the domains is organised on the level of the matrix-vector multiplication in the BICGSTAB2 solver. The preconditioning is then local on each domain. The data exchange uses MPI (Message Passing Interface) on the computers with distributed memory. On shared-memory computers the code runs also parallel by applying OpenMP. For more details on the numerical procedures and parallelisation the reader is referred to [6,7].

3. Adaptive Turbulence Model

Classical turbulence models, which are usually applied for solving engineering flow problems, model the whole turbulent spectrum. They usually show excessive viscous behaviour and very often damp down unsteady motion quite early. Therefore they are not completely successful for some flow cases.

For resolving at least one part of turbulence spectrum and by this getting more precise picture of the flow behaviour, VLES is used. Depending on the type of the flow and grid size applied model should automatically adjust to one of the modelling approaches (schematically shown in Fig. 3.1). Thus an adaptive model is developed. Its advantage is that with increasing computer power it can be afforded that a larger part of spectrum is resolved (due to a finer computational grid). As a result the accuracy of the calculation improves.

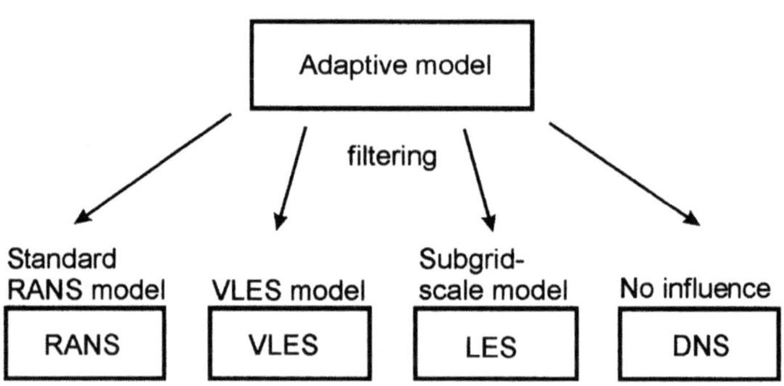

Fig. 3.1 Adjustment for adaptive model

For distinguishing resolved and modelled turbulence spectrum, the adaptive model uses a filtering technique. There are several of them described in the literature [e.g. 2, 5, 11], but the applied one is similar to Willems [13]. Employed nomenclature can be seen in Fig. 3.2.

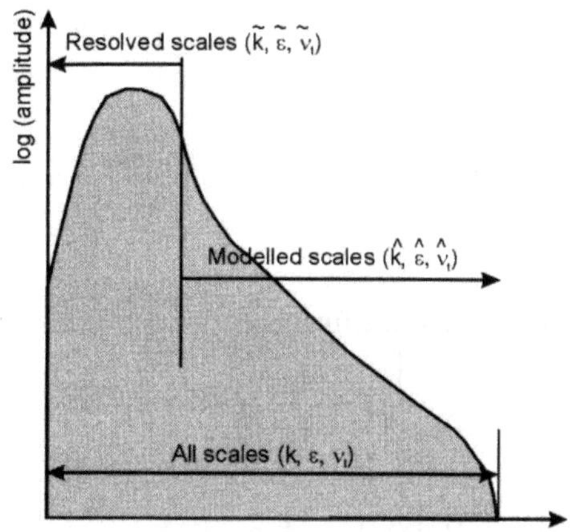

Fig. 3.2 Distinguishing of turbulence spectrum by VLES

As already stated, the basis of the adaptive model is the k-ε model of Chen and Kim [1]. It is chosen due to its simplicity and capacity to better handle unsteady flows. Its transport equations for k and ε are given as

$$\frac{\partial k}{\partial t}+U_j\frac{\partial k}{\partial x_j}=\frac{\partial}{\partial x_j}\left[\left(\nu+\frac{\nu_t}{\sigma_k}\right)\frac{\partial k}{\partial x_j}\right]+P_k-\varepsilon \tag{3.1}$$

$$\frac{\partial \varepsilon}{\partial t}+U_j\frac{\partial \varepsilon}{\partial x_j}=\frac{\partial}{\partial x_j}\left[\left(\nu+\frac{\nu_t}{\sigma_k}\right)\frac{\partial \varepsilon}{\partial x_j}\right]+c_{1\varepsilon}\frac{\varepsilon}{k}P_k-c_{2\varepsilon}\frac{\varepsilon^2}{k}+\underbrace{c_{3\varepsilon}\left[\frac{P_k}{k}\right]\cdot P_k}_{\text{additional term}}. \tag{3.2}$$

Additionally they need to be filtered. According to the Kolmogorov theory it can be assumed that the dissipation rate is equal for all scales. This leads to

$$\varepsilon=\hat{\varepsilon}. \tag{3.3}$$

Turbulent kinetic energy still needs a filtering

$$\hat{k}=k\cdot\left[1-f\left(\frac{\Delta}{L}\right)\right]. \tag{3.4}$$

As a suitable filter

$$f=\begin{cases}0 & \text{for} \quad \Delta\geq L\\ 1-\left(\frac{\Delta}{L}\right)^{2/3} & \text{for} \quad L>\Delta\end{cases} \tag{3.5}$$

is applied where

$$\Delta = \alpha \cdot \max \begin{cases} |u| \cdot \Delta t \\ h_{\max} \end{cases} \quad \text{with} \quad h_{\max} = \begin{cases} \sqrt{\Delta V} & \text{for 2D} \\ \sqrt[3]{\Delta V} & \text{for 3D} \end{cases} \tag{3.6}$$

with a model constant $\alpha\,(\alpha = 2\text{–}5)$. ΔV is the volume of the local element, u is the local velocity and Δt is the time step. The Kolmogorov scale L is given as

$$L = \frac{k^{3/2}}{\varepsilon}. \tag{3.7}$$

The filtering procedure leads to the final equations

$$\frac{\partial k}{\partial t} + U_j \frac{\partial k}{\partial x_j} = \frac{\partial}{\partial x_j} \left[\left(\nu + \frac{\hat{\nu}_t}{\sigma_k} \right) \frac{\partial k}{\partial x_j} \right] + \hat{P}_k - \varepsilon \tag{3.8}$$

$$\frac{\partial \varepsilon}{\partial t} + U_j \frac{\partial \varepsilon}{\partial x_j} = \frac{\partial}{\partial x_j} \left[\left(\nu + \frac{\hat{\nu}_t}{\sigma_k} \right) \frac{\partial \varepsilon}{\partial x_j} \right] + c_{1\varepsilon} \frac{\varepsilon}{k} \hat{P}_k - c_{2\varepsilon} \frac{\varepsilon^2}{k} + c_{3\varepsilon} \left[\frac{\hat{P}_k}{k} \right] \cdot \hat{P}_k \tag{3.9}$$

with the production term

$$\hat{P}_k = \hat{\nu}_t \left(\frac{\partial U_i}{\partial x_j} + \frac{\partial U_j}{\partial x_i} \right) \frac{\partial U_i}{\partial x_j}. \tag{3.10}$$

For more details of the model and its characteristics the reader is referred to [10].

The simulation of the vortex shedding behind the trailing edge (Fig. 3.3), which can be considered as a convenient test case for CFD computations, shows very often difficulties when URANS is applied. Unsteady RANS with the standard k-ε model usually leads to a steady state solution (see Fig. 3.4). The vortex shedding and its unsteadiness are suppressed by the too diffusive turbulence model. More sophisticated turbulence model i.e. extended k-ε of Chen and Kim is less diffusive and therefore vortex shedding is gained. Their comparison is shown in Fig. 3.4.

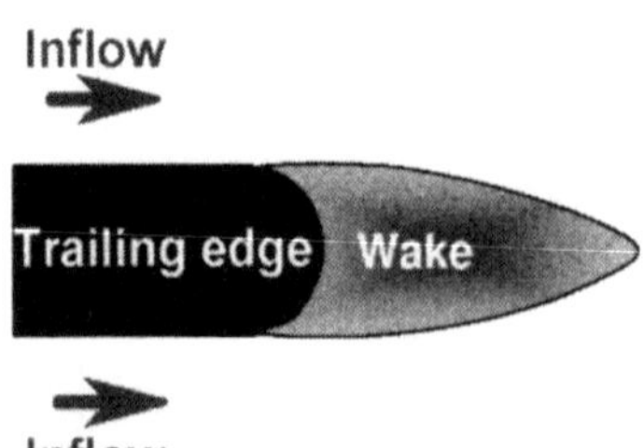

Fig. 3.3 Flow behind trailing edge

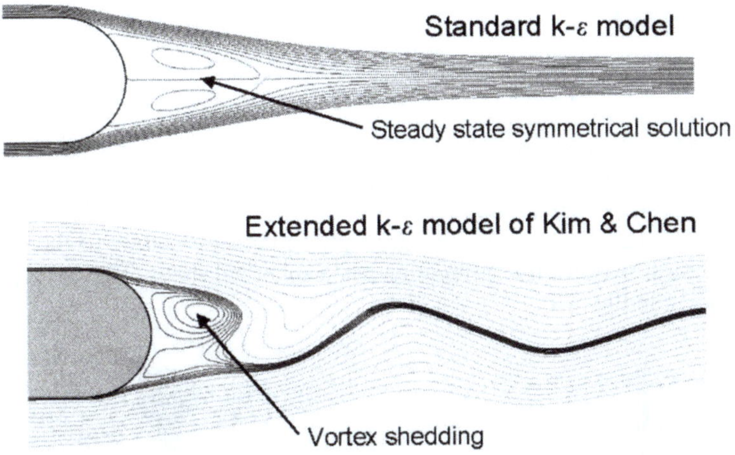

Fig. 3.4 Streamlines of the vortex shedding behind the trailing edge

Further simulation with VLES method provides slightly improved results (see Fig. 3.5). In comparison to k-ε model of Chen and Kim, it proves to be less damping in the downstream flow behind trailing edge and agrees better with experimental data.

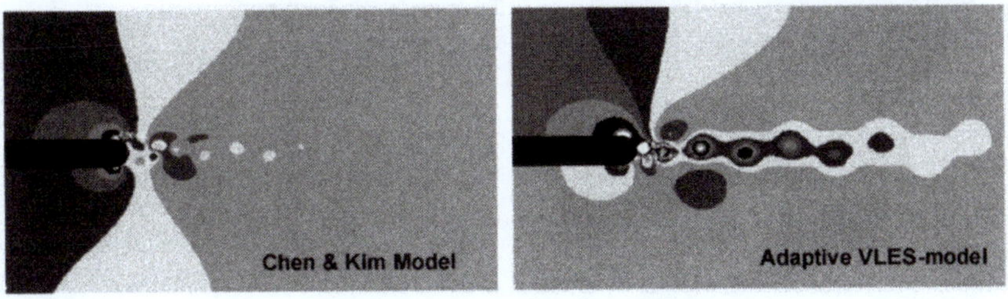

Fig. 3.5 Pressure distribution by vortex shedding behind the trailing edge, comparison of extended k-ε model of Chen and Kim and adaptive VLES model

4. Application

4.1. Problem Description

As a practical application the flow in a pipe trifurcation of a water power plant is presented. The complete water passage consists of the upper reservoir, channel, surge tank, penstock, trifurcation and three turbines. The trifurcation distributes the water from the penstock into the three pipe branches leading to the turbines. Due to structural reasons, the trifurcation has a spherical shape. The water passage is partly shown in Fig. 4.1.

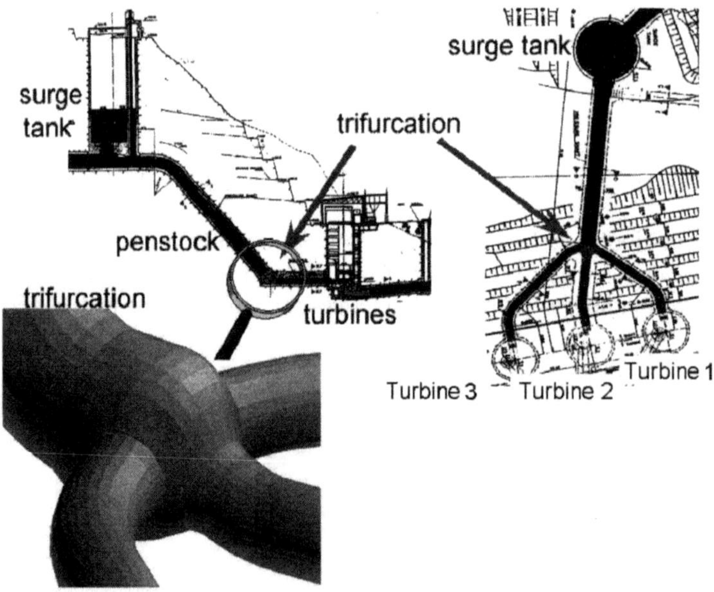

Fig. 4.1 Water passage with trifurcation

During the power plant exploitation severe power oscillations were encountered at the outer turbines (1 and 3) in the range of ± 10% of nominal power. Vortex instability was discovered as a cause of these fluctuations. The vortex is formed in the sphere, appearing (starting) at the top and extending into a one of the outer branches. After a certain period it changes its behaviour and extends into opposite outer branch staying in it for a certain time period. Then the vortex jumps back again. In Fig. 4.2 are schematically shown the two vortex positions. This unstable vortex motion is not periodic and due to its strong swirling flow produces a very high losses in the branch in which it is located. These losses reduce the head of the turbine and consequently the power output. For better understanding and analysis of this flow phenomenon , a computer simulation is performed and results are compared with available model test measurements.

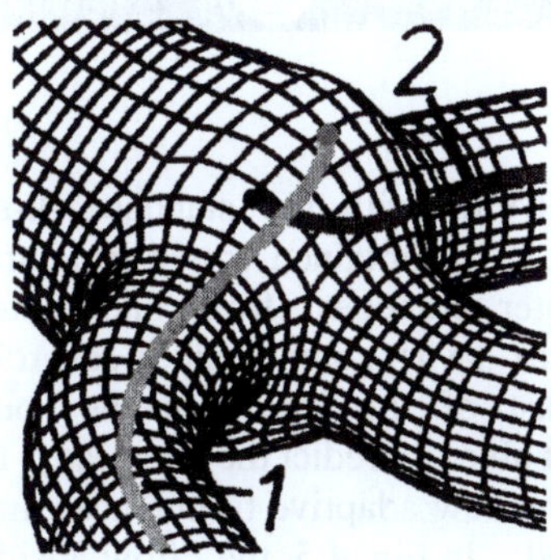

Fig. 4.2 Positions of the unsteady vortex structure

4.2. Unsteady Vortex Simulation

Computational domain is discretised by approximately 500 000 elements and decomposed into 32 partitions (see Fig. 4.3) for parallel computing. Computations are then run on CRAY T3E.

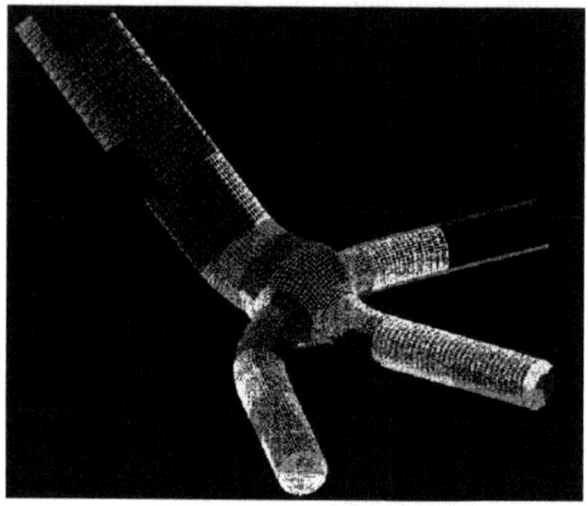

Fig. 4.3. Computational grid

Simulation applying URANS with the standard k-ε turbulence model leads to a steady state solution. The obtained vortex structure is shown in Fig. 4.4. It extends through both outer branches (from one outer branch to the other) and is fully stable. Vortex swirl component is severely underpredicted which leads to an poor forecast of the losses in the outer branches. It clearly shows that URANS is not able to predict the unsteady flow phenomenon.

Applying VLES with the new adaptive turbulence model, this unstable vortex movement is predicted. In Fig. 4.5 the flow inside branch 1 at a certain time step is shown . The vortex is represented by an iso-pressure surface and instantaneous streamlines. It can be seen that the vortex appears at the top of the sphere and extends into branch 1.

After some time (see Fig. 4.6), vortex "jumps" to the opposite branch. Since the geometry is not completely symmetric, the vortex stays longer in branch 3 than in branch 1. It is observed in the simulation as well as in the model test.

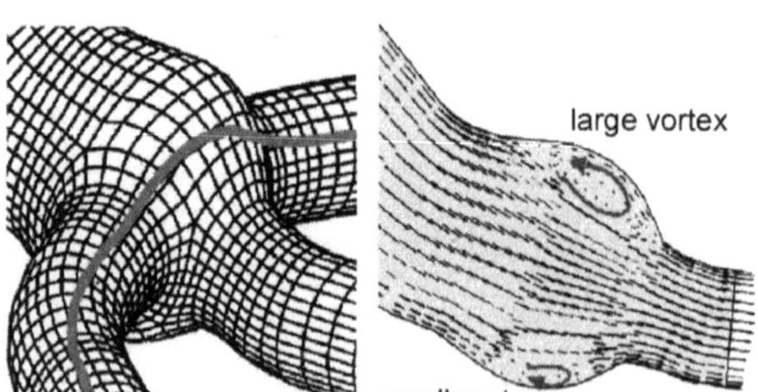

Fig. 4.4 Vortex structure obtained by URANS with standard k-ε turbulence model

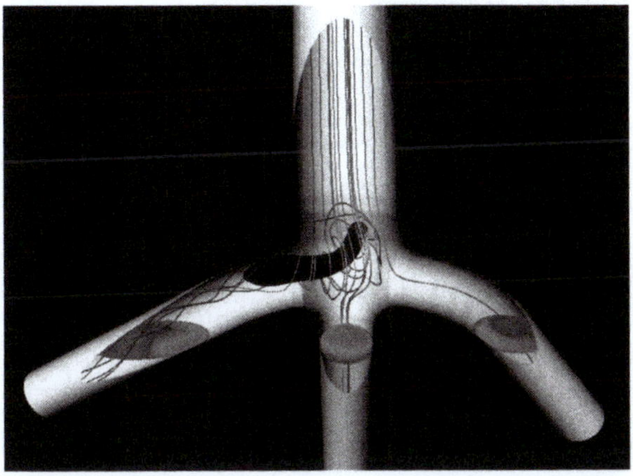

Fig. 4.5 Flow inside the trifurcation – vortex position in the branch 1

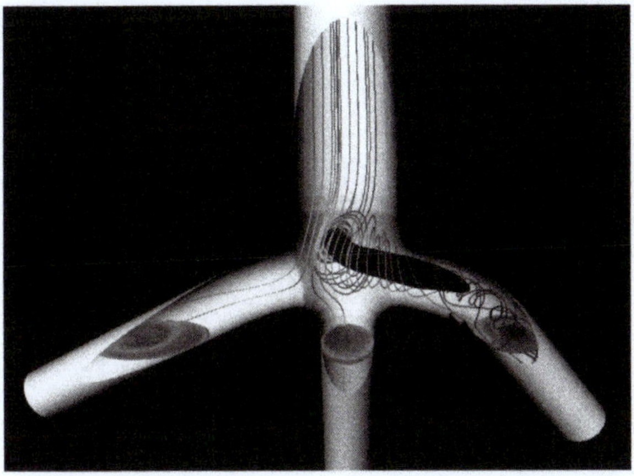

Fig. 4.6 Flow inside the trifurcation – vortex position in the branch 3

Due to the strong swirl at the inlet of the branch in which the vortex is located, the losses inside this branch are much higher compared to the other two. Therefore, the discharge through this branch is reduced. In Fig. 4.7 the discharge characteristics through all three branches are shown and the low discharge corresponds to the vortex presence. It is obvious that the discharges through two outer branches vary successively, while the discharge in the middle branch shows much smaller oscillations.

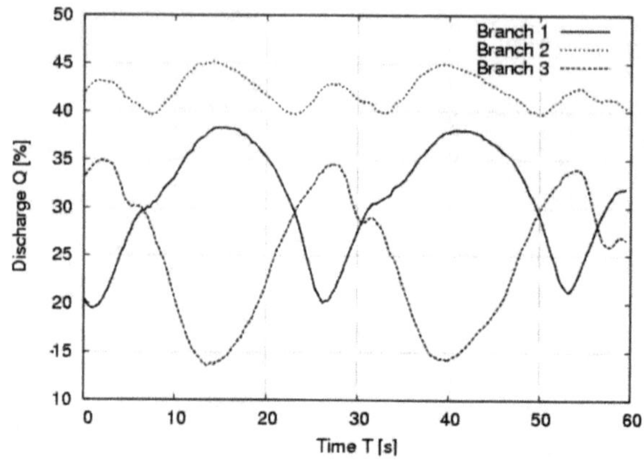

Fig. 4.7 Discharge through the three branches

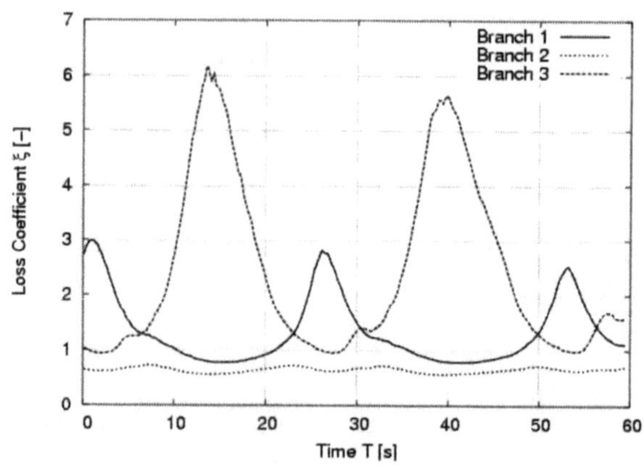

Fig. 4.8 Loss coefficients of the three branches

In the reality turbines are located at the outlet of each branch. Therefore the discharge variation is rather small since the flow rate through the different branches is prescribed by the turbines. In the simulation, however, a free outflow boundary condition is applied which leads to the higher discharge variations. For comparison with the experiment, loss coefficients for each branch are calculated. They are shown in Fig. 4.8. Branch 3 shows the highest values and has a strong fluctuation. The loss coefficient of the middle branch is nearly constant.

4.3. Model Tests

Model tests were carried out by ASTRÖ in Graz, Austria. For more details the reader is referred to [4]. In Fig. 4.9 the test rig is shown , looking from the penstock down to the trifurcation. For easier observation of the vortex motion, trifurcation was made of acrylic glass.

Fig. 4.10 shows the view from the top of the trifurcation. One can see that the vortex appears at the top and extends into a one of the outer branches. Due to the pressure decrease small cavitation bubbles shows up in the vortex centre without disturbing the flow structure. It makes vortex visible. In the model tests vortex also moved from one branch to the other , as described above.

Loss coefficients for each branch were calculated from the pressure and discharge measurements. They are shown in the Fig. 4.11.

Fig. 4.9 The test rig at ASTRÖ, Graz

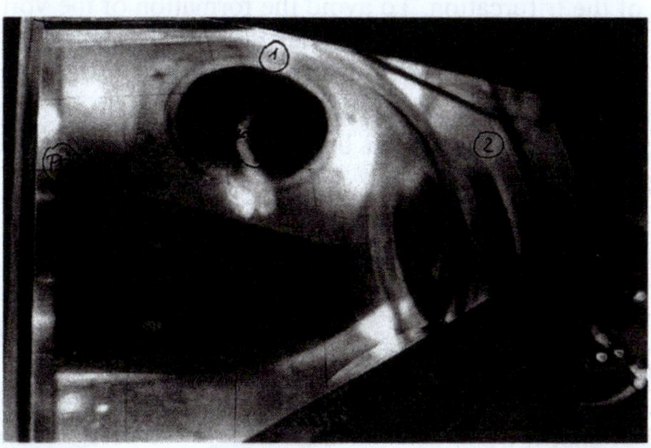

Fig. 4.10 Vortex inside the trifurcation – measurements by ASTRÖ

Comparing the measured loss coefficients with one gained by simulation, it can be seen that the maximum values are still underpredicted, although general flow tendency and quantitative prediction fit in with measurement data reasonably well. For the branch 3 the computational simulation shows a maximum of 6, while in the measurements the value 8 is obtained. The values for branch 1 are clearly lower (approximately 3 in the simulation and 4 in the measurements).

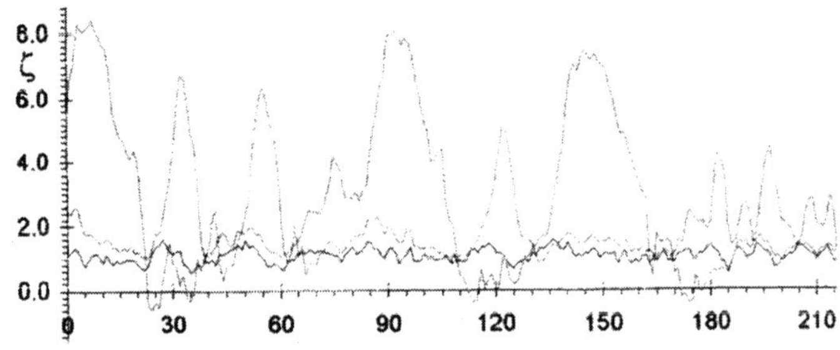

Fig. 4.11 Measured loss coefficients for all three branches

This underprediction of the loss coefficient is assumed to be primarily due to the rather coarse grid and secondly due to a strong anisotropic turbulent behaviour which cannot be accurately predicted by the turbulence model based on the eddy viscosity assumption. Therefore it is intended to develop and adaptive algebraic Reynolds-stress model in the future.

4.4. Problem Solution

In order to solve the oscillation problem in the hydro power plant, it was proposed to change the shape of the trifurcation. To avoid the formation of the vortex, upper and lower parts of the sphere are cut off in flat plates. This modified shape is shown in the Fig. 4.12. In the meantime this modification was made and the power oscillations have disappeared. As a result the power output has increased significantly since the losses in the trifurcation have decreased. Furthermore, no unsteady vortex has been noticed.

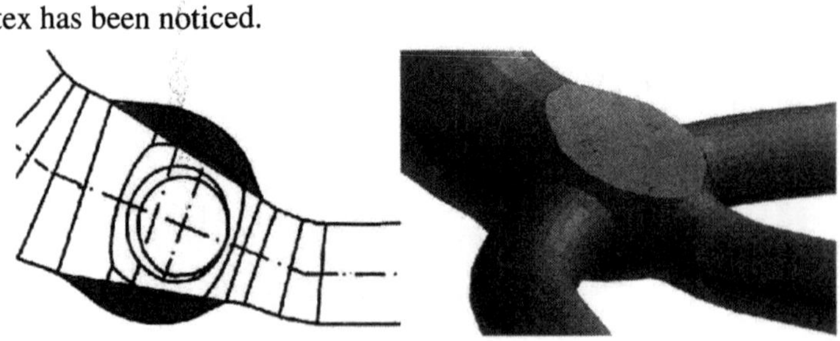

Fig. 4.12 Modified trifurcation shape

5. Conclusions

An adaptive turbulence model for Very Large Eddy Simulation is presented. It is based on the extended k-ε model of Chen and Kim. Introducing a filtering technique the new turbulence model distinguishes between numerically resolved and unresolved part of the flow.

With the help of this new model the unstable vortex motion in a pipe trifurcation is calculated. Using classical RANS method and common turbulence models this flow phenomenon cannot be predicted. Applying VLES with adaptive turbulence model unsteady vortex motion was obtained and the results agree reasonably well with measurement data. However, the vortex swirl is still underpredicted. A further improvement is expected by applying an adaptive Reynolds-stress model to take into account the anisotropy of the turbulence in swirling flows.

References

[1] Chen YS, Kim SW (1987) Computation of turbulent flows using an extended k-ε turbulence closure model, NASA CR-179204

[2] Constantinescu GS, Squires KD (2000) LES and DES Investigations of Turbulent flow over a Sphere, AIAA-2000-0540

[3] De Langhe C, Dick E (2002) Very-Large-Eddy Simulation of Turbulent Flows, http://allserv.rug.ac.be/~edick/

[4] Hoffmann H, Roswora RR, Egger A (2000) Rectification of Marsyangdi Trifurcation, In: Hydro Vision 2000 Conference Technical Papers, HCI Publications, Inc., Charlotte

[5] Magnato F, Gabi M (2000) A new adaptive turbulence model for unsteady flow fields in rotating machinery, In: Proceedings of the 8th International Symposium on Transport Phenomena and Dynamics of Rotating Machinery (ISROMAC 8)

[6] Maihöfer M (2002) Effiziente Verfahren zur Berechnung dreidimensionaler Strömungen mit nichtpassenden Gittern, Ph.D. thesis, University of Stuttgart

[7] Maihöfer M, Ruprecht A (2003) A Local Grid Refinement Algorithm on Modern High-Performance Computers, In: Proceedings of Parallel CFD 2003, Elsavier, Amsterdam

[8] Pope SB (2000) Turbulent flows, Cambridge University Press, Cambridge

[9] Ruprecht A (1989) Finite Elemente zur Berechnung dreidimensionaler turbulenter Strömungen in komplexen Geometrien, Ph.D. thesis, University of Stuttgart

[10] Ruprecht A (2003) Numerische Strömungssimulation am Beispiel hydraulischer Strömungsmaschinen, Habilitation thesis, University of Stuttgart

[11] Spalart PR, Jou WH, Strelets M, Allmaras SR (1997) Comments on the Feasibility of LES for Wings, and on Hybrid RANS/LES Approach, In: Liu C, Liu Z (eds) Advances in DNS/LES, Greyden Press, Columbus

[12] Van der Vorst HA (1994) Recent Developments in Hybrid CG Methods, In: Gentzsch W, Harms U (eds) High-Performance Computing and Networking, vol. 2: Networking and Tools, Lecture Notes in Computer Science, 797, Springer, Berlin, Heidelberg, New York , pp 174–183
[13] Willems W, Peters N (1997) Large Eddy Simulation of a Turbulent Mixing Layer Using a New Two-Level-Turbulence Model, In: Proceedings of the 11th Symposium on Turbulent Shear Flows, Grenoble
[14] Zienkiewicz OC, Vilotte JP, Toyoshima S, Nakazawa S (1985) Iterative method for constrained and mixed finite approximation. An inexpensive improvement of FEM performance. Comput Methods Appl Mech Eng 51:3-29

A new stabilized finite element method for advection-diffusion-reaction equations using quadratic elements

Alessandro CORSINI*

Franco RISPOLI

Andrea SANTORIELLO

Department of Mechanics and Aeronautics
University of Rome "La Sapienza"

*corresponding author: Via Eudossiana, 18, I00184 Roma, Italy
Tel.: +39 06 44585231, Fax: +39 06 4881759
Email: corsini@dma.ing.uniroma1.it

Abstract

A new stabilized *FEM* formulation for advective-diffusive-reactive problems is presented. The new method, called *Spotted Petrov-Galekin* (*SPG*), is a quadratic *Petrov-Galerkin* formulation based on two perturbations of the weight function: the first one is a generalized *SUPG* operator, the second one a nodal spot-like controlling operator for reactive instabilities. The formulation covers all the combinations of advective and reactive effects, associated to the dimensionless element Peclet and reaction numbers. After an introduction to the method, we present some numerical applications, addressing the reliability of *SPG* in turbomachinery *CFD*.

1. Introduction

In this work we focus on the numerical solution of advective-reactive-diffusive problems using the Finite Element Method (FEM) on quadratic spaces of approximation. Here, diffusion, advection and reaction refer to those terms in the partial differential equations (PDEs) involving second, first and zero order derivatives of the unknowns. This family of equations, that governs several phenomena of industrial interest, is here discussed because of its importance in the modelling of turbomachinery fluid dynamics.

Several sources of oscillations affect the solution of PDEs in fluid dynamics if standard schemes are used (e.g. central finite differences or Galerkin finite elements). In the finite element framework a number of stabilized formulations has been proposed during the last two decades as remedial strategies. Most of them were based on a Petrov-Galerkin (PG) approach, where the stabilization is achieved preserving the Euler-Lagrange conditions consistency by adding a perturbation to the Galerkin weights (such as SUPG [1-4], or PSPG [3], or Discontinuity Capturing [5] schemes).

Moving towards the turbomachinery CFD, an additional origin of instabilities stems from the reaction or zero order derivative terms. These terms are usually related to the rotation of turbomachinery frame of reference (e.g. in the modelling of Coriolis forces), but it is worth noting that they appear also in the turbulence closure scale determining equations (e.g. one or two equations eddy viscosity models (EVM) and second moment closure models). Local oscillations arise near boundaries or solution discontinuities and it is not possible to obtain a global stability estimate in the H^1 norm, though it could be evaluated in L^2 [6].

To the best of the authors' knowledge, in the open literature only equal order PG formulations have been developed to control advective-diffusive-reactive flow problems. Most of them deal with scalar equations (e.g. (SU+C)PG in [7]) or linear reactive operators [8]. Few works concern with reactive problems pertinent to real turbomachinery fluid dynamics [6].

In this viewpoint, the present work addresses the definition of a new Petrov-Galerkin like stabilization scheme for the reactive flow limit, formulated on a quadratic finite element space of approximation. The use of a higher order stabilized formulation, though its complexity due to the non-negligibility of second order derivatives, guarantees the best compromise between solution stability and accuracy [9]. In particular, the authors propose a stabilized formulation that performs well both in the advection and in the reaction dominated case. The new method is called *Spotted Petrov-Galerkin* (SPG) and possesses some distinctive features. For advection-diffusion problems it behaves like a SUPG method, whereas in the reactive-diffusive limit the space invariant problem is controlled by a perturbation able to give rise to spot-like weight function, symmetric and concentrated around each nodal position. In intermediate situations, the scheme combines the perturbation integrals using tuning or *upwind* coefficients that depend on element Peclet and reaction numbers.

The remainder of the paper is organised as follows. In Sect. 2 the SPG formulation is presented for linear scalar advective-diffusive-reactive equation. The extension of the formulation to multi-dimensional case is discussed, and the family of weights for Q2 element is shown. In Sect. 3 are commented the reactivity features of general PDE that models the budget of turbulent determining-scale variables. Finally in Sect. 4 the performance of SPG are assessed on three test cases against solutions provided by SUPG, Streamline Upwind and Galerkin methods.

2. Finite Element scalar advective-diffusive-reactive problem

Let take the general linear scalar advective-reactive-diffusive problem statement on the closed domain $\underline{\Omega}$ for the unknown ϕ:

$$\begin{aligned} & u_j\phi_{,j} - k\phi_{,jj} + c\phi = f \qquad (j = 1,nsd) \\ & \phi(\Gamma_g) = \phi_g \\ & \phi_{,n}(\Gamma_h) = \theta_n \end{aligned} \tag{1}$$

where *nsd* is the number of space dimensions, $k > 0$ is the constant diffusivity, u_j are the velocity components, $c \geq 0$ is the reaction coefficient, f the source term. The boundary conditions are specified along $\Gamma = \underline{\Gamma}_h \cup \underline{\Gamma}_g$ ($\underline{\Gamma}_h$ and $\underline{\Gamma}_g$ are closed, disjoint subsets of Γ), including Dirichlet (ϕ_g) and Neumann conditions (θ_n), and $\boldsymbol{n}$ is the unit outward normal vector.

Given a finite element partition of the original closed domain $\underline{\Omega}$ into elements Ω_e, $e = 1, nel$ (*nel* number of elements) such that

$$\bigcup_e \Omega_e = \Omega \text{ and } \bigcap_e \Omega_e = \varnothing \tag{2.1}$$

Consider the definition of interior boundary as

$$\Gamma_{int} = \bigcup_e \Gamma_e - \Gamma \tag{2.2}$$

Let define the finite dimensional spaces of trial and weight functions respectively as:

$$\begin{aligned} S^h = \{ & \phi^h \mid \phi^h \in H^{1h}(\Omega), \phi^h = \phi_g \text{ on } \Gamma_g, \\ & \phi_g \in H^{(1/2)h}(\Gamma_g) \} \end{aligned} \tag{3.1}$$

$$W^h = \left\{ w^h \mid w^h \in H_0^{1h}, w^h = 0 \text{ on } \Gamma_g \right\} \tag{3.2}$$

where $H^{1h}(\Omega)$ and $H_0^{1h}(\Omega)$ are the Sobolev spaces for the continuous pair of finite element functions, $H^{(1/2)h}(\Gamma_g)$ is their restriction to the domain boundary, and the superscript h denotes the characteristic length scale of the domain discretization. The Galerkin weak formulation of problem (1) reads as follows:

$$\sum_{e=1}^{nel}\int_{\Omega_e} w^h u_j \phi_{,j}^h + \sum_{e=1}^{nel}\int_{\Omega_e} w_{,j}^h k \phi_{,j}^h d\Omega +$$

$$\sum_{e=1}^{nel}\int_{\Omega_e} w^h c \phi^h d\Omega = \sum_{e=1}^{nel}\int_{\Omega_e} w^h f d\Omega + \int_{\Gamma_h} w^h k \frac{\partial \phi^h}{\partial n} d\Gamma \tag{4}$$

2.1 Galerkin formulation in one-dimension

Let consider the ordinary differential equation obtained from (1) when $nsd=1$, and source term $f = 0$. The problem statement reads as:

$$u\frac{d\phi}{dx} - k\frac{d^2\phi}{dx^2} + c\phi = 0 \tag{5}$$

The discretization of (5), for constant physical properties, using the Galerkin method on a quadratic space of interpolation, with uniform element of length h, gives rise to the following difference equations:

$$\phi_{i-1}\,[-4 - 2Pe + r/10] + \phi_i\,[8 + r\,4/5] + \phi_{i+1}\,[-4 + 2\,Pe + r/10] = 0 \tag{6}$$

for (i) element central node, and

$$\phi_{i-2}\,[1 + Pe - r/10] + \phi_{i-1}\,[-8 - 4Pe + r/5] + \phi_i\,[14 + r\,4/5] +$$
$$\phi_{i+1}\,[-8 + 4Pe + r/5] + \phi_{i+2}\,[1 - Pe - r/10] = 0 \tag{7}$$

for (i-2, i, i+2) element extreme nodes.

In the above equations, the magnitudes of advection or reaction versus diffusion are, respectively, given by element Peclet number $Pe = \|u\| h/2k$, and element reaction number $r = c\,h^2/k$.

In order to have a first insight on the instability origin of the reaction dominated case, let now consider the null advection limit of the studied problem, focusing for instance on element central nodes. In this case Eq. (6) turns to the following expression:

$$\phi_{i-1}\,[-4 + r/10] + \phi_i\,[8 + r\,4/5] + \phi_{i+1}\,[-4 + r/10] = 0 \tag{8}$$

The solutions a_G of the characteristic equation associated to (8), the so-called Galerkin nodal amplification factors [8], purely depend on the magnitude of reaction:

$$a_G = \frac{-(8 + \frac{4}{5}r) \pm \sqrt{(8 + \frac{4}{5}r)^2 - 4(-4 + \frac{r}{10})^2}}{2(-4 + \frac{r}{10})} \tag{9}$$

The dependence of a_G from r shows that the exact solution exponential behavior is preserved only with $r < 40$. This circumstance confirms the need of designing a stabilized scheme with built-in component to preclude oscillatory behavior in reaction dominated cases.

2.2 SPG formulation

The stabilized SPG formulation has been designed for quadratic element on the basis of 1D advection-diffusion-reaction problem (5). The rationale for the SPG scheme, aimed at the generalization to higher order finite elements of the concept presented by Idelsohn et al. in [7], is to build the PG weight functions as:

$$\tilde{w}_i = w_i + \hat{w}_i, \qquad \text{with} \qquad \hat{w}_i = \boldsymbol{\tau}_i^{sQ2} \cdot \boldsymbol{\pi}_i^{sQ2} \tag{10}$$

where the perturbation $\hat{w}_i$ to Galerkin weights w_i could be seen as the inner product of element node-vector intrinsic times ($\boldsymbol{\tau}_i^{sQ2} = \left\{\tau_i^{adv}, \tau_i^{rea}\right\}$) and perturbation operators ($\boldsymbol{\pi}_i^{sQ2} = \left\{\pi_i^{adv}, \pi_i^{rea}\right\}$). The components of these two vectors contribute to the nodal SPG weight perturbation acting, respectively, on the advective and reactive dominated flow instabilities. Thus the inner product could be seen as the sum of two perturbations; the first is formally similar to a quadratic SUPG one and reads as:

$$P_{1i} = \tau_i^{adv}\pi_i^{adv}, \text{ with: } \tau_i^{adv} = \frac{h}{2\|u\|}\zeta_i^{adv}(Pe, r) \text{ and } \pi_i^{adv} = u_k w_{i,k} \tag{11}$$

The latter is a spot-like perturbation defined as:

$$P_{2i} = \tau_i^{rea}\pi_i^{rea}, \text{ with: } \tau_i^{rea} = \frac{r}{c}\zeta_i^{rea}(Pe, r) \text{ and } \pi_i^{rea} = \frac{k}{h^2} w_i^{SPG} \tag{12}$$

In Eqs. (11) and (12) ζ_i^{adv} and ζ_i^{rea} are the stabilization tuning coefficients. The second perturbation has been designed accordingly to the ideal weight for the *pure reaction* limit ($r \to \infty$), thus preserving the weight symmetry and attempting to concentrate $\tilde{w}_i$ (about) around each nodal position in a Dirac's delta-like fashion [7]. In order to prevent from singularities, the authors have proposed a negative definite even order polynomial for w_i^{SPG}, that for a one-dimensional master element with $|\xi| \leq 1$ and dimension $l_\xi = 2$, is defined as:

$$w_i^{SPG}(\xi) = -\frac{C_{SPG}}{l_\xi^6}\left[\xi^6 - \frac{l_\xi^2}{2}\xi^4 + \frac{l_\xi^4}{16}\xi^2\right] \tag{13}$$

The sixth order expansion is the lowest one able to fulfil the set of conditions on null perturbation and first derivative values on central and extreme nodal positions. The C_{SPG} is a coefficient chosen to module the P_{2i} perturbation intensity at $\xi = \pm\, l_\xi /4$. Fig.1 shows the resulting weight behavior in case of null advection for central ($\xi = 0$) and extreme ($\xi = -1,\ +1$) node.

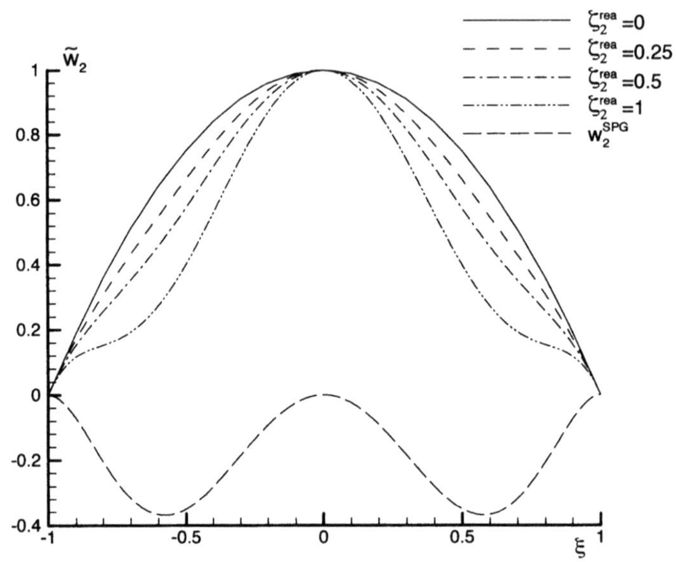

a)

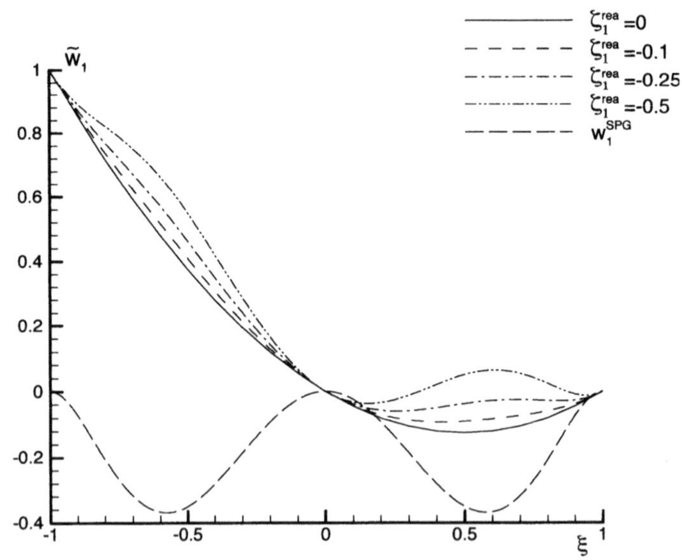

b)

Fig. 1. 1-D null advection: PG weights (**a**) $\tilde{w}_2$ for central node ($\xi = 0$) and (**b**) $\tilde{w}_1$ for extreme node ($\xi = -1$)

The stabilization tuning coefficients $\zeta_i^{adv}(Pe,r)$ and $\zeta_i^{rea}(Pe,r)$, that appear in the advective and reactive intrinsic time scales, are obtained by imposing the super-convergence condition on the numerical solution. Provided that different equations could be obtained for extreme and central nodes, similarly to other SUPG Q2 formulation [4] it is possible to find two optimal perturbations on the basis of the discrete nodal difference equations. In particular, Figs. 2 and 3 show how ζ_i^{adv} and ζ_i^{rea} behave with different combinations of *Pe* and *r* on central ($\xi = 0$) and extreme ($\xi = -1, +1$) nodes.

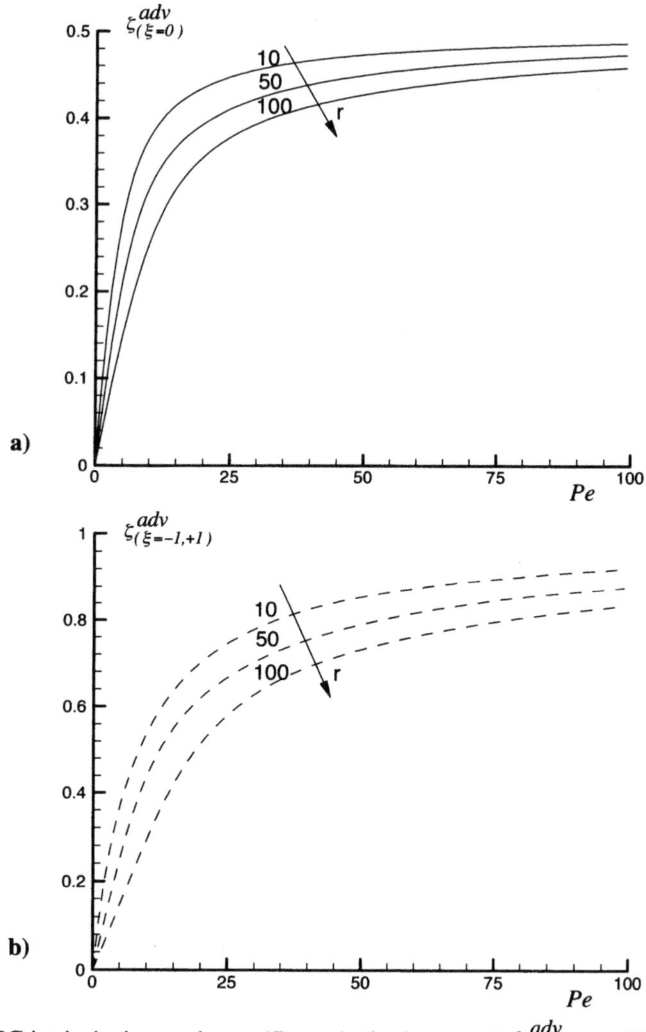

Fig. 2. SPG intrinsic time scales on 1D quadratic element: **a)** $\zeta^{adv}_{(\xi=0)}$, and **b)** $\zeta^{adv}_{(\xi=-1,+1)}$

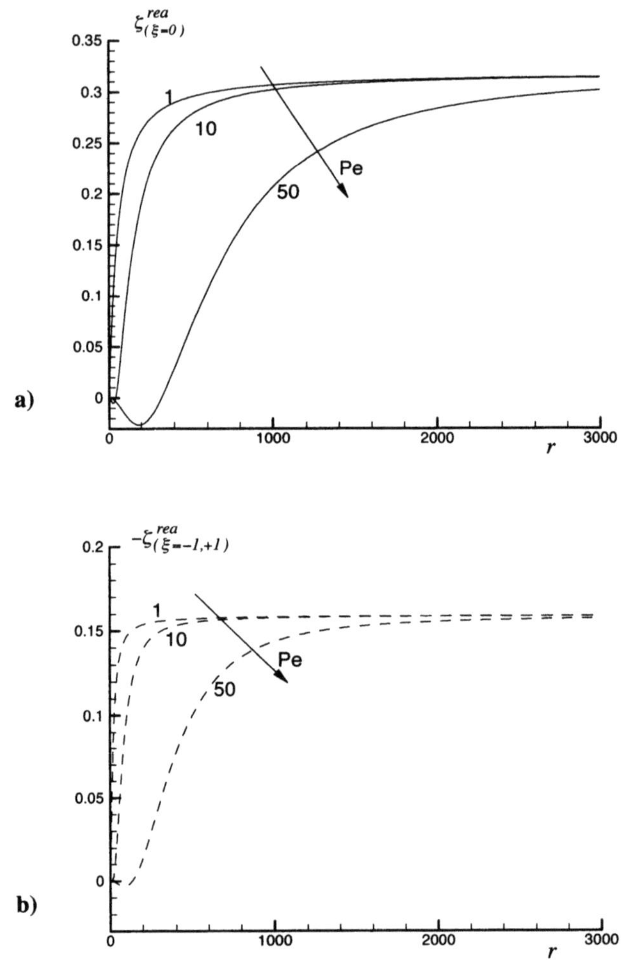

Fig. 3. SPG intrinsic time scales on 1D quadratic element: **a)** $\zeta^{rea}_{(\xi=0)}$, and **b)** $\zeta^{rea}_{(\xi=-1,+1)}$

From the comparison of Figs. 2 and 3 we can infer that generally the effect of ζ_i^{rea} grows up with *r* and, on the contrary, is attenuated by increasing the local advection magnitude, whereas opposite effects of *Pe* and *r* can be observed on ζ_i^{adv}. Concerning the limit behaviour of the stabilization tuning coefficients, it is also worth noting that for null advection $\zeta_1^{rea} = \zeta_3^{rea} = \zeta^{rea}(\xi = -1,+1)$ have a threshold value $-\zeta^{rea}_{min}(\xi = -1,+1) = 0.5 \cdot \zeta^{rea}_{max}(\xi = 0)$ and for null reaction $\zeta_1^{adv} = \zeta_3^{adv} = \zeta^{adv}(\xi = -1,+1)$ have threshold $\zeta^{adv}_{max}(\xi = -1,+1) = 2 \cdot \zeta^{adv}_{max}(\xi = 0)$.

2.3 Extension of SPG formulation to multi-dimensional case

The multi-dimensional extension of the SPG stabilization requires a space prolongation of second perturbation function $P_{2i} = \tau_i^{rea} \pi_i^{rea}$, dealing with the shape of the sixth order perturbation operator as well as the tuning functions.

The 2D SPG function is designed to preserve its 1D requirement in the reaction limit, that is the isotropic concentration of the perturbed weight $\tilde{w}_i$ around each nodal position. To this end, we designed a Cartesian product between the 1D counterparts of the second perturbation function, where the w_i^{SPG} spots are moved in the element portion closer to the corresponding nodes. The concept is depicted in Fig. 4 that shows the w_i^{SPG} 2D shapes in the master element reference. It is thus possible to maintain the continuity of w_i^{SPG} on the inter-element boundary.

As far as the 2D intrinsic time scales τ_i^{rea} are concerned, they have been obtained using a pure positional criterion based on master element image. For instance, with respect to 1D functions, $\tau^{rea}(\xi = 0)$ is used for the central node, $\tau^{rea}(\xi = -1, +1)$ on corner nodes and their linear combination with equal coefficients for mid-side nodes.

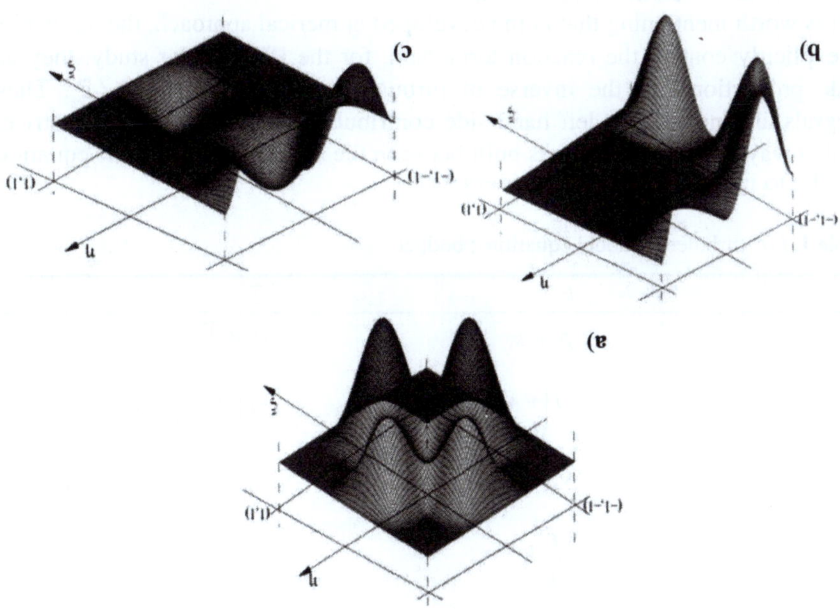

Fig. 4. w_i^{SPG} function in 2D logic coordinates (ξ, η): **a**) central node (*0, 0*), **b**) corner node (*-1, -1*), and **c**) mid-side node (*-1, 0*)

3. Problem statement for EVM

In this section is briefly commented the suitability of SPG formulation for determining-scale equations in the first moment turbulence closures used as the modelling base-line for turbomachinery CFD. Let consider the general steady-state advective-diffusive-reactive equation used to model the budget of a turbulent variable ϕ (i.e. ϕ could represent turbulent kinetic energy k, its dissipation rate ε, &c.):

$$F_{\phi a} + F_{\phi d} = P_{\phi} + \varepsilon_{\phi} + S_{\phi} \tag{14}$$

here: F_{ϕ_a} are the convective fluxes, F_{ϕ_d} are the diffusive fluxes, P_{ϕ} is the production, ε_{ϕ} is the dissipation (or destruction) term, and S_{ϕ} is the term containing the near wall extra sources. When the standard k-ε model proposed by Launder and Sharma (1974) [10] is used, the budget structure (14), gives rise to the terms sketched in Table 1.

In Table 1: $\nu_t = c_{\mu} f_{\mu} k^2 / \tilde{\varepsilon}$ is the scalar eddy viscosity; f_1, f_2, f_{μ} are damping functions; $c_1, c_2, c_{\mu}, \sigma_k, \sigma_{\varepsilon}$ are empirical constants. The extra term D accounts for the dissipation value on solid boundaries, and E is the buffer-layer source to correct the near wall dissipation behaviour.

It is worth mentioning that in the developed numerical approach, the dissipation ε_{ϕ}, explicitly contain the reaction terms and, for the EVM under study, they are made proportional to the inverse of turbulence time scale $\tau = k / \tilde{\varepsilon}$. These integrals are included as left hand-side contributions to the coefficient matrices. By that way a strong coupling is built between the k-ε determining-scale equations in order to improve the solver convergence.

Table 1. k-ε turbulence model equations; budget

	k	$\tilde{\varepsilon}$
F_{ϕ_a}	$\rho\ u_j k_{,j}$	$\rho\ u_j\ \tilde{\varepsilon}_{,j}$
F_{ϕ_d}	$-[\left(\nu + \frac{\nu_t}{\sigma_k}\right) k_{,j}]_{,j}$	$-[\left(\nu + \frac{\nu_t}{\sigma_k}\right) \tilde{\varepsilon}_{,j}]_{,j}$
P_{ϕ}	$\nu_t [u_{i,j} + u_{j,i}] u_{i,j}$	$c_1 f_1 \frac{\tilde{\varepsilon}}{k} \nu_t [u_{i,j} + u_{j,i}] u_{i,j}$
ε_{ϕ}	$-\frac{\tilde{\varepsilon}}{k} k$	$-c_2 f_2 \frac{\tilde{\varepsilon}}{k} \tilde{\varepsilon}$
S_{ϕ}	$D = 2\ \nu (\partial \sqrt{k} / \partial x_i)^2$	$E = 2\ \nu\ \nu_t (\partial^2 u_i / \partial x_j \partial x_k)^2$

4. Numerical examples

In this section we assess the numerical performance of the proposed SPG formulation for model problems and for configurations pertinent to turbomachinery fluid dynamics. In these validation studies the improvement of the SPG are discussed with respect to the classical stabilization schemes, such as the SUPG or Streamline Upwind. It is remarkable that, since all the stabilization schemes usually share the optimal property in 1D, all the investigated test cases violate one of the super-convergent conditions (i.e. non-uniform mesh, multidimensional domain, non-linear equations and problems with source terms).

4.1 Scalar advective-diffusive-reactive equation on a square domain

The first test cases (labeled TC1 and TC2) concern with the numerical solution of the linear scalar advective-diffusive-reactive model problem (1), in a unit square domain. The mesh is uniform with 10×10 quadratic elements, thus consisting of 441 nodes. The complete problem statements are resumed in Fig. 5, for both TC1 and TC2. The known velocity field u is assumed to have a parabolic profile (e.g. $u(x,y) = 2y - y^2$, $v(x,y) = 0$), with maximum value equal to 1. The coefficients are: $k = 10^{-5}$, $c = 5 \times 10^2$. The maxima for dimensionless numbers are: $Pe = o(10^3)$ and $r = o(10^5)$, based on a pure geometrical element characteristic length scale.

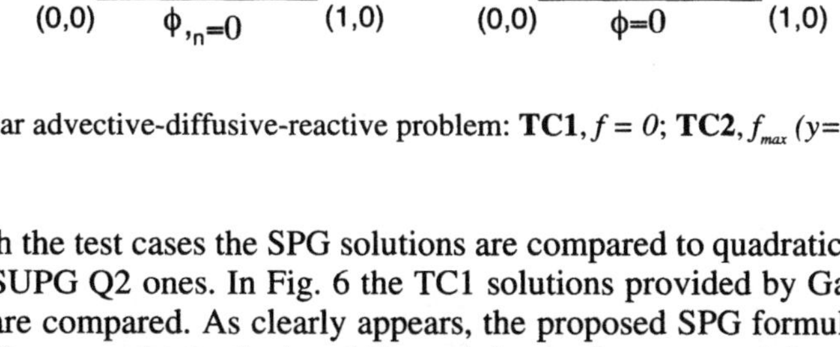

Fig. 5. Scalar advective-diffusive-reactive problem: **TC1,** $f = 0$; **TC2,** $f_{max}\ (y=0.1) = 50$

For both the test cases the SPG solutions are compared to quadratic Galerkin (G Q2), and SUPG Q2 ones. In Fig. 6 the TC1 solutions provided by Galerkin G Q2 and SPG are compared. As clearly appears, the proposed SPG formulation is able of controlling completely the instability origins in the near- and far-wall regions. Fig. 7 shows the ϕ streamwise profiles predicted by G Q2, SUPG Q2 and SPG schemes at $y = 0.05$ where reaction dominates.

Fig. 7 shows that the PG-like solutions are both able of predicting smooth ϕ profiles, thus improving the G Q2 oscillatory behaviour. Nonetheless, the SUPG Q2 returns an over-diffused layer close to the Dirichlet boundary with respect to

SPG solution that predicts a sharp but continuous solution layer. As far as the TC2 case with non uniform source f is concerned, in Fig. 8 the SUPG Q2 and SPG solutions are compared. It is worth noting that the source integral has been approximated linearly, according to Q2 element optimal conditions [4]. Moreover, Fig. 9 shows the ϕ streamwise profiles predicted by G Q2, SUPG Q2 and SPG schemes at $y = 0.05$ where the reactive effect is combined with a positive gradient of the source.

The comparison between stabilized PG schemes, confirms that the SPG is able to totally recover a non-oscillatory solution, also where the sharp streamwise solution layer develops under the effect of a non uniform source.

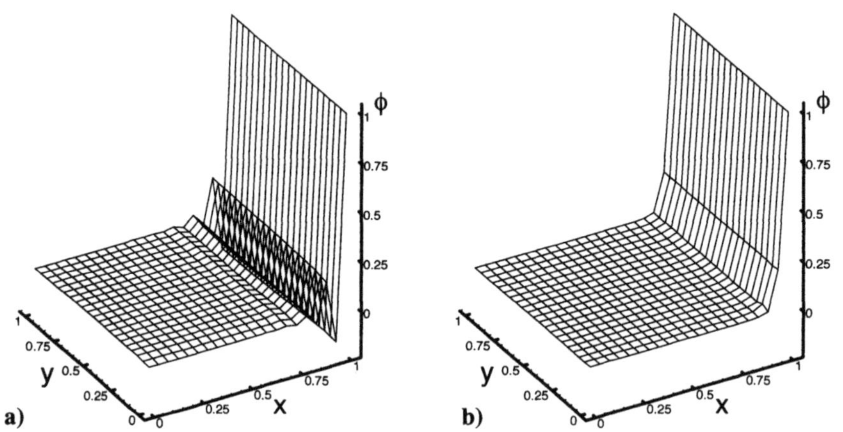

Fig. 6. TC1 comparison of solution: **a)** G Q2, **b)** SPG

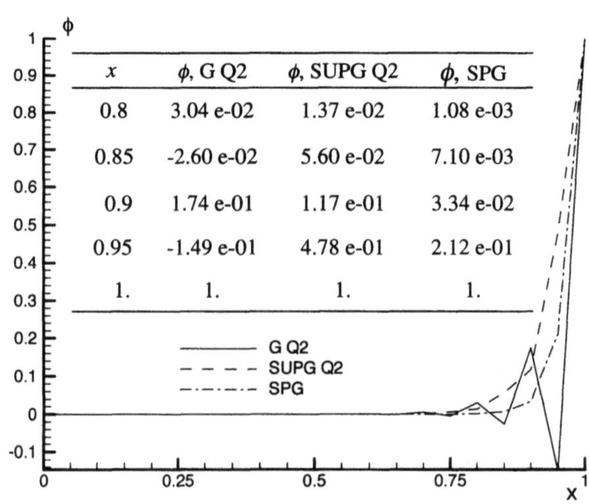

x	ϕ, G Q2	ϕ, SUPG Q2	ϕ, SPG
0.8	3.04 e-02	1.37 e-02	1.08 e-03
0.85	-2.60 e-02	5.60 e-02	7.10 e-03
0.9	1.74 e-01	1.17 e-01	3.34 e-02
0.95	-1.49 e-01	4.78 e-01	2.12 e-01
1.	1.	1.	1.

Fig. 7. TC1 comparison of streamwise ϕ profiles

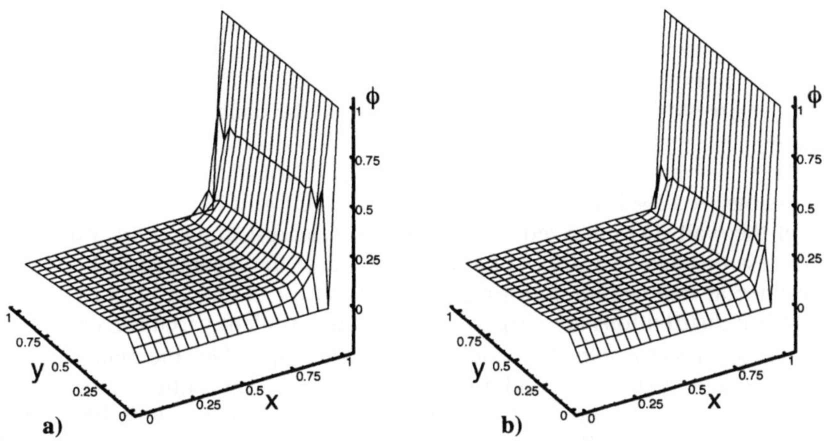

Fig. 8. TC2 comparison: **a**) SUPG Q2, **b**) SPG

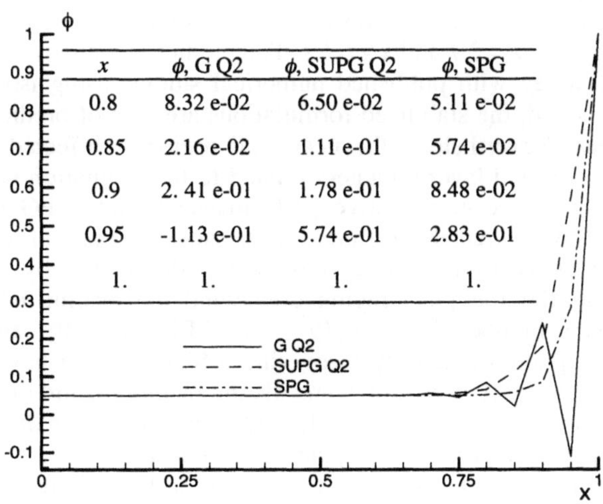

x	ϕ, G Q2	ϕ, SUPG Q2	ϕ, SPG
0.8	8.32 e-02	6.50 e-02	5.11 e-02
0.85	2.16 e-02	1.11 e-01	5.74 e-02
0.9	2. 41 e-01	1.78 e-01	8.48 e-02
0.95	-1.13 e-01	5.74 e-01	2.83 e-01
1.	1.	1.	1.

Fig. 9. TC2 comparison of streamwise ϕ profiles

4.2 Semi-circular leading edge

The last test case concerns with the prediction of the boundary layer development on a flat plate, with a semi-circular leading edge. The leading edge configuration is that proposed by ERCOFTAC Special Interest Group on Transition in 1991 (labelled T3L). The experimental data have been provided by Palikaras et al. [11], for the zero pressure gradient configuration. The Reynolds number, based on inlet velocity and leading edge radius (e_r = *5 mm*) is equal to *1660*. The free stream turbulence intensity (*TI*) at the inlet is set to *7%*, and the chosen dissipation length l_ε is *18 mm*.

The flow is assumed two-dimensional with constant temperature and incompressible. A 12681 nodes block-structured (H-O) grid has been used. In the vicinity of the wall (O-connected region) the node nearest to the solid wall has a dimensionless distance $y^+ = 1.0$. At the inlet section of the computational domain, the experimental free-stream uniform profile is used for the mean velocity (u = 5 *m/s*). Uniform distributions are also imposed for the turbulent variables, computed on the basis of *TI* and l_ε. No-slip conditions are then applied on the plate surface, and homogeneous Neumann conditions are imposed at the outlet section. The SUPG and SPG formulations have been used on Q2Q1 elements, with PSPG-like relaxation of incompressibility constraint. The turbulence closure is the standard k-ε model [10], in its near wall extension. A GMRes(50) solver has been used with convergence thresholds for error R_{res} and solution R_{sol} residuals set to 10^{-6}. The convergence histories are first compared in Fig. 10, clearly showing the faster convergence of SPG. Figs. 11 and 12 show the streamwise turbulence intensity and velocity profiles computed in two stations close to the leading edge stagnation point, respectively at $x/e_r=2.4$ and $x/e_r=3.2$. The predicted velocity and turbulence intensity profiles agree with published numerical studies using isotropic EVMs (e.g. [10]). To this end, the stabilized formulations are able of predicting well the free-stream values. Nevertheless, the near wall region is affected by an over-prediction of *TI* level and layer thickness related to the stagnation point anomaly [12]. With respect to the comparative performance between PG schemes, the solutions do not show appreciable differences, as shown in Figs. 11 and 12.

This evidence confirms that in a flow region dominated by the effect of advection, such as the flat plate region downwind the leading edge, here the maximum reaction number is of $o(10^2)$, the SPG correctly annihilates its sensitivity to the equation reactivity recovering a SUPG-like behaviour. Moving upwind in the stagnating flow region, some distinguishing feature of the proposed SPG could be found. To this end in Figs. 13 and 14 are compared the computed *TI* and turbulence time scale profiles along the stagnation streamline.

The comparison of turbulence intensity profiles, in Fig. 13, demonstrates that the SPG is able of reducing the *TI* peak (TI_{SPG} =*28%* instead TI_{SUPG} = *36%*). This suggests that approaching a null-advection region, the SPG partially corrects one of the well recognised drawback of linear two-equation EVMs, the so-called stagnation point anomaly. Fig. 14 shows that this remarkable feature of SPG formulation is related to the capacity of controlling the over-prediction of turbulence time scale in the pure reactive-diffusive flow limit [13]. In this condition, the SPG perturbation activates a dependence of weights from the predicted turbulence scale intensity, which sensitises the residual projection basis.

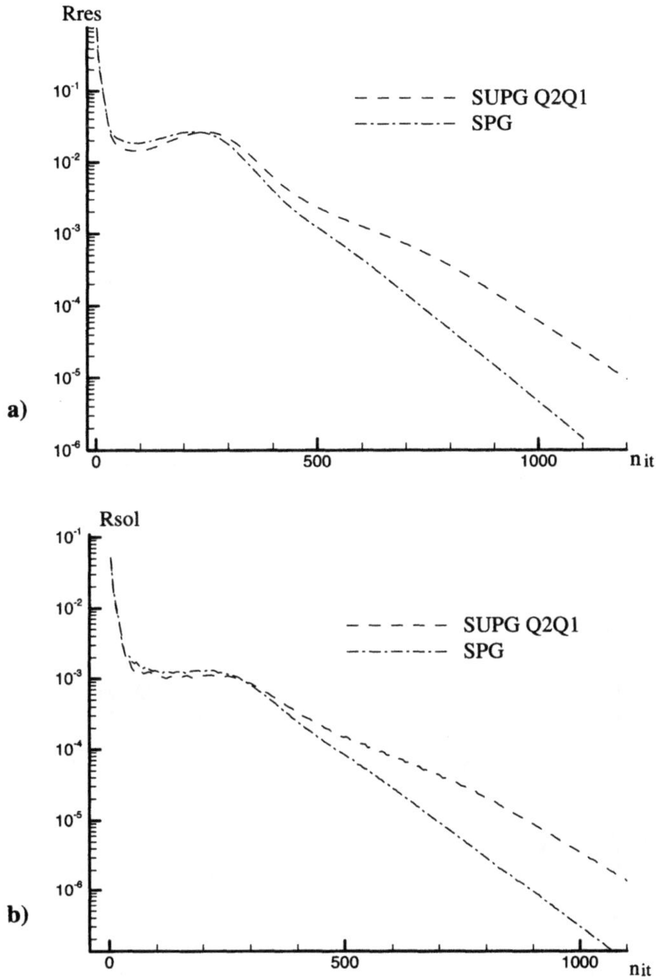

Fig. 10. T3L convergence histories for PG formulations: (**a**) *Rres* and (**b**) *Rsol*

a)

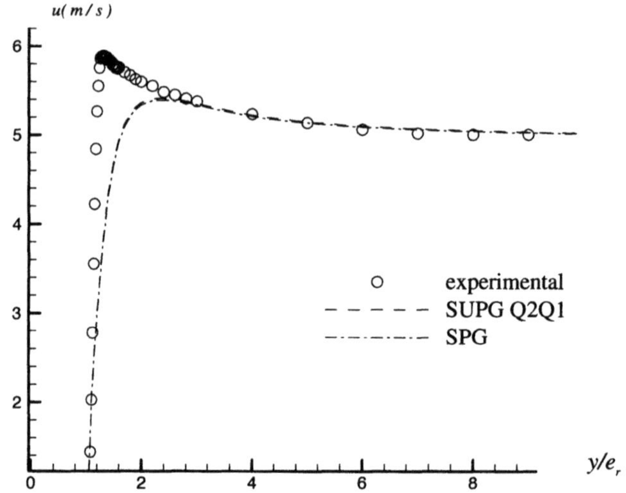

b)

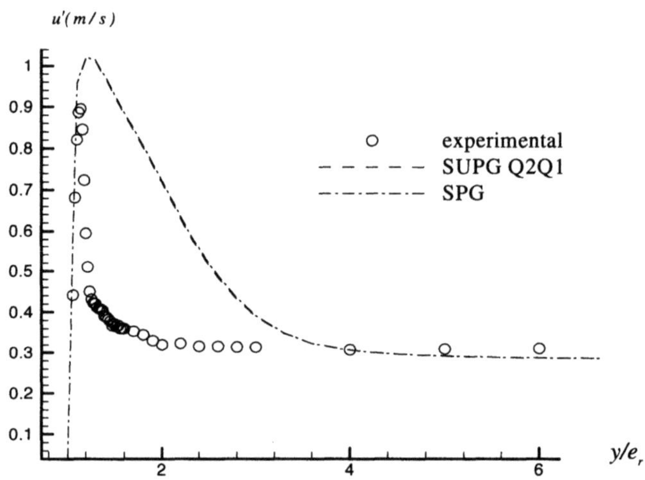

Fig. 11. Comparison of streamwise: velocity *u (m/s)* (**a**) and turbulence intensity *u′(m/s)* (**b**) at $x/e_r = 2.4$

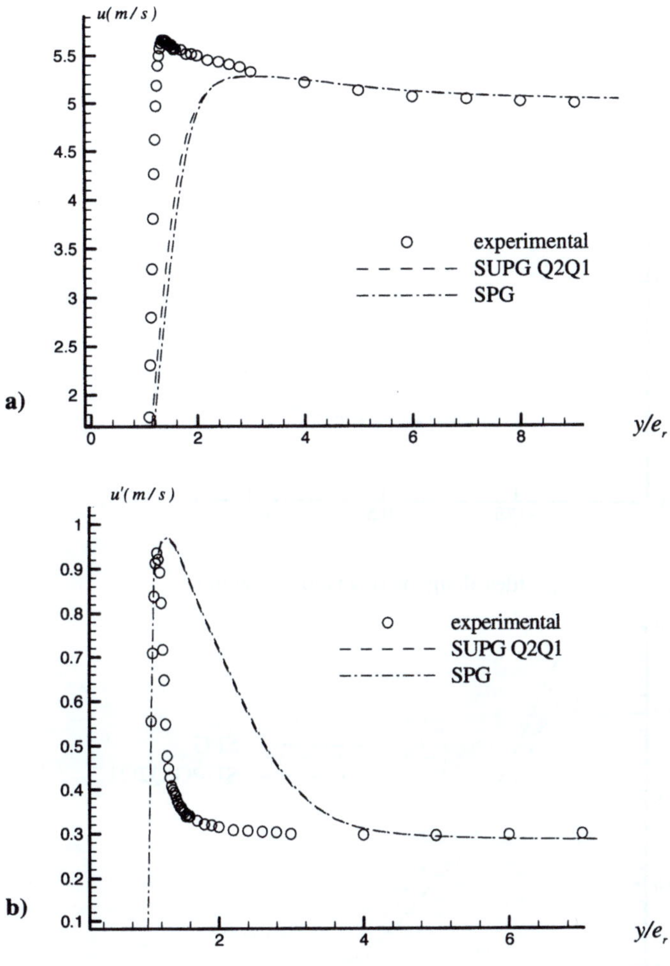

Fig. 12. Comparison of streamwise: velocity u *(m/s)* (**a**) and turbulence intensity u'*(m/s)* (**b**) at $x/e_r = 3.2$

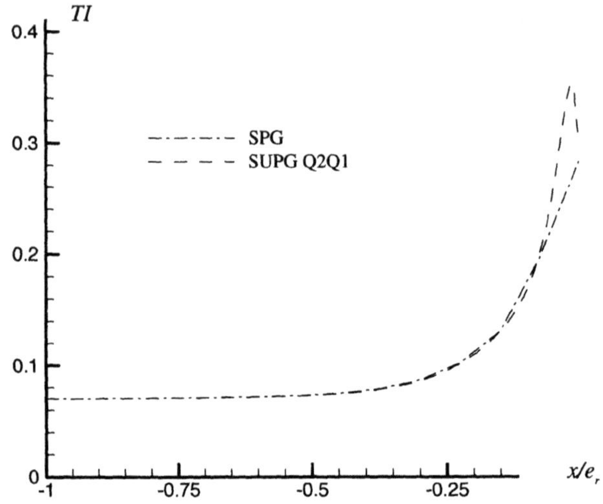

Fig. 13. Comparison of *TI* profiles along the stagnation streamline

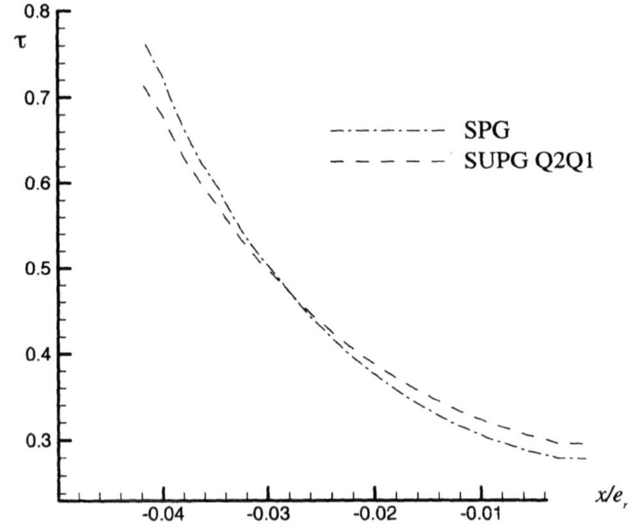

Fig. 14. Comparison of turbulence time scale τ profiles along the stagnation streamline

5. Summary

The paper has investigated the predicting capabilities of a FEM stabilized formulation developed for the purpose of solving advective-diffusive-reactive problems. This scheme, called SPG, addresses the use of a perturbation to the weight function composed by two contributions. The first is a SUPG like operator and is used to overcome instabilities due to advective or skew-simmetric terms, whereas the second operator is a symmetric one aiming at precluding oscillations due to reactive terms. The FEM formulation has been obtained by means of one-dimensional nodal exactness, but has been tested in several more complex examples that violate the super convergence conditions (e.g. 2D and non-linear problems). In this respect, the SPG method demonstrates its suitability in solving the typical equations of turbulence EVMs, with specific application to turbomachinery flows.

Acknowledgments

The authors acknowledge MIUR under the projects COFIN 2001, and MIUR-Ateneo 2001.

References

1. Hughes TJR, Brooks AN (1982) Streamline Upwind/Petrov-Galerkin formulations for convection dominated flows with particular emphasis on the incompressible Navier-Stokes equations. Comp Meth Appl Mech Eng 32
2. Hughes TJR, Brooks AN (1982) A theoretical framework for Petrov-Galerkin methods with discontinuous weighting functions: application to the Streamline-Upwind procedure. J Wiley&s
3. Tezduyar TE, Mittal S, Ray SE, Shih R (1992) Incompressible flow computations with stabilized bilinear and linear equal-order-interpolation velocity-pressure elements. Comp Meth Appl Mech Eng 95
4. Codina R, Oñate E, Cervera M (1992) The intrinsic time for the streamline upwind/Petrov-Galerkin formulation using quadratic elements. Comp Meth Appl Mech Eng 94
5. Hughes TJR, Mallet M, Mizukami A (1986) A new finite element formulation for computational fluid dynamics: II. Beyond SUPG. Comp Meth Appl Mech Eng 54
6. Codina R (2001) A stabilized finite element method for generalized stationary incompressible flows. Comp Meth Appl Mech Eng 190
7. Idelsohn S, Nigro N, Storti M, Buscaglia G (1996) A Petrov-Galerkin formulation for advection-reaction-diffusion problems. Comp Meth Appl Mech Eng 136
8. Harari I, Hughes TJR (1994) Stabilized finite element methods for steady advection-diffusion with production. Comp Meth Appl Mech Eng 115
9. Borello D, Corsini A, Rispoli F (2003) A finite element overlapping scheme for turbomachinery flows on parallel platforms. Computers & Fluids 32
10. Launder BE, Sharma BI (1974) Application of the energy dissipation model of turbulence to the calculation of flow near a spinning disc. Lett in Heat and Mass Transfer

11. Palikaras A, Yakinthos K, Goulas A (2002) Transition on a flat plate with a semi-circular leading edge under uniform and positive shear free-stream flow. Int J Heat Fluid Flow 23
12. Durbin PA (1996) On the k-ε stagnation point anomaly. Int J of Heat and Fluid Flow 17

FLUID MACHINERY

Nonuniform Flow in a High Speed Compressor Due to Asymmetric Tip Clearance

Hyun Suk JOO, Graduate Student
School of Mechanical and Aerospace Engineering, Seoul National University, Seoul 151-742, Korea

Seung Jin SONG, Associate Professor
School of Mechanical and Aerospace Engineering, Seoul National University, Seoul 151-742, Korea

Abstract This paper presents an analytical model to predict flow response in a high speed compressor stage caused by asymmetric tip clearance distribution. The model assumes inviscid, compressible flow, and uses basic conservation equations. The stage is modelled as an actuator disc, and the analysis is done in the meridional and radial planes. First, an axisymmetric meridional flow model for tip clearance and passage flows is developed. The model takes into consideration the detached shock that occurs in the rotor passage at the design point. This shock model is used to calculate the change in flow variables. Then, this model is perturbed to obtain non-axisymmetric flow quantities in the radial plane.
Key Words: axial compressor, tip clearance, compressibility

Nomenclature

c	absolute flow velocity; blade chord
f	force
H	total enthalpy; annulus height
h	enthalpy
M	Mach number
M_t	tangential Mach number
p	pressure
Q	strength of shear layer
q	nondimensional vorticity strength; local mass flux
R	mean compressor radius
s	blade pitch
t	radial tip clearance
U	compressor rotational speed at the mean radius

W relative flow velocity
X direction along the rotor offset
x axial direction
Y direction perpendicular to the rotor offset
y azimuthal direction
z radial direction
α absolute flow angle
β relative flow angle
Δ thickness of underturned layer downstream of actuator disc
ϕ flow coefficient ($= c_x/U$)
λ nondimensional mass fraction of underturned flow
θ azimuthal angle measured in the direction of rotation from the minimum gap location
ρ density
ω angular velocity of rotor shaft rotation
ψ meridional stream function

Subscripts

m mean
r relative component
1 rotor inlet
2 rotor exit
3 stator exit
$\perp$ meridional component

Superscripts

$-$ main passage flow
$+$ tip clearance flow
$'$ nonaxisymmetric perturbation
— azimuthal mean, or axisymmetric value
^ complex amplitude

1. Introduction

Tip clearance asymmetry can have many causes such as rotor shaft whirling, bending and deformation of components, and the asymmetry degrades both aerodynamic and structural performance. Horlock and Greitzer [5] 1983 and Cho [1] and Song 2000 developed actuator disc models to examine the effects of tip clearance asymmetry in compressors. The former examined effects on the flow field and performance while the latter focused on rotordynamic effects. Also, Graf [4] et al. 1998 and Enrich [2] 1993 developed parallel compressor models to examine tip clearance asymmetry's influence on aerodynamic and structural stability, respectively.

However, such studies have been concentrated on incompressible flows, and models for compressible flows are still lacking. Therefore, this investigation aims to understand the flow field in a high speed compressor caused by nonaxisymmetric rotor tip clearance. The scope of current investigation is limited to the effects of static tip clearance asymmetry in a single stage high speed compressor.

2. Model Description

A two-step process is used to solve for the flow through a compressor stage. The first part is an axisymmetric, two-dimensional analysis in the meridional plane, and the second part is a nonaxisymmetric, two-dimensional, radial plane analysis. The model assumes the flow to be inviscid and compressible. The compressor geometry is assumed to be two dimensional at the mean radius values.

2.1 Concentric Model

This analysis examines radial redistribution of flow due to axisymmetric rotor tip clearances. The modeled tip clearance flow is calculated considering only axisymmetric tip vortex sheet without viscous and vortex shedding effects. Only a brief summary is given here, and more details can be found in Joo [6] 2003.

Upon going through the stage, the flow splits into tip clearance and main passage flows (Fig. 1). Stream "+" is associated with the rotor tip clearance, and Stream "-" is called the main passage flow.

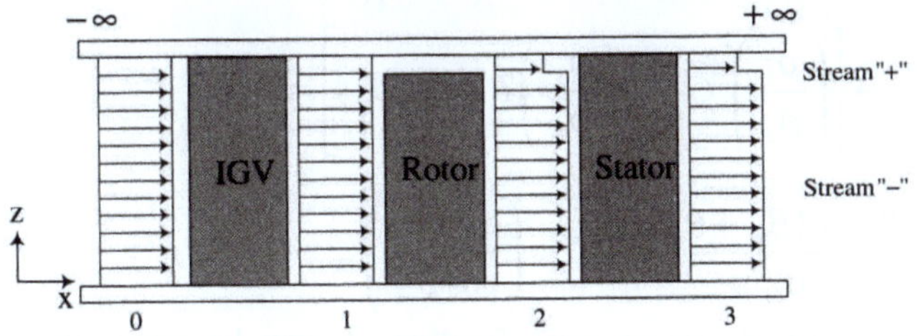

Fig. 1. Schematic view of a compressor stage

The actuator disc in blade scale model consists of an IGV row, a rotor row, and a stator row. The IGV and stator blade rows have full span blades while the rotor has a tip clearance. Axial, tangential, and radial directions are denoted by x, y, and z, respectively. Upstream of IGV is referred to as Station 0. Station 1 is the inlet to the rotor and Station 2 is the rotor exit. Downstream of the stator row is called Station 3.

The downstream vorticity between the under-turned and passage streams can be determined as

$$\omega_{y3} = \frac{\rho_3}{\rho_1}\frac{\partial H_{\perp 3}}{\partial \psi} \tag{1}$$

where $H_\perp$ is enthalpy based on the meridional velocity and $\psi(x, z)$ is the stream function. To focus on tip clearance effects, the coordinate system is transformed to the streamline coordinate from the z coordinate. Then, the equation for ψ becomes

$$\text{Upstream } (x < 0)\ \nabla_\perp^2 \psi = 0 \tag{2}$$

$$\text{Downstream } (x > 0)\ \nabla_\perp^2 \psi = Q\delta(\psi - \psi_{tip}) \tag{3}$$

where $Q = \int_i^j \omega_y d\psi = H_{\perp 3}^i - H_{\perp 3}^j$ is the strength of azimuthal vorticity ω_y between streams i and j, and δ is Dirac' delta function.

A quadratic equation for Q as a function of the blade geometry, mass fraction of the underturned stream, and density ratios is shown below.

$$\begin{aligned}
&\left[(1-\lambda)^2 G + (1-2\lambda)\tan^2\alpha_3\right]\left(\frac{q}{2}\right)^2 \\
&+2\left[\frac{1}{\phi}\left(\frac{\rho_1}{\rho_2}\right)\left(\frac{\rho_3}{\rho_1}\right)(\lambda\tan\beta_2 + (1-\lambda)T)\right. \\
&\left.+((1-\lambda)G + \tan^2\alpha_3) + 2\left(\frac{\rho_1}{\rho_3}\right)\right]\left(\frac{q}{2}\right) \\
&+\left[\left(\frac{\rho_1}{\rho_3}\right)^2 G + \frac{2}{\phi}\left(\frac{\rho_1}{\rho_2}\right)(T - \tan\beta_2)\right] = 0
\end{aligned} \tag{4}$$

where $T = \dfrac{\cos\theta \cdot \sin(\beta_m + \theta)}{\cos\beta_m}$ and $G = \left(\dfrac{\sin\theta}{\cos\beta_m}\right)$.

Next, the underturned stream mass fraction λ can be determined from the given tip clearance as

$$\lambda = \frac{4\left(\frac{t}{H}\right)}{\left(1 - \frac{q}{2}\frac{\rho_1}{\rho_3}\right) + \sqrt{\left(1 - \frac{q}{2}\frac{\rho_1}{\rho_3}\right)^2 + 4\frac{q}{2}\frac{\rho_1}{\rho_3}\frac{t}{H}}} \tag{5}$$

The density ratios are obtained from the shock relation. In most high speed compressors, a detached shock occurs in the rotor passage at the design point, and, thus the rotor passage flow is unchoked. Freeman and Cumpsty [3] 1992 developed a detached shock model considering a control volume extending from the inlet region to the maximum blade thickness location. But in our model the entire blade passage is considered as the control volume (Fig. 2)

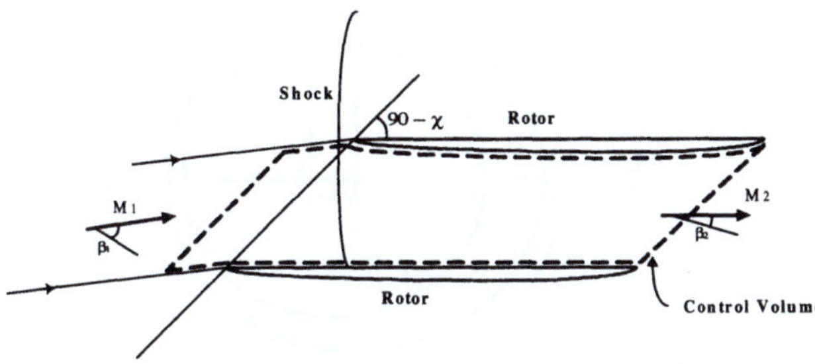

Fig. 2. Extended control volume considering the entire rotor passage

In the direction of the blade chord, continuity mass and momentum are conserved. Also, the relative total enthalpy is constant. Thus, the density ratio across the rotor is easily obtained as

$$\frac{\rho_2}{\rho_1}=\frac{M_{r1}}{M_{r2}}\left(\frac{1+\frac{\gamma-1}{2}M_{r2}^2}{1+\frac{\gamma-1}{2}M_{r2}^2}\right)^{\frac{1}{2}}\frac{\cos\beta_1}{\cos\beta_2} \tag{6}$$

where M_{r1} and M_{r2} are the rotor inlet and exit relative Mach numbers, respectively.

In the stator, there is no shock, and the total enthalpy remains constant. Thus, the density ratio across the stator is obtained as

$$\left(\frac{\rho_3}{\rho_2}\right)^{\gamma-1}+\frac{\gamma-1}{2}M_2^2\left(\frac{\rho_3}{\rho_2}\right)^{-2}\left(\frac{\cos\alpha_2}{\cos\alpha_3}\right)^2=1+\frac{\gamma-1}{2}M_2^2 \tag{7}$$

Here, M_2 is the absolute Mach number at the rotor exit. Finally, the system of Eqs. (4), (5), (6) and (7) are solved simultaneously.

2.2. Eccentric Model

This section presents the radius scale model. The rotor offset, *e*, is assumed to be much smaller than the blade span. Therefore, the tip clearance distribution is given by

$$t = \bar{t} + \mathrm{Re}\left[\hat{e}e^{iy/R}\right] \tag{8}$$

where $\bar{t}$ is the mean rotor tip gap, and y is the distance from the maximum tip gap in the azimuthal direction (Fig. 3)

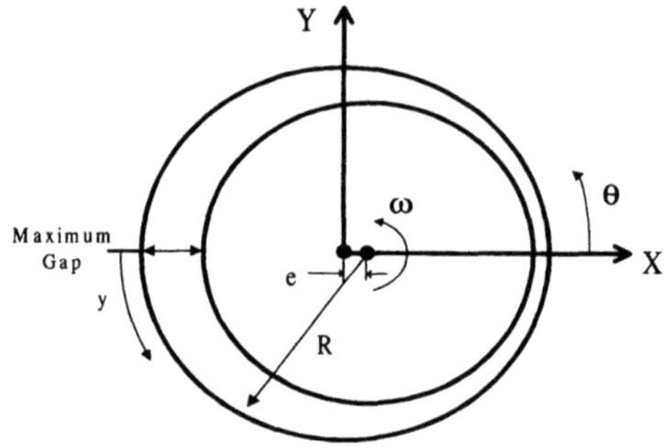

Fig. 3. Coordinate system for the eccentric analysis

The flow far upstream is uniform and steady, and, thus, irrotational. Near upstream of the stage at $x = 0-$, the assumed axial velocity is

$$c_x(0-, y, t) = \mathrm{Re}[c_{x(-\infty)} + \hat{K}_0 e^{\alpha x + i\frac{y}{R}}] \tag{9}$$

where $\hat{K}_0$ is the complex amplitude of axial velocity perturbation as the flow approaches the disc. Far upsteam, $c_{x(-\infty)} = \bar{c}_{x0}$

With these conditions, linearization of the equations of conservation of mass, x-momentum, y-momentum, and energy yields

$$c_x = c_{x(-\infty)} + \mathrm{Re}[\hat{K}_0 e^{\frac{\alpha_c}{R}x + i\frac{y}{R}}] \tag{10}$$

$$c_y = \mathrm{Re}[\frac{i}{\alpha_c}\hat{K}_0 e^{\frac{\alpha_c}{R}x + i\frac{y}{R}}] \tag{11}$$

$$h_x = h_{x(-\infty)} + \mathrm{Re}[-c_{x(-\infty)}\hat{k}_0 e^{\frac{\alpha_c}{R}x + i\frac{y}{R}}] \tag{12}$$

where $\alpha_c = \dfrac{\sqrt{a^4 - a^2 c_{x(-\infty)}{}^2}}{a^2 - c_{x(-\infty)}{}^2}$.

Downstream of the stage, the flow consists of two regions – the tip clearance and the main passage flows. The mass conservation equation for each region can be written as

$$\frac{\partial(\rho\Delta)}{\partial t}+\frac{\partial(\rho c_x^+\Delta)}{\partial x}+\frac{\partial(\rho c_y^+\Delta)}{\partial y}=0 \tag{13}$$

$$\frac{\partial(\rho(H-\Delta))}{\partial t}+\frac{\partial(\rho c_x^+(H-\Delta))}{\partial x}+\frac{\partial(\rho c_y^+(H-\Delta))}{\partial y}=0 \tag{14}$$

The momentum equations, implicitly containing the energy equation, can be written as

$$\frac{\partial\vec{c}^{\pm}}{\partial t}+(\vec{c}^{\pm}\cdot\nabla)\vec{c}^{\pm}+\nabla h=0 \tag{15}$$

where all vectors are two-dimensional in *(x, y)*.

Assuming $e << H$, $c_x^+, c_y^+, c_x^-, c_y^-$, and h can be expressed as

$$\Delta=\overline{\Delta}+\Delta' \tag{16}$$

where

$$\Delta'=\mathrm{Re}[\hat{\Delta}e^{\alpha x+i\frac{y}{R}}] \tag{17}$$

is a small perturbation about the mean. A homogeneous set of equations can be obtained by substituting for each flow variable and linearizing.

$$\begin{bmatrix} \alpha\overline{\Delta} & i\overline{\Delta}/R & 0 & 0 & A & A\overline{\Delta}M_t^2 \\ 0 & 0 & (H-\overline{\Delta})\alpha & i(H-\overline{\Delta})/R & -B & (H-\overline{\Delta})BM_t^2 \\ A & 0 & 0 & 0 & 0 & \alpha \\ 0 & A & 0 & 0 & 0 & i/R \\ 0 & 0 & B & 0 & 0 & \alpha \\ 0 & 0 & 0 & B & 0 & i/R \end{bmatrix}\begin{bmatrix} \hat{c}_x^+ \\ \hat{c}_y^+ \\ \hat{c}_x^- \\ \hat{c}_y^- \\ \hat{\Delta} \\ \hat{h} \end{bmatrix}=0 \tag{18}$$

Here, $A=\alpha\overline{c}_x^+ + i\frac{\overline{c}_y^+}{R}$, $B=\alpha\overline{c}_x^- + i\frac{\overline{c}_y^-}{R}$, and $M_t=\frac{U}{a}$.

Upon determining eigenvalues α_i and eigenvectors E_i, the nontrivial homogenous solution to the system of equations above can be written as

$$\left(\hat{c}_x^+,\hat{c}_y^+,\hat{c}_x^-,\hat{c}_y^-,\hat{\Delta},\hat{h}\right)=\sum_{i=1}^{5}\hat{K}_iE_i \tag{19}$$

where the complex constants $\hat{K}_i$'s have to be determined.

The eccentric model analysis relates the flow variables at $x=0-$ to those at $x=0+$, equivalent to far upstream (Station 0) and far downstream (Station 3), respectively, in the concentric model. According to the concentric model analysis, the nondimensional flow variables depend on the local nondimensional tip clearance, *t/H*, flow coefficient, ϕ, and upstream Mach number, M_1. If both rotational speed and blade geometry are known, the upstream Mach number can be obtained. Thus, the flow variables can be said to be functions of the local tip clearance and flow coefficient.

The downstream and upstream perturbation quantities are determined from the concentric model results as shown below.

$$\begin{bmatrix} \hat{c}_x^+ \\ \hat{c}_y^+ \\ \hat{c}_x^- \\ \hat{c}_y^- \\ \hat{\Delta} \\ \hat{h} \end{bmatrix}_{x=0+} = \left[\hat{\phi}\frac{\partial}{\partial \phi} + \left(\frac{e}{H}\right)\frac{\partial}{\partial \frac{t}{H}} \right] \begin{bmatrix} c_{x4}^+ \\ c_{y4}^+ \\ c_{x4}^- \\ c_{x4}^+ \\ \Delta \\ h_4 \end{bmatrix} = \sum_{i=1}^{6} \hat{K}_i E_i \tag{20}$$

The tangential force exerted on the compressor by the fluid per azimuthal length f_y can be determined as

$$f_y = \lambda q\left(c_{y1} - c_{y2}^+\right) + (1-\lambda)q\left(c_{y1} - c_{y2}^-\right) \tag{21}$$

Then, its perturbations is given as

$$\begin{aligned} f_y' &= \bar{\lambda}\bar{q}\left(\bar{c}_{y1} - \bar{c}_{y2}^+\right)\left[\frac{\lambda'}{\bar{\lambda}} + \frac{q'}{\bar{q}} + \frac{c_{y1}' - c_{y2}^{+\prime}}{\bar{c}_{y1} - \bar{c}_{y2}^+}\right] \\ &+ \left(1-\bar{\lambda}\right)\bar{q}\left(\bar{c}_{y1} - \bar{c}_{y2}^-\right)\left[-\frac{\lambda'}{1-\bar{\lambda}} + \frac{q'}{\bar{q}} + \frac{c_{y1}' - c_{y2}^{-\prime}}{\bar{c}_{y1} - \bar{c}_{y2}^-}\right] \end{aligned} \tag{22}$$

3. Model Predictions

This section presents the model predictions for a test compressor. First, the radial flow redistribution induced by axisymmetric rotor tip clearance is presented. Second, the azimuthal flow redistribution due to non-axisymmetric tip clearances is brought out.

The test compressor is a Lewis single-stage compressor described in Reid [8] and Moore [7] 1978, 1980. This compressor, called Stage 37, consists of rotor and stator rows. Table 1 shows the specifications of this compressor. Input upstream stagnation conditions are P_{t1}=101.4 kPa and T_{t1}=288.2 K. Also, the value of upstream Mach number and flow coefficient is 0.661 and 0.453, respectively. The tip clearance value is assumed to be 2% of the annulus height.

Table 1. Lewis singe-stage compressor specifications at the design point

Parameter	Value
t/H	0.02
s/c	1.471
α_1	0°
β_1	56.53°
α_2	45.31°
β_2	38.87°
α_3	2.54°

3.1. Concentric Model Predictions

Figures 4a and 4b show the radial profiles of relative axial and tangential velocities at the rotor exit. z/H=1 is the casing and z/H= 0 is the hub. The radially uniform upstream flow has split into two streams – the clearance stream and the passage stream. Due to the density increase across the rotor without area variation, the axial velocities at station 2 are smaller than those upstream of the rotor. Furthermore, relative to the passage stream, the clearance stream has a lower axial velocity (Fig. 4a) and a lower tangential velocity in the direction of rotation (Fig. 4b). Thus, the clearance flow is underturned, and this can be seen in Fig. 4c which shows the radial profile of the relative flow angle at the rotor exit. The angle is defined to be positive in the direction of rotation. These effects are due to the kinematics of the tip clearance flow similar to the incompressible case.

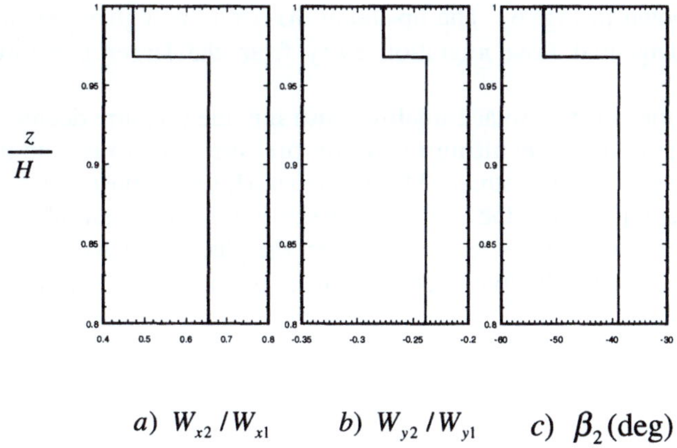

Fig. 4. Radial distribution of (a) relative axial velocity; (b) relative tangential velocity; (c) relative flow angle

Next, the predictions' sensitivity to tip clearance is analyzed. Fig. 5 shows a plot of the leakage flow vs. tip clearance for various Mach numbers. As the tip clearance increases, the mass flow increases proportionally. Velocity magnitudes and flow angles are not significantly affected; only the thickness of the under-turned layer changes. Also, as seen in Fig. 5, the Mach number does not have much influence on the leakage flow mass fraction.

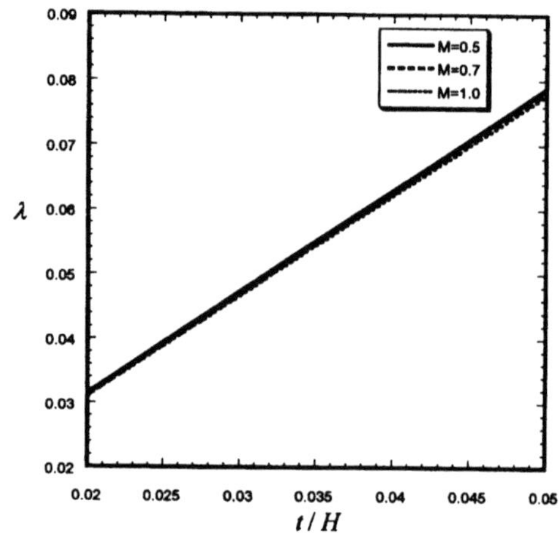

Fig. 5. Leakage mass flow amount vs. tip clearance for M=0.5, 0.7, and 1.0

3.2. Eccentric Model Predictions

First, the plots of predicted upstream axial velocity perturbation vs. azimuthal location θ are shown in Fig. 6. The upstream mass flux is higher near minimum gap due to a tangential flow migration away from the larger gap toward the smaller gap.

Then, due to the constant total enthalpy constraint, the pressure decreases when flow accelerates, and the resulting upstream pressure pattern is seen in Fig. 7. To better see the effects of upstream Mach number M_1 on the upstream flow field, the magnitudes and phases of the non-uniformities in C_{x0} and P_0 are plotted vs. M_1 in Figs. 8 and 9, respectively. In both cases, the phase remains relatively constant. However, the perturbation magnitudes increase at increasing rates as M_1 increases.

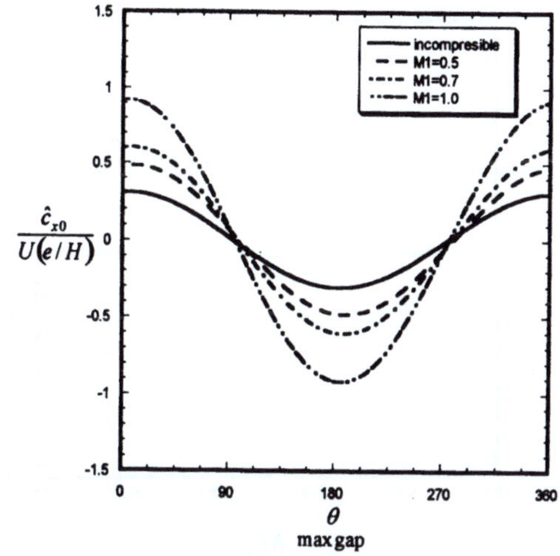

Fig. 6. Predicted upstream axial velocity perturbation versus azimuthal angle for M_1= 0.5, 0.7, and 1.0

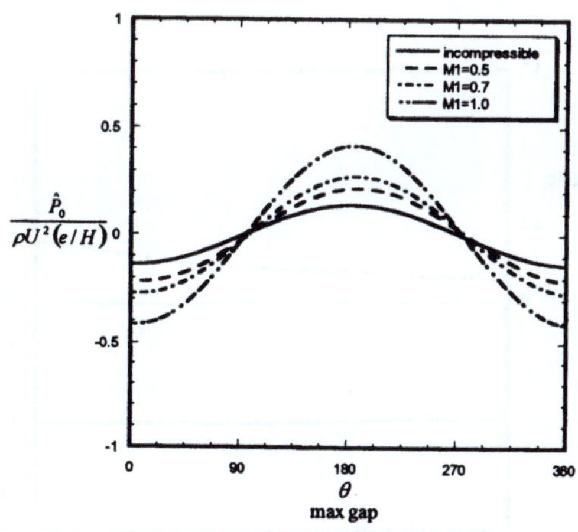

Fig. 7. Predicted upstream pressure perturbation versus azimuthal angle for M_1= 0.5, 0.7, and 1.0

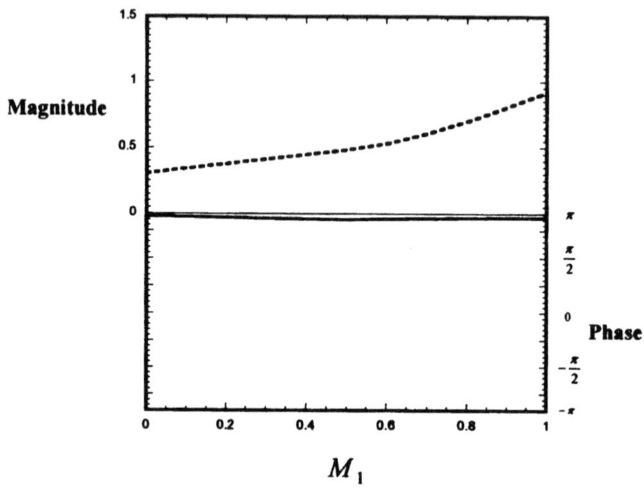

Fig. 8. Phase and magnitude of upstream axial velocity versus upstream Mach number M_1

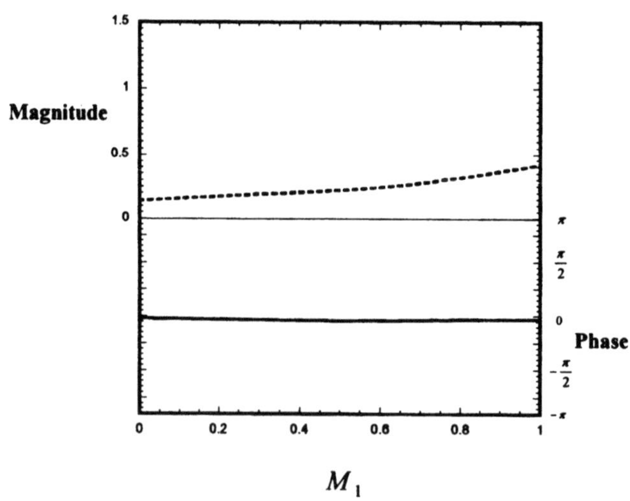

Fig. 9. Phase and magnitude of upstream pressure perturbation versus upstream Mach number M_1

Figure 10 shows the plots of tangential force distribution versus θ for various Mach numbers. Since the force on the compressor by the fluid acts in a direction opposite to the direction of rotation, the mean value of tangential force is negative.

Therefore, the compressor rotor blade is loaded less near the maximum gap. As M_l increases, the magnitude of f_y' increases while the phase remains approximately constant (Fig. 11). Such dependence arises because f_y' is primarily influenced by the perturbation in leakage mass fraction λ', and the magnitude of λ' increases with M_l as shown in Fig. 12.

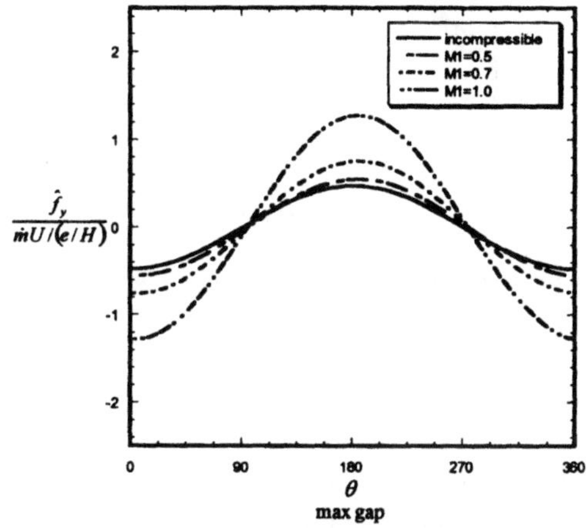

Fig. 10. Predicted rotor blade loading perturbation versus azimuthal angle for M_l= 0.5, 0.7, and 1.0

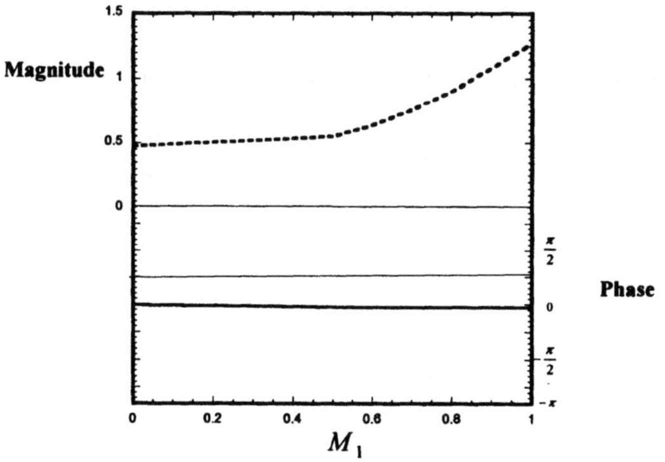

Fig. 11. Phase and magnitude of rotor blade loading perturbation versus upstream Mach number

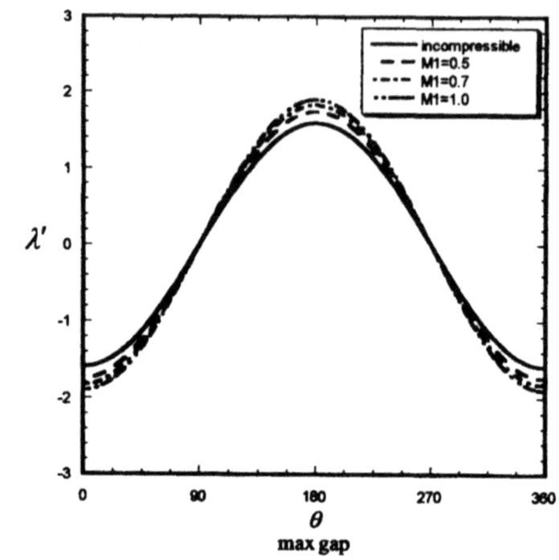

Fig. 12. Leakage mass fraction perturbation versus azimuthal angle for M_1= 0.5, 0.7, and 1.0

4. Conclusions

The intent of this study is to gain physical understanding of rotor tip clearance effects in high speed compressors. The following conclusions can be made.

1. An analytical model for tip clearance effects in high speed axial compressors has been developed.
2. In compressible flow, the tip clearance flow has axial and tangential momentum defects relative to the passage flow.
3. Tip clearance variation mainly affects the tip clearance flow, mass fraction which increases linearly with tip clearance.
4. With eccentricity, the upstream flow migrates azimuthally toward the smaller gap.
5. In both incompressible and compressible flows, rotor blade loading has its maximum near $\theta \approx 180°$, but the magnitude is larger in compressible flow.

References

[1] Cho S H, Song S J (2000) Non-uniform Flow in a Compressor Due to Asymmetric Tip Clearance. ASME Journal of Turbomachinery 122 : 751–760

[2] Enrich F F (1993) Rotor Whirl Forces Induced by the Tip Clearance Effect in Axial Flow Compressor. ASME Journal of Vibration and Acoustics 115 : 509–515

[3] Freeman C, Cumpsty N A (1992) Method of the Prediction of Supersonic Compressor Blade Performance. Journal of Propulsion and Power 8 : 199–208

[4] Graf M B, Wong T S, Greitzer E M, Marble F E, Tan C S, Shin H W, Wisler D C (1998) Effects of Nonaxisymmetric Tip Clearance on Axial Comprsssor Performance and Stability. ASME Journal of Turbomachinery 120 : 648–661

[5] Horlock J H, Greitzer E M (1983) NonUniform Flows in Axial Compressors Due to Tip Clearance Variation. Proceeding of the Institution of Mechanical Engineering 197C : 173–178

[6] Joo H S (2003) Nonuniform Flow in a High Speed Compressor Due to Asymmetric Tip Clearance. M.S. Thesis, Seoul National University

[7] Moore R D, Reid L (1980) Performance of Single-Stage Axial Flow Transonic Compressor with Rotor and Stator Aspect Ratios of 1.19 and 1.26, Respectively, and With Design Pressure Ratio of 2.05. NASA TP1659

[8] Reid L, Moore R D (1978) Design and Overall Performance of Four Highly Loaded, HighSpeed Inlet Stages for an Advanced High-Pressure-Ratio Core Compressor. NASA TP1337

Deterministic Unsteady Vorticity Field in a Driven Axisymmetric Cavity Flow

Daniel RUSCH, PhD student
Institute of Fluid Dynamics, Swiss Federal Institute of Technology, Zurich

Anestis I. KALFAS, senior scientist[1],
Reza S. ABHARI, professor
Axel PFAU, PhD student,
Joël SCHLIENGER, PhD student,
Turbomachinery Laboratory, Swiss Federal Institute of Technology, Zurich

Abstract This paper introduces a new data visualization technique for the evaluation of 3D unsteady data using the various terms of the deterministic unsteady vorticity transport equation. The toroidal vortex residing in the inlet cavity of an axial turbine rotor labyrinth is discussed using the proposed technique. Especially secondary flow effects and the effect of unsteadiness with respect to its contribution to loss generation were investigated. The analysis has allowed further insight in flow physics. It turned out that the rotational acceleration of the vortex shows a phase shift of one quarter of blade passing period relative to the vortex strength.

Nomenclature

C	[-]	circumferential position, θ / p_{blade}
f_{bp}	[Hz	blade passing frequency
h	[J/kg]	specific enthalpy
p	[Pa]	static pressure
p_{blade}	[deg]	blade pitch, 360°/42
r	[m]	radial position
R	[-]	radial height, $(r - r_{Hub})/(r_{Tip} - r_{Hub})$
s	[$J/(kgK)$]	specific entropy

[1] Corresponding author:
ETH Zentrum, CH-8092 Zurich, Switzerland
Tel.: +41 1 632 68 31, Fax: +41 1 632 11 00, Email: anestis.kalfas@ethz.ch

t	$[s]$	time
t_{bp}	$[s]$	blade passing period, $1/f_{bp}$
T	$[K]$	temperature
$\vec{v}$	$[m/s]$	velocity vector
v_{∞}	$[m/s]$	undisturbed velocity
$v_{\perp}$	$[m/s]$	velocity component perpendicular to $\vec{\omega}$
$v_{\Uparrow}$	$[m/s]$	velocity component parallel to $\vec{\omega}$
v_z	$[m/s]$	axial velocity component
z	$[mm]$	axial direction parallel to machine axis
$\hat{\alpha}$	$[1/s^2]$	rotational acceleration
θ	$[deg]$	angular position
ρ	$[kg/m^3]$	density
σ	$[m]$	coordinate along a vortex line
$\Delta\sigma$	$[m]$	coordinate increment along a vortex line
τ	$[N/m^2]$	stress tensor
$\vec{\omega}$	$[1/s]$	vorticity vector
ω_{θ}	$[1/s]$	circumferential component of $\vec{\omega}$
$\hat{\omega}$	$[1/s]$	rotational speed
Y	[-]	non dimensional time,

1. Introduction

One of the key issues in modern turbomachinery design is the improvement of component efficiency. The aerodynamic efficiency of a turbine stage is drastically reduced due to secondary flows. Langston [2] estimates the contribution of the secondary loss to be up to 50% of the total aerodynamic loss.

Secondary flow control plays a key role in the strive for more efficient and environmentally friendly turbomachines. This goal can only be reached through improved understanding of the secondary flow development and interactions. There have been a number of investigations focusing on the total pressure loss coefficients, secondary velocity [1] and - in more recent papers - the vorticity distribution [9].

Considering the terms of the vorticity equation such as the vortex stretching and the time derivative term, a deeper insight in flow physics is gained. The present work is aiming to provide a novel technique of analytical study of complex flow fields involving vortical structures.

2. Experimental Setup

The data presented in this paper were taken form a two-stage axial research turbine, which is described in detail in Sell et al. [7]. All measurements were taken at the same operating point using a single sensor fast response aerodynamic pressure probe (FRAP) [3, 6]. The raw data is phase-lock averaged using an accurate rotor trigger and finally reduced to the requested time resolved flow parameters such as the 3D, unsteady velocity vector, as well as the total and static pressure.

The location of the measurement volume is indicated in Fig. 2.1. The resolved volume covers both the cavity and a part of the main flow region. The coordinate system is defined as follows: C denotes the non-dimensional circumferential and R the non-dimensional radial position. According to the definition, R=1 represents the tip radius. The z axis denotes the axial direction as indicated in Fig. 2.1. A local axial coordinate axis is introduced denoting the first measurement plane with z=0. The last of the 5 measurement planes is found at z=7.5mm.

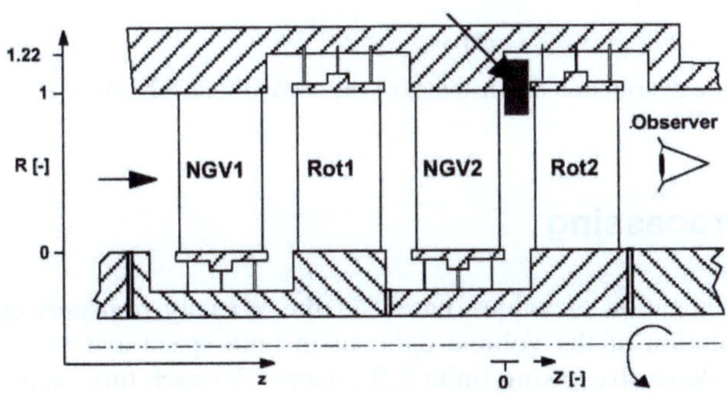

Fig. 2.1. Geometry overview. The box indicates the location of the measurement volume.

As an example, how secondary flow effects the cavity flow situation, the time averaged total pressure distribution downstream the second nozzle guide vane (NGV2) is depicted in Fig. 2.2. The data range covers both, main and cavity flow in radial direction and 1.1 pitches in circumferential direction. The hub is located at R=0 and the tip at R=1, respectively. In order to visualize the periodicity of the flow, the results are copied three times in circumferential direction. The view is from a downstream position into the upstream direction. The flow region of low total pressure behind the blade shows the typical wake-loss core structure in the main flow. Both the passage vortices at the hub as well as at the tip can be clearly identified. These loss cores are created by the passage vortices, which entrain all incoming boundary layer fluid and move it to the suction side of the wake. The cavity flow regions at hub and tip are characterized with significantly lower total pressure. The interface between main and cavity flow both at hub and tip shows a wavy structure.

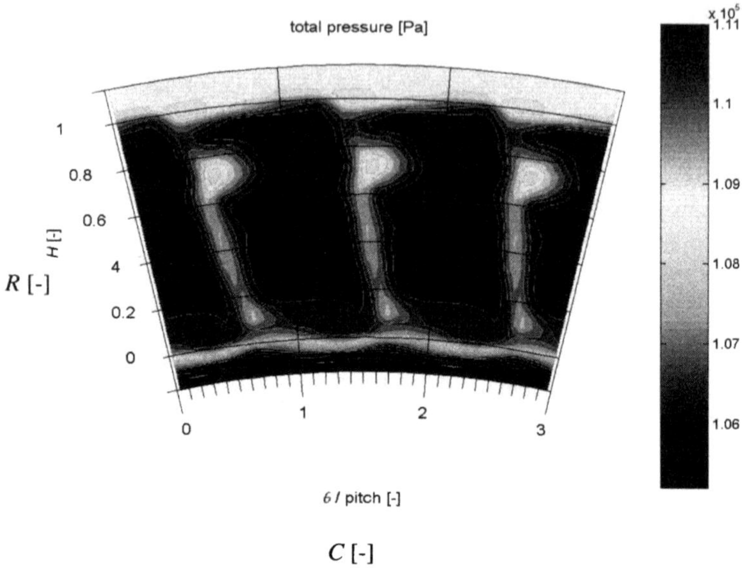

Fig. 2.2. Total pressure distribution downstream of the second stator (NGV2).

3. Data Postprocessing

The new flow visualization technique bases on the vorticity transport equation Eq. (3.2). The resolution of the volume data set in both space and time is high enough to evaluate derivatives using finite differences. For each time step, the unsteady vorticity field is evaluated directly using the definition

$$\omega = \nabla \times \vec{v}\,. \tag{3.1}$$

Having both, the full three dimensional time resolved vorticity and velocity field, the evaluation of the substantial derivative within the compressible unsteady vorticity transport equation Eq. (3.2) [8] is straight forward using its definition Eq. (3.3).

$$\frac{D}{Dt}\left(\frac{\vec{\omega}}{\rho}\right) = \frac{1}{\rho}\vec{\omega}\cdot\nabla\vec{v} + \frac{1}{\rho^3}\nabla\rho\times\nabla p + \frac{1}{\rho}\nabla\times\left(\frac{1}{\rho}\nabla\cdot\tau\right) \tag{3.2}$$

$$\frac{D}{Dt}\left(\frac{\vec{\omega}}{\rho}\right) = \frac{\partial}{\partial t}\left(\frac{\vec{\omega}}{\rho}\right) + (\vec{v}\cdot\nabla)\left(\frac{\vec{\omega}}{\rho}\right) \tag{3.3}$$

For visualization reasons, both terms on the right hand side of Eq. (3.3) were evaluated and discussed in [5] as they allow to flip between the Lagrangian and the Eulerian point of view and describe the acceleration of the vortices. Besides the acceleration term, the new flow visualization technique bases especially on the vortex stretching term

$$\frac{1}{\rho}\vec{\omega}\cdot\nabla\vec{v}\,, \tag{3.4}$$

which describes the vortex tilting and stretching. A discussion of this term and its application to flow field visualization will be done in the next section.

From the application point of view, the baroclinic generation term

$$\frac{1}{\rho^3}\nabla\rho\times\nabla p \tag{3.5}$$

cannot be derived form measurements. The missing information is the unsteady temperature distribution. Together with the static pressure distribution, the time resolved density field could be evaluated. For the evaluation of the other terms in Eq. (3.2), the lack of accurate density information has only a minor influence, since the values of the temperature distribution are within a relatively narrow band.

Nevertheless, it is worthwhile to theoretically discuss the baroclinic vorticity generation term as it can be related to the entropy distribution. Taking the curl of

$$Tds = dh - \frac{1}{\rho}dp \quad \text{or } T\nabla s = \nabla h - \frac{1}{\rho}\nabla p \tag{3.6}$$

using $\nabla\times\nabla h \equiv 0$, Eq. (3.6) can be written as

$$\nabla\times(T\nabla s) = \nabla\times\nabla h - \nabla\times\left(\frac{1}{\rho}\nabla p\right) = -\nabla\times\left(\frac{1}{\rho}\nabla p\right) \tag{3.7}$$

Using Eq. (3.7) and the vector identity (Wilcox [10])

$$\nabla\times\left(\frac{1}{\rho}\nabla p\right) = \frac{1}{\rho}\nabla\times\nabla p + \nabla\frac{1}{\rho}\times\nabla p = \nabla\frac{1}{\rho}\times\nabla p \tag{3.8}$$

the following relation can be derived:

$$\frac{1}{\rho^3}\nabla\rho\times\nabla p = \frac{1}{\rho}\nabla T\times\nabla s \tag{3.9}$$

In addition to the ideal gas law, Eq. (3.9) represents three coupled differential equations for the time resolved entropy. However, the time resolved temperature measurement is missing for the presented data set.

4. Results

In this section, the toroidal vortex located at the inlet cavity of a turbine rotor labyrinth seal - as discussed by Pfau et al. [4] - is under investigation applying both standard as well as the new proposed visualization technique using the vortex stretching term (3.4). The cavity vortex is driven by the main flow due to viscous interaction. The behavior of this vortex is of interest, as it affects the leakage mass flow over the shroud and redirects cooling air in cooled turbines.

4.1. Total pressure distribution

A useful property to study loss generation is the steady total pressure. The corresponding distribution downstream of the second stator is given in Fig. 4.1. Five radial cuts and one axial cut, which is located at z=0, were used to intersect the measurement volume. In addition, streamlines were added in order to visualize the flow.

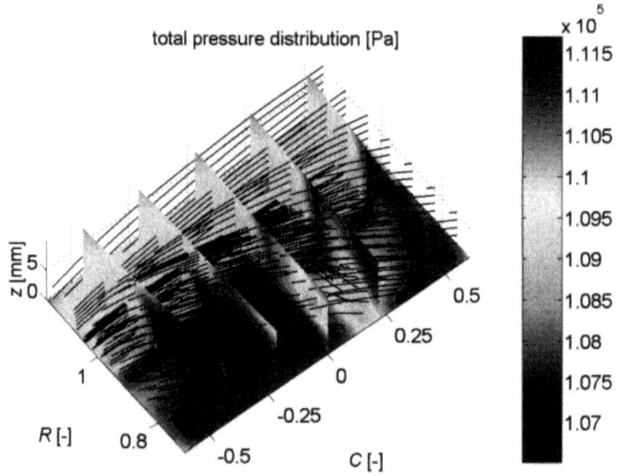

Fig. 4.1. Total pressure distribution behind a stator visualized using radial slices. The black lines represent streamlines.

From the total pressure distribution, the volume under investigation can be split into two regions, one being the main flow region (R<1) which is represented by high total pressure and the other being the cavity flow region (R>1), which is characterized by low total pressure. The motion of the main flow shows a large circumferential component as expected downstream of swirl generating stator. The location of low total pressure within the main flow represents the loss core as found in Fig. 2.2.

Within the cavity, the presence of the toroidal vortex aligned in circumferential direction can be verified by observing the streamlines which show very small axial motion and even back flow at the outer part. A qualitative description of the flow situation within the cavity is depicted in Fig. 4.2. The axial component of the velocity $v_z(r)$ is constant in the main flow region and then decays while moving into the cavity. At the centre of the ring vortex, the sign changes and finally the fluid comes to rest due to the non-slip condition at the cavity wall.

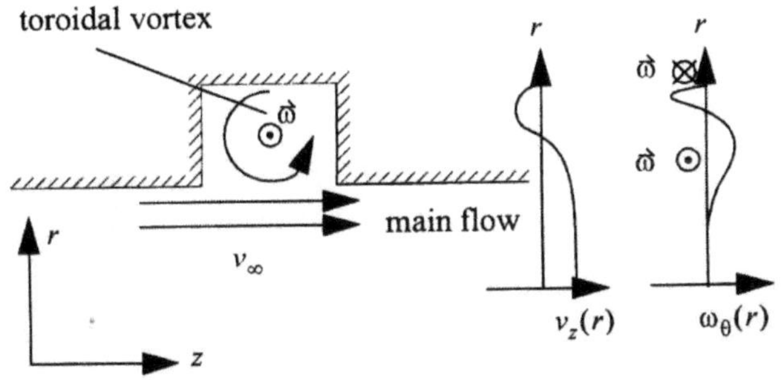

Fig. 4.2. Generic cavity flow structure in an (r,z) cut: expected axial velocity and circumferential vorticity distribution.

4.2. Static pressure and vorticity distribution

In this section, the unsteady flow structure in the measurement volume is investigated. At first, the time resolved static pressure distribution is given in Fig. 4.3. The different plots are equidistant in time and represent the flow during one blade passing period.

Vortices create a local minimum of static pressure in their centre of rotation. The vortex gets stronger if the static pressure in the centre is decreased. The toroidal vortex located in the cavity (R>1) can be identified with this reflection in the plot as a region of low static pressure. This zone of the flow is pointed out in the middle radial cut of Fig. 4.3 using a black circle. Looking at this region for a blade passing period, it can be seen that the level of the static pressure doesn't remain constant. It shows a cyclic change where the pressure level within this middle radial cut reaches its minimum between $Y = 1/6$ and $Y = 2/6$. The cyclic change is correlated to the blade passing period.

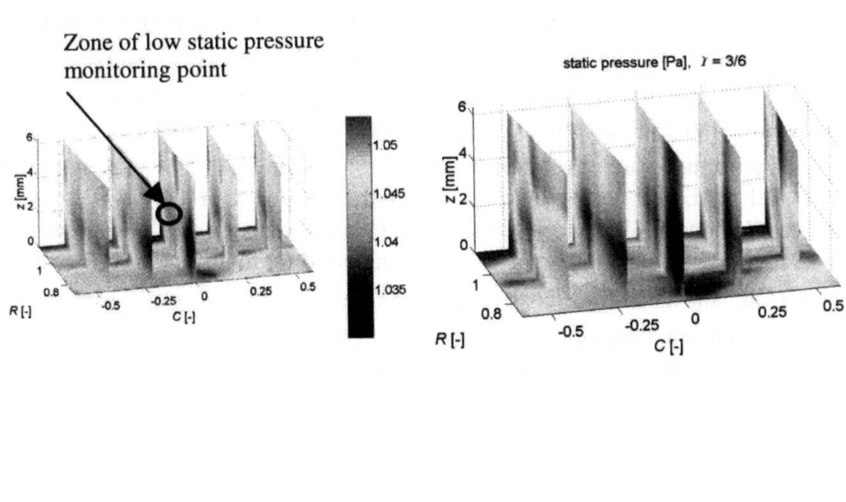

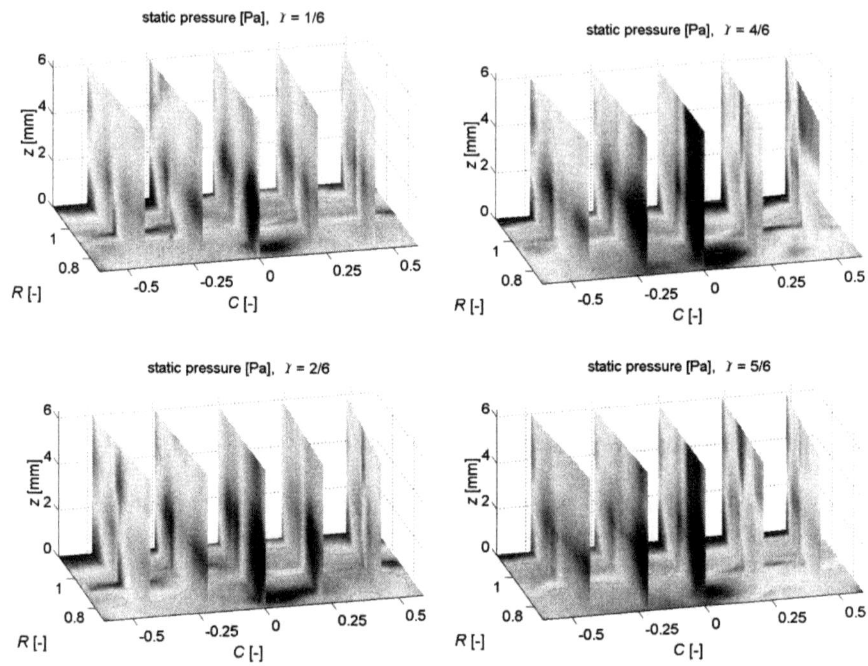

Fig. 4.3. Unsteady static pressure distribution downstream of the second stator: One blade passing period.

Fig. 4.4 shows the circumferential component of the vorticity vector ω_θ. Considering the cavity flow model depicted in Fig. 4.2, it can be stated that ω_θ changes while going radially outward through the cavity: As ω_θ is proportional to the slope of the axial velocity v_z, it vanishes in the main flow region, where v_z is about constant. In the tip zone, v_z decays and therefore ω_θ gets positive and stays positive as long as the slope of the axial velocity profile shows the same sign. At maximum negative axial velocity, ω_θ is zero. At this radial position, ω_θ changes its sign and remains negative until the cavity wall is reached. The presented idealized radial distribution of the circumferential component of the vorticity vector can also be found in the unsteady measurements presented in Fig. 4.4, considering one of the radial cuts. ω_θ vanishes in the undisturbed main flow. Going radially outward, ω_θ shows at first positive and then negative values as discussed. Moreover, the transient behavior is as expected from the previous reflections on the static pressure distribution: Looking again at the middle radial cut, a maximum value is reached between $\Upsilon = 1/6$ and $\Upsilon = 2/6$.

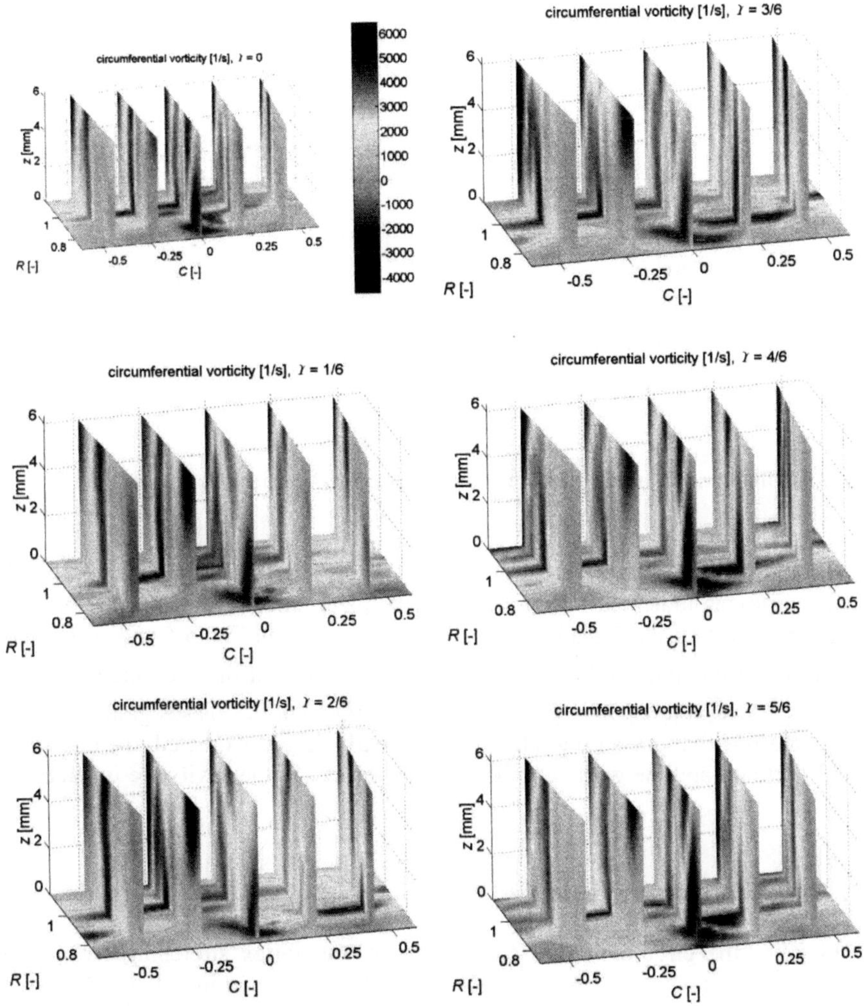

Fig. 4.4. Unsteady circumferential vorticity component downstream of the second stator: One blade passing period.

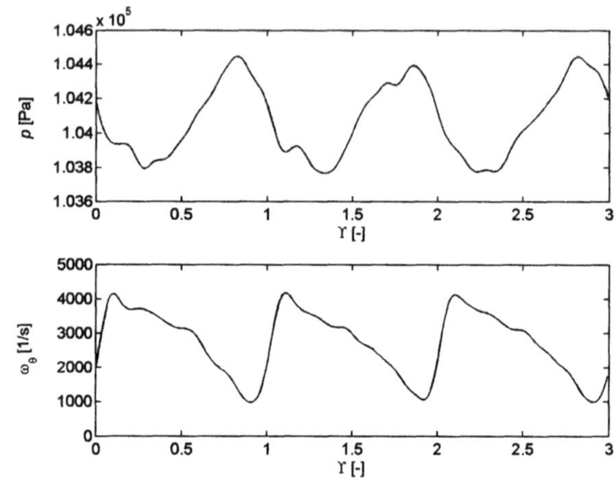

Fig. 4.5. Time series of static pressure and circumferential vorticity component.

A monitor point was selected at the middle radial slice within the core of the cavity vortex as indicated in Fig. 4.3 with the black circle. The behavior of p and ω_θ at this point during 3 blade passing periods is given in Fig. 4.5 and allows a direct transient comparison between p and ω_θ. It is evident that ω_θ increases by as much as 100% in a negative pressure gradient zone and decreases gradually by as much in a positive pressure gradient area. Clearly and as expected, the circumferential vorticity and the gradient of the static pressure in the cavity are inversely related.

4.3. Vortex stretching distribution

In Fig. 4.6, the circumferential component of the vortex stretching term is depicted. Consider a vortex line in a velocity field as indicated in Fig. 4.7 According to the definition, vortex lines are tangential to the local vorticity vector $\vec{\omega}$. The local velocity vector $\vec{v}$ can be split into a component tangential $v_{\Uparrow}$ and a component normal $v_{\perp}$ to the vortex line. Neglecting the density, term (3.4) can be written as

$$\vec{\omega} \cdot \nabla \vec{v} = |\vec{\omega}| \frac{\partial}{\partial \sigma} \vec{v} = |\vec{\omega}| \frac{\partial}{\partial \sigma} \vec{v}_{\Uparrow} + |\vec{\omega}| \frac{\partial}{\partial \sigma} \vec{v}_{\perp} \tag{4.1}$$

where σ denotes the coordinate along the vortex line. Considering the derivative of the first term in Eq. (4.1) to be positive, which corresponds to the case that

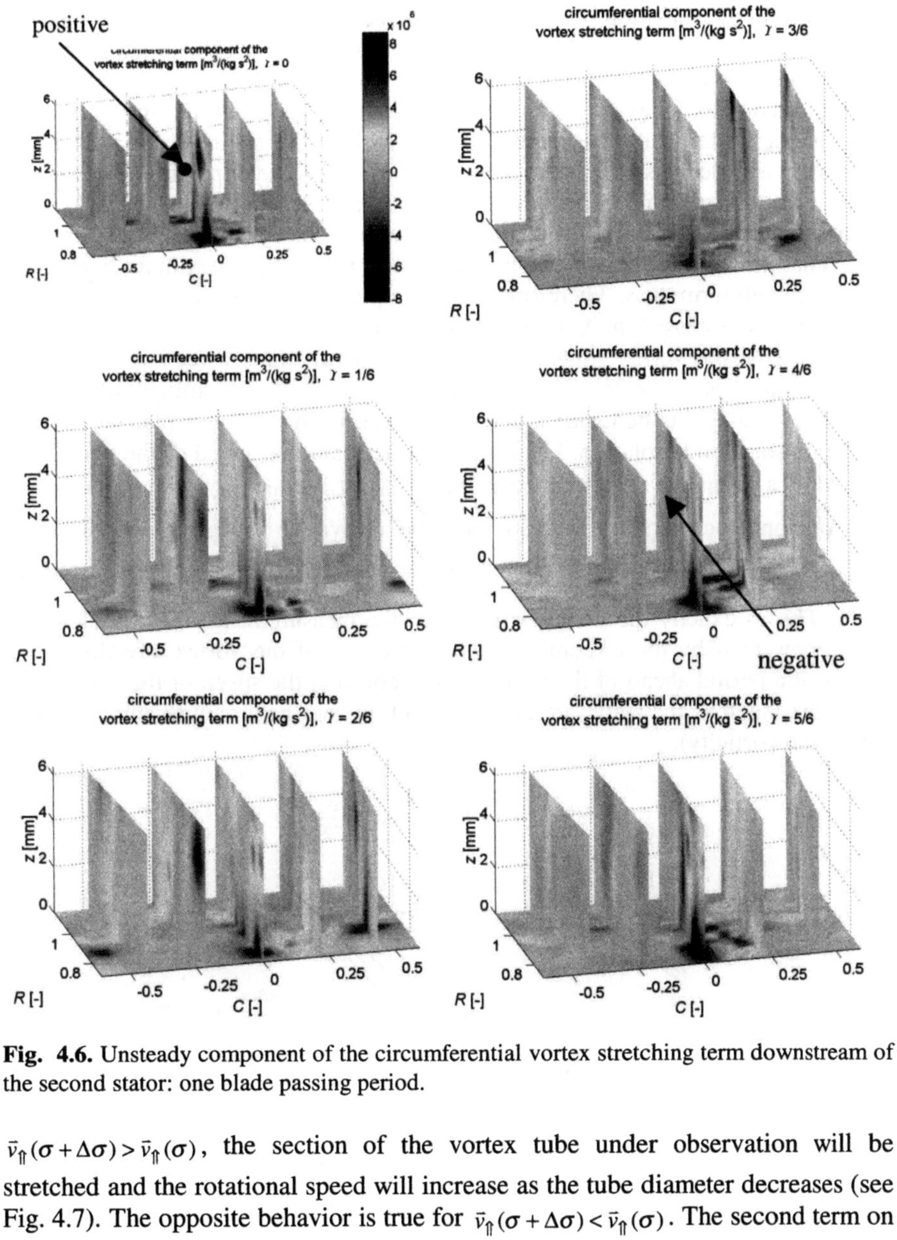

Fig. 4.6. Unsteady component of the circumferential vortex stretching term downstream of the second stator: one blade passing period.

$\bar{v}_{\Uparrow}(\sigma+\Delta\sigma) > \bar{v}_{\Uparrow}(\sigma)$, the section of the vortex tube under observation will be stretched and the rotational speed will increase as the tube diameter decreases (see Fig. 4.7). The opposite behavior is true for $\bar{v}_{\Uparrow}(\sigma+\Delta\sigma) < \bar{v}_{\Uparrow}(\sigma)$. The second term on the right hand side of Eq. (4.1) is responsible for the vortex tilting as can be seen by considering $\bar{v}_{\perp}(\sigma+\Delta\sigma) \neq \bar{v}_{\perp}(\sigma)$. Going back to Fig. 4.6, one can see that the circumferential component of the vortex stretching term undergoes a cyclic change as well. At $\Upsilon = 0$ the stretching term is positive, whereas it is negative at $\Upsilon = 3/6$, looking again at the middle radial cut. As mentioned before, a positive component in the direction of the vortex line represents a stretching of the vortex and a negative one squeezes the vortex, if the vortex is aligned with the stretching term vector. In case of the toroidal vortex, the vortex is aligned in circumferential direction and the vortex stretching term will therefore primarily stretch (rather than tilt) the vortex and thus increase the strength of the circumferential vorticity component and decrease the static pressure in its core. The opposite is true for the negative vortex stretching term, which is consistent with the findings up to now. It has to be stated, that the vortex stretching term is ahead of the vortex strength term. That means that the vortex stretching term reaches its maximum before the vortex strength term does. Therefore, the spin up process of the vortex needs some time, which is discussed in the following paragraph.

At $Y = 3/6$, the vortex stretching term in circumferential direction is negative and thus squeezes the vortex. As it can be seen in Fig. 4.5, the strength of the vortex will decrease. The behavior found in the experiment can be explained by considering a solid body rotation. Let $\hat{\omega}$ denote the rotational speed of the vortex expressed as a harmonic function according to $\hat{\omega} = \overline{\omega} + \omega' \sin(\omega_{bp} t)$ with $\omega_{bp} = 2\pi f_{bp}$. The rotational acceleration is then the time derivative of the rotational speed $\hat{\alpha} = \frac{\partial}{\partial t}\hat{\omega} = \omega_{bp}\omega' \cos(\omega_{bp} t)$ and is thus a quarter of a period ahead of the rotational speed. This is exactly the behavior found in the measurements: The acceleration term represented by the circumferential component of the vortex stretching is a quarter of a period ahead of the terms which represent the speed or the strength of the vortex (such as the static pressure distribution and the circumferential component of the vorticity).

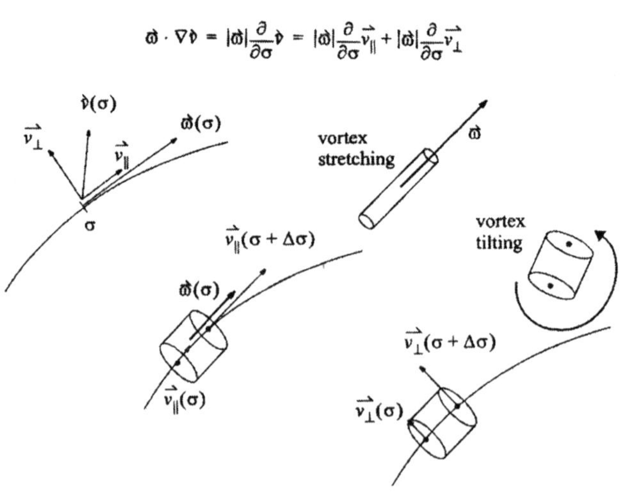

Fig. 4.7. Influence of the velocity field on the vorticity.

5. Conclusions

Nowadays, sophisticated probe technology and high computational power make it possible to analyze complex flow in four dimensions, i.e. space and time. Considering the huge amount of data, intuitive visualization is the key for further processing and analyzing data as well as to locate interesting features. It allows to gain insight in physical mechanisms and promotes flow understanding.

The visualization of the vortex stretching term of the deterministic unsteady vorticity transport equation has proven to be a powerful tool. It makes it possible to locate complex features in the flow, such as the toroidal vortex and to associate different properties with each other. In the example under investigation, the stretching and squeezing of the vortex line around the annulus could be connected with the transient behavior of the static pressure and the circumferential component of the vorticity vector distribution.

The baroclinic vorticity generation term was related to the gradient of the entropy. For the evaluation of the entropy differences, the temperature distribution has to be time resolved. Hence, the time resolved temperature measurement is a very important research area for the future as it would allow to calculate entropy differences.

Acknowledgments

The flow measurements in the turbine were supported by the German Federal Ministry of Economy (BMWI) under file numbers 0327060D and 0327060F. The authors gratefully acknowledge AG Turbo, Alstom Power and Rolls-Royce Germany for their support and permission to publish this paper.

References

[1] Chaluvadi V.S.P., Kalfas A., Hodson H.P., Ohyama H., Watanabe E. (2003) Blade Row Interaction in a High Pressure Steam Turbine. J. of Turbomachinery: Vol 125: pp 14-25.

[2] Langston L.S. (2001) Secondary Flows in Axial Turbines - A Review. New York Academy of Sciences.

[3] Pfau A., Schlienger J., Kalfas A.I., Abhari R.S. (2003) Unsteady, 3-dimensional flow measurement using a miniature virtual 4 sensor Fast Response Aerodynamic Probe (FRAP). GT2003-38128, ASME Turbo Expo, 16-19 June 2003, Atlanta.

[4] Pfau A., Schlienger J., Rusch D., Kalfas A.I., Abhari R.S. (2003) Unsteady Flow Interactions within the Inlet Cavity of aTurbine Rotor Tip Labyrinth Seal. GT2003-38271, ASME Turbo Expo, 16-19 June 2003, Atlanta.

[5] Rusch D. (2002) Evaluation of the steady and unsteady vorticity field from experimental data. Semester Thesis, Turbomachinery Laboratory, ETH Zurich.

[6] Schlienger J., Pfau A., Kalfas A.I., Abhari R.S. (2003) Effects of Labyrinth Seal Variation on Multistage Axial Turbine Flow, Proceedings of ASME Turbo Expo. GT-2003-38270, June 16-19, WCC, Atlanta, Georgia, USA.
[7] Sell M., Schlienger J., Pfau A., Treiber M., Abhari R.S. (2001) The 2-stage axial turbine test facility LISA. Proceedings of the ASME Turbo Expo, 2001-GT-492, June 4-7, New Orleans, Louisiana.
[8] Thompson P.A. (1972) Compressible Fluid Dynamics. McGraw-Hill, New York.
[9] Treiber M., Abhari R.S., Sell M. (2002) Flow physics and vortex evolution in annular turbine cascades. ASME Turbo Expo 2002, GT-2002-30540.
[10] Wilcox D.C. (1997) Basic Fluid Mechanics. DCW Industries, Inc., California.

An Application of the Buffer Layer Technique to Computations of Flow in Turbomachinery

Franco MAGAGNATO*, Jaroslaw RACHWALSKI, Martin GABI
Institute of Fluid Machinery University of Karlsruhe

* Corresponding author: Kaiserstr. 12, D-76128 Karlsruhe, Germany
Tel.: (+49 721) 608 3813, Fax: (+49 721) 608 3529,
Email: magagnato@ism.uka.de

Abstract Computations of turbomachinery are usually carried out with truncated domains. To eliminate the undesirable effect of spurious reflections, a non-reflecting boundary approach must be employed at boundaries. The buffer layer technique, proposed by Freund [4], was chosen. This approach is designed for external flow. The investigations were carried out in order to verify if the approach is suitable for internal flow. The flow through a VKI-turbine cascade was considered in order to examine this boundary condition. These investigations together with comparison to experimental data are presented in the paper. The paper describes mainly difficulties of using this kind of approach.
Key Words: buffer layer technique, non-reflecting boundary condition, turbomachinery

Nomenclature

a	*[m/s]*	speed of sound
c	*[m]*	chord length
d	*[m]*	normal distance to the boundary
d_A	*[m]*	thickness of the buffer layer consequently for A
d_σ	*[m]*	thickness of the buffer layer consequently for σ
g	*[m]*	pitch height
h	*[m]*	blade height
t	*[s]*	time
u	*[m/s]*	physical convective velocity
x	*[m]/[mm]*	catresian coordinate/coordinatealong the chord
y	*[mm]*	distance to wall
y^+		non-dimensional distance to the wall
$\mathbf{A}$	*[m/s]*	artificial convective velocity
$A=\lvert\mathbf{A}\rvert$		
A_0	*[m/s]*	maximal values of A

C_D		drag coefficient
C_L		lift coefficient
Ma		mach number
γ	*[°]*	stagger angle
σ	*[1/s]*	damping coefficient
σ_0	*[1/s]*	maximal values of σ
Φ		generic variable
Φ_{target}		quiescent target state

1. Introduction

Computational Fluid Dynamics (CFD) is intensively used today in the field of fluid machinery in research and development as well as in the daily design business. One of the most complicated flow field in fluid dynamic practice is observed in turbomachinery. Usually steady-state simulations are applied in industry. However, recently it has become more and more important to understand details of an instantaneous flow structure.

Since turbomachinery computations are often performed on truncated domains, the solution scheme should be used in conjunction with non-reflecting boundary conditions. This is because the standard numerical boundary conditions produce unphysical reflections of disturbances, which travel down to an inlet/outlet boundary of the computational domain. Besides avoiding the spurious reflection, one can encounter difficulties with setting appropriate values at boundaries e.g. in the case of shear flows. Vortical and acoustical waves must pass freely through boundaries with minimal reflections.

According to recent papers, one can find some different approaches of non-reflecting boundaries. The linear treatment of the boundaries and/or the governing equations is widely used by many researches. *Rieman Variables Condition, Gile's Conditions* [1] or *Quasi 3-D non-reflecting boundary condition* [2] are the well-known approaches. This kind of treatments are reviewed by Hayder and Turkel in [3]. In most of the cases these boundary conditions are not working well enough with shear flows. The difficulties are discussed in [3] and [4]. Rowley and Colonius presented discretely non-reflecting boundary condition for the linear hyperbolic systems in [5]. They derived the boundary condition which is based on three different approaches: radiation boundary condition [6], perfectly matched layer [7,8] and the decompositions of the solution into Fourier/Laplace modes [9].

It is possible to derive boundary conditions that are non-reflecting to arbitrarily order of accuracy for equations of motions linearized about uniform mean flow. For non-linear equations, especially at outflow boundaries, the interaction of disturbance and mean flow gradients severely limits the accuracy of any linear boundary conditions [10]. As a remedy for this situation a buffer layer (zone) technique has been suggested.

The boundary condition proposed by Freund [4] is based on the buffer layer (zone) technique. The idea lies in combining two different basic zonal approaches. The first one comes from Ta'asan and Nark [11], who enforce a supersonic flow in the boundary layers by adding a convective term to the governing equations. The second one is taken from Berenger [7] and Hu [8]. It is based on adding a damping term to the equations. The damping term drives the solution in the buffer layers toward a quiescent target state. The boundary condition treatment proposed by Freund was successfully used by Collonius et al. [12], Rowley et al. [13] and Boersma and Lele [14].

This technique has been implemented into our code (called SPARC) [15] and tested with some test cases, like a pressure impulse (1D) and plane waves inclined at an angle to the boundary (2D) that are convected out of the domain. The computations showed an improvement (the reduction of reflections) of about 97% [16].

The buffer layer technique, proposed by Freund, is designed for external flow. The investigations were carried out due to verify if the approach is suitable for internal flow. The flow through a VKI-turbine cascade was considered in order to examine this boundary condition. This investigations together with comparison to experimental data are going to be presented in this paper.

2. The Code Description

SPARC [15] is compressible and parallelized. It solves 3D steady and unsteady Reynolds Averaged Navier-Stokes (RANS) equations as well as LES. It is a block structured, Finite-Volume code using non-orthogonal grids and full multigrid methods. The code uses the cell-centred approach. For steady computations an explicit Runge-Kutta scheme is implemented, while for unsteady flows the explicit Runge-Kutta and the implicit dual time stepping schemes are used. Several turbulence models like algebraic, one-equation, linear and non-linear two-equation models are implemented into the code.

3. Theoretical Model

The buffer layer technique according to Freund is based on the construction of the modified equations in the buffer layer by changing the domain of dependence for the problem, or equivalently by changing the dispersion relation for the equations near the boundary. Therefore a layer at the inlet/outlet boundary was introduced as well as additional convective and dissipative terms were added into the compressible Navier-Stokes equations. The generic simplified one-dimensional equation shows the basic idea:

$$\frac{\partial \Phi}{\partial t} + A(x)\frac{\partial \Phi}{\partial x} + \frac{\partial (u\Phi)}{\partial x} = -\sigma(x)[\Phi - \Phi_{\text{target}}]\,.$$

The additional convective term (with the artificial convective velocity) changes locally the character of the equations from elliptic to the hyperbolic type. In the inlet buffer layer it causes stopping of the disturbances which travel upstream, while in the outlet buffer layer it convects out any disturbances that travel downstream. The disturbances which are stopped at the inlet layer are slowly damped in consequence of the impact of the additional dissipative term. The terms act only in the buffer layers, so the Navier-Stokes equations remain unchanged in the computational domain. While the artificial velocity and the damping coefficient are equal to zero in the computational domain they are smoothly increased in the buffer layers. The change has to be done in a smooth way in order to eliminate reflection caused by the introduced inhomogeneity. Values are calculated according to the following equations:

$$A_{(d)} = A_0 (d / d_A)^{\beta}, \qquad \sigma_{(d)} = \sigma_0 (1 + \tanh(l \cdot d))$$

where $\beta = 3$, l is adjusted to $\sigma(d_\sigma)/\sigma_0 \approx 10^{-6}$.

This kind of non-reflecting boundary allows one to set time-dependent inlet-values, because any change at the inlet, accelerated in the buffer layer by the artificial convection, is almost unaffected by the damping term.

4. Implementation

The set of the equations solved by the code were extended by adding the previously described terms. The eigenvalues for the modified equations were derived. For this, only the artificial convective terms were taken into account. According to the new eigenvalues the computation of the time step was altered.

A field of values of the artificial velocity and the damping coefficient is calculated once at the beginning of a computation. This is done in a way that the values are non-zero only up to a given distance from the numerical boundary. This defines the buffer layer. A direction of **A** can be chosen either normal to the boundary or arbitrarily in the Cartesian system.

Only the variables which are enforced at the numerical boundary (the incoming characteristic) are damped. This is different compared to Freund's approach where all variables are driven to a quiescent target state.

The dissipative term is coded in such a way that for $\sigma_0 = 1$ a deviation is removed completely during one time step.

5. Description of Test Case

The VKI turbine cascade was used to test the considered non-reflecting condition. The cascade was designed by Sieverding and Citatelli [17] to investigate time varying wakes behind turbine blades. A description of the cascade can be found in **Table 1**.

Experimental data for comparison with numerical results were taken from Ubaldi et al. [18]. In the numerical computations the boundary conditions were adjusted in a manner to obtain at the outlet the isentropic Reynolds number $Re_{2is}=1.6\cdot10^6$ (which is based on the chord length) and the isentropic Mach number $Ma_{2is}=0.23$. The turbulence level was measured upstream of the cascade and was $Tu=3\%$.

Table 1. Cascade characteristics.

chord length – c	0.3m
pitch-to-chord ratio – g/c	0.7
aspect ratio – h/c	1.0
stagger angle (with axial direction) – γ	49.83°

A geometry of the mesh used in the current investigations is shown in **Figure 1**. 2-dimensional simulations are carried out in a very fine mesh, which consists of about *72700* control volumes for one blade. No wall-law is used in the turbulence model, therefore y^+ at the wall is kept at about *1* and the boundary layer is resolved with about 30 control volumes. For the upstream boundary condition the total pressure condition was applied. For the downstream boundary condition the static pressure and the non-reflecting boundary condition were used. Only one blade has been calculated using periodic boundary conditions to simulate other blades. The non-linear Craft/Launder/Suga turbulence model was used in the calculations.

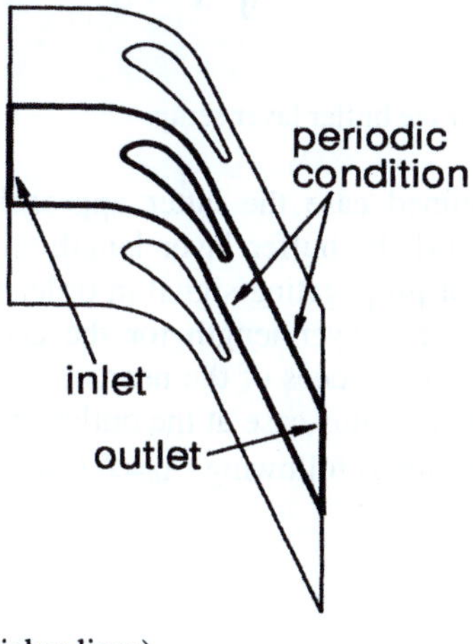

Fig. 1. The mesh geometry (thicker lines).

The non-reflecting boundary condition was applied in a way that part of the computational domain at the outlet was utilized as the buffer layer (see **Figure 2**). The thick lines show the thickness of the buffer layers used for acceleration (continuous line) and damping (dashed line). Unfortunately the thickness of the buffer layers and the parameters (the artificial velocity and the damping coefficient) have to be adjusted for every considered test case. This is the disadvantage of this approach. An indispensable condition for applying this kind of boundary condition for internal flows is knowledge of a quiescent/target/average state of a flow field in the buffer layer region. Gradients of damped variables can not be in the buffer layer because they will be changed in an unpredictable manner.

There are two possibilities of choosing variables that are going to be damped. One can damp all variables [4], but this is troublesome in an implementation as well as for use (one must know quiescent state of all variables). The other possibility is more straightforward, because a known parameter applied at a boundary (for example static pressure) is damped in the buffer layer, and other are extrapolated.

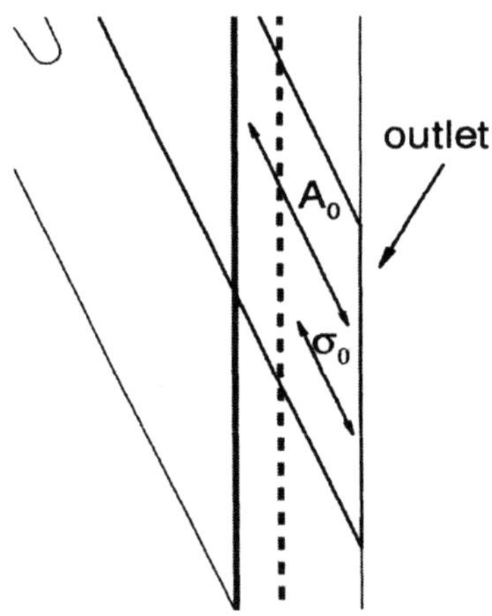

Fig. 2. Schematic diagram of the buffer layer setup.

In the currently examined case the latter approach was used. According to Ta'san [11] and Freund [4] the buffer layer length should be approximately *2.5* times of the wavelength of propagating signal in order to obtain reflections of less than one percent. The buffer layer length for the acceleration was $d_A=0.114m$. This fulfils the condition for vortices of the near wake. It has been found [16] that for this values of velocities (in this case at the outlet) the other parameters describing the buffer layer can be set as following: $A_0=1.15a$, $d_\sigma=0.3d_A$, $\sigma_0=0.5$.

6. Results

In order to provide an insight into the high turbulent flow in the wake, the magnitude of mach number *Ma* is presented in **Figure 3**.

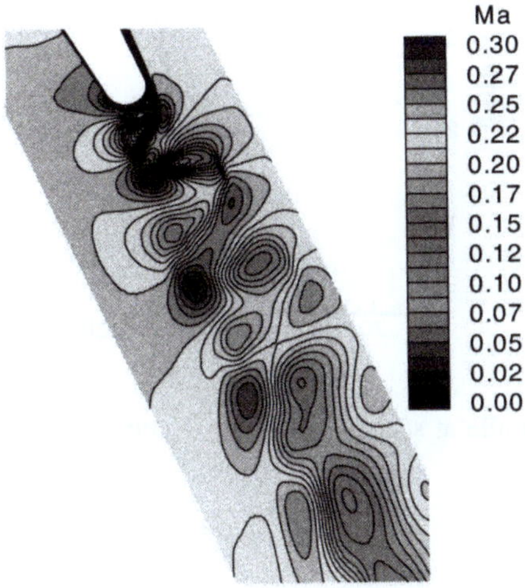

Fig. 3. Mach number (Ma) magnitude.

Velocity profiles at the boundary layer have been measured in the experiment [18]. In **Figure 4** a comparison of velocity profiles is showed. The difference of the profiles with and without the non-reflecting boundary condition was negligible. Therefore only one profile obtained from the numerical simulation, together with the experimental results, is presented at the plot. The agreement with experiment is satisfactory.

The frequency measured in the experiment of the vortex shedding was *1700Hz*. In the calculations the obtained frequencies were *1443Hz* with the reflecting condition and *1451Hz* with the non-reflecting condition. Here the discrepancy lies in an improper turbulence modelling. An investigation of this problem as well as the dependency of the vortex shedding frequency and the boundary layer profile were reported in [19,20].

The mean isentropic Mach number on the blade (presented in **Figure 5**) also shows negligible differences between the calculations performed with and without the non-reflecting condition.

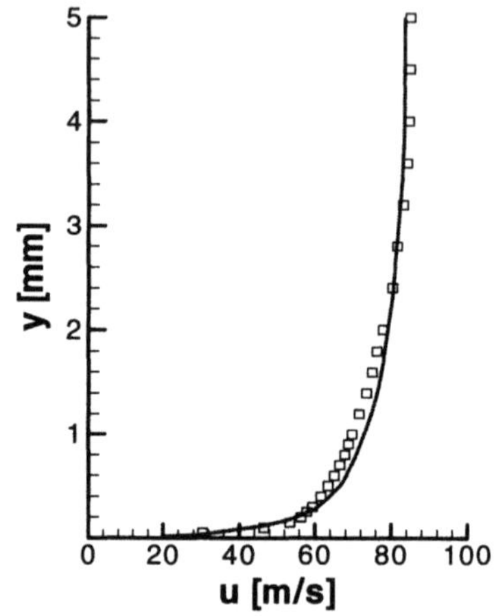

Fig. 4. Mean velocity profile at s/smax=0.95 (line – numerical results, squares -- experimental data).

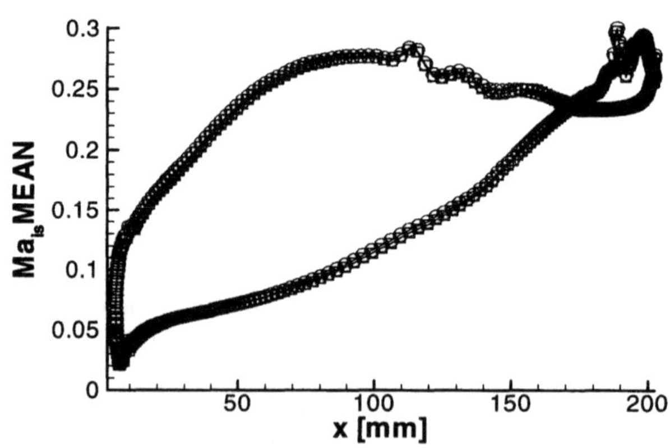

Fig. 5. Mean isentropic Mach number on the blade (squares – reflecting, circles – non-reflecting)

The comparison of the mean flow quantities (see Figs. 4 and 5) drives us to the conclusion that the buffer layer doesn't seem to work in the desired manner. Is seems to be not thick enough. This means, in order to force it to work properly, it is necessary to generate a longer mesh, because it is not allowed to set the buffer layer closer to the blade. On the other side it would be also the best to keep the same (equidistant) point distribution for the extended part of the mesh. This is necessary to resolve outgoing disturbances as well as to keep smoothness of an alteration of the boundary layer coefficients. This "smoothness" is also an important condition in order to minimise spurious reflections. Unfortunately farther from the cascade the disturbances dissipate and become larger what needs also a thicker buffer layer.

Analysing the lift C_L (see **Figure 6**) and the drag C_D coefficients one can see that the vortex shedding disappears periodically and then it is recreated. This seems to be caused by reflections of disturbances of longer wavelength. The same behaviour is observed for both cases: with and without the non-reflecting condition. This confirms the previous suggestion that the non-reflecting boundary condition has almost no effect. Additionally one can observe an unforeseen effect of a bigger C_L amplitude (see Fig. 5, right side). This is caused probably by reduced reflections of higher frequency disturbances.

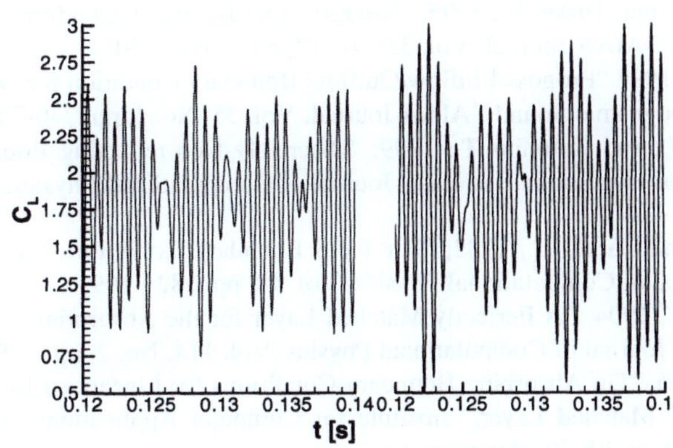

Fig. 6. The lift coefficient on the blade (left – reflecting condition, right – non-reflecting condition).

7. Summary

The boundary condition, based on two different zonal approaches, was proposed by Freund and is dedicated especially to external flows. It gives good results for external flows and is quite straightforward to implement as well as to use.

The investigations were carried out in order to verify if the approach is suitable for internal flow. The flow through VKI turbine cascade was used to verify the non-reflecting boundary condition. In this case this approach present more difficulties. One must know a quiescent/target state of the flow where the buffer layer is going to be applied as well as to be sure that a quiescent state contain negligible gradients of a damping variable. For the purpose of a proper grid generation and setting right parameters of the buffer layer the flow field must be known.

The investigations showed the above mentioned difficulties. The non-reflecting boundary condition could not stop spurious reflections due to an insufficient size of the buffer layer and probably the buffer layer was applied too close to the blade.

References

[1] Giles, M. B., 1988, "Non-Reflecting Boundary Conditions for Euler Equations," Rep. MIT, Computational Fluid Dynamics Laboratory TR-88-1.

[2] Saxer, A. P., 1992, "A Numerical Analysis of 3-D Inviscid Stator/Rotor Interactions Using Non-Reflecting Boundary Conditions," PhD Thesis, MIT Gas Turbine Laboratory.

[3] Hayder M. E., and Turkel E., 1995, "Nonreflecting Boundary Conditions for Jet Flow Computations,“ AIAA Journal, Vol. 33, No. 12, pp. 2264-2270.

[4] Freund J. B., 1997 "Proposed Inflow/Outflow Boundary Condition for Direct Computation of Aerodynamic Sound," AIAA Journal, Vol. 35, No. 4, pp. 740-742.

[5] Rowley C. W., and Colonius T., 1999, "Discretely Non-reflecting Boundary Conditions for Linear Hyperbolic Systems," Journal of Computational Physics, Vol. 157, pp. 500 - 538.

[6] Bayliss A., and Turkel E., 1982, "Far Field Boundary Conditions for Compressible Flows," Journal of Computational Physics, Vol. 48, pp. 182 - 199.

[7] Berenger J. P., 1994, "A Perfectly Matched Layer for the Absorption of Electromagnetic Waves," Journal of Computational Physics, Vol. 114, No. 2, pp. 185-200.

[8] Hu F. Q., 1995, "On Absorbing Boundary Conditions for Linearized Euler Equations by a Perfectly Matched Layer," Institute for Computer Applications in Science and Engineering, Rept. 95-70, Hampton, Va.

[9] Engquist B., and Majda A., 1977, "Absorbing Boundary Conditions for the Numerical Simulation of Waves," Mathematics of Computation, Vol. 31, No. 139, pp. 629 - 651.

[10] Colonius T., Lele S. K., and Moin P., 1993, "Boundary Conditions for Direct Computation of Aerodynamic Sound," AIAA Journal, Vol. 31, pp. 1574 - 1582.

[11] Ta'asan S., and Nark D. M., 1995, "An Absorbing Buffer Zone Technique for Acoustic Wave Propagation," AIAA Paper 95-0146.

[12] Colonius T., Mohseni K., Freund J. B., and Moin P., 1998, "Evaluation of Noise Radiation Mechanisms in a Turbulent Jet," Proceeding of the Summer Program, Center for Turbulence Research, pp. 159-167.

[13] Rowley C. W., Colonius T., and Basu A. J., 2002, "On Self-Sustained Oscillations in Two-Dimensional Compressible Flow Over Rectangular Cavities," Journal of Fluid Mechanics, Vol. 455, pp. 315-346.

[14] Boersma B. J., and Lele S. K., 1999, "Large Eddy Simulation of Compressible Turbulent Jets," Center for Turbulence Research, Annual Research Briefs, pp. 365-377.

[15] Magagnato F., 1998, "*KAPPA* – Karlsruhe Parallel Program for Aerodynamics," TASK quarterly, Vol. 2, No. 2, pp. 215-270.
[16] Rachwalski J., Magagnato F., and Gabi M., 2002, "The Buffer Layer Technique Applied To Transonic Flow Calculations," Proceeding IUTAM Symposium Transsonicum IV, to be published.
[17] Cicatelli G., and Sieverding C. H., 1996, "The Effects of Vortex Shedding on the Unsteady Pressure Distribution around the Trailing Edge of a Turbine Cascade," ASME paper 96-GT-359.
[18] Ubaldi M., Zunino P., Campora U., and Ghiglione A., 1996, "Detailed Velocity and Turbulence Measurements of Profile Boundary Layer in a Large Scale Turbine Cascade," ASME paper 96-GT-042.
[19] Magagnato F., 1999, "Unsteady Flow Past a Turbine Blade Using Non-linear Two-equation Turbulence Models," Proceeding IMechE C557/139.
[20] Magagnato F., Gabi M., 2002, "A New Adaptive Turbulence Model for Unsteady Flow in Rotating Machinery," International Journal of Rotating Machinery, Vol. 8, No. 3, pp. 175-183.

CFD-analysis of Secondary Flows and Pressure Losses in a NASA Transonic Turbine Cascade

V.D. GORIATCHEV[1], N.G. IVANOV[2], E.M. SMIRNOV[2], V.V. RIS[2]
[1] Department of Mathematics, Tver State Technical University, 170026 Russia
[2] Department of Aerodynamics, St.-Petersburg State Polytechnic University, 195251, Russia

Abstract Results of numerical simulation of highly three-dimensional turbulent flow and pressure losses in a transonic turbine cascade are presented. Computations have been carried out using the one-equation Spalart-Allmaras turbulence model. Predictive capabilities of an academic CFD code and a commercial one, both of second-order accuracy, are analyzed in comparison with measurements in a linear cascade at the NASA GRC transonic turbine blade cascade facility.
Key Words: numerical simulation, secondary flow, transonic turbine cascade

Nomenclature

C_x	[m]	blade axial chord
$C_{p,t}$	[-]	total pressure coefficient
M	[-]	Mach number
P	[Pa]	pressure
Re	[-]	Reynolds number, $Re = \rho U_{\text{in}} C_x / \mu$
s	[m]	blade span
U	[m/s]	total velocity
x	[m]	chordwise (axial) direction
y	[m]	pitchwise direction
y^+	[-]	normalized distance to a wall
z	[m]	spanwise direction
α	[degrees]	pitch angle (x-y plane)
μ	[$kg/s{\cdot}m$]	dynamic viscosity
ρ	[kg/m^3]	density
ex		exit freestream value
in		inlet freestream value
p		computational point nearest to a solid wall
x		axial
$'$		total conditions

1. Introduction

Accurate prediction of turbine cascade flow and pressure losses remains a challenging task despite a lot of work in this area. No doubt that justified conclusions on verification of Computational Fluid Dynamics (CFD) codes should result from comparisons of calculated data with measurements at discrete points rather than comparing results for overall performance of a turbine cascade.

For highly three-dimensional (3D) turbulent flow in a linear cascade with a large turning angle, benchmark-quality well-documented data for verification of CFD codes are available from measurements at the NASA Glenn Research Center transonic turbine blade cascade [1, 2].

A brief description of the experimental facility and data available is as follows. The transonic turbine blade linear cascade consists of 11 passages. A flow field with strongly pronounced 3D phenomena inside and after the blade passages was obtained by allowing the endwall turbulent boundary layer to develop (under low freestream turbulence conditions) in a long inlet section upstream of the cascade. Blade and endwall static pressures were measured at four combinations of the inlet Reynolds number, *Re*, and the isentropic exit Mach number, M_{ex}. Flow field probe measurements were carried out at $Re = 1.0\times10^6$ and $M_{ex} = 1.3$. These measurements covered three blade pitches and extended from near the endwall to just above the mid-span plane. Experimental results were presented using a combination of 3-hole boundary-layer and 5-hole pitch/yaw probe data. To define the cascade inlet flow conditions, aerodynamic probe measurements were made at a section located one axial chord upstream of the blade leading edge plane. Turbulence measurements were performed with a constant temperature hot wire anemometer. Additional details are given by Giel et al. [1].

The present contribution covers CFD results obtained for the NASA GRC transonic turbine cascade with the commercial CFD package FLUENT 5.5, and with the 3D Navier-Stokes code SINF being under development at the Department of Aerodynamics of the St.-Petersburg State Polytechnic University. Data computed are compared thoroughly with the static pressure and flow-field probe measurement results. A special attention is paid to analysis of secondary flows developing in the cascade.

2. Problem definition

The geometry of the linear cascade is that available from the NASA GRC CD-ROM database arranged by Giel and Gaugler [2]. The computational domain is illustrated in Figure 1. Table 1 covers cascade dimensions and basic flow parameters at the design inlet flow angle.

We treat the fluid (air) as a perfect gas with the specific heat ratio $\gamma = 1.4$. The governing equations are the Reynolds-averaged Navier-Stokes equations and the energy equation written for the total enthalpy. A power-law is adopted to account for the dependency of the dynamic viscosity on temperature, $\mu \sim T^{0.76}$. The one-equation Spalart-Allmaras turbulence model [3] with no wall functions is used to close the governing equation system.

To define proper boundary conditions at the computational domain inlet section placed one axial chord upstream of the blade leading edge, the two-dimensional (2D) turbulent flow developing in a parallel-plate channel was computed first, assuming the adiabatic wall conditions. The inlet value of the turbulent to molecular viscosity ratio, $\nu_{t,\,in}/\nu$, was taken as 11.5 that approximately corresponded to the measured freestream turbulence parameters. In the 2D flow computed, a section was chosen that corresponded to the boundary layer thickness of 3.2 cm. Flow field data at this section were used to define the total temperature, total pressure, velocity vector angle and eddy viscosity distributions over the inlet plane of the 3D blade cascade computational domain.

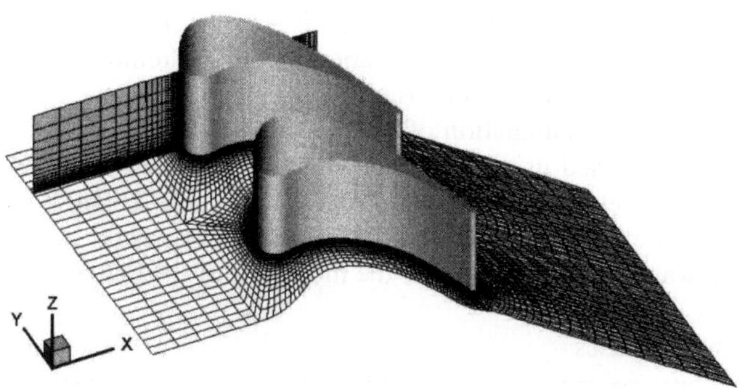

Fig. 1. View of the blade cascade and a three-block computational grid

Table 1. Cascade dimensions and flow parameters

Axial chord, C_x, cm	12.70
Pitch, cm	13.00
Span, cm	15.24
True chord, cm	18.42
Stagger angle, degrees	41.54
Inlet flow angle, degrees	63.6
Design flow turning, degrees	136
Prandtl number, Pr	0.72
Inlet Reynolds number, Re	1.0×10^6
Inlet Mach number, M_{in}	0.38
Exit Mach number, M_{ex}	1.32
Inlet boundary layer thickness, cm	3.2
Inlet turbulence intensity, %	0.25
Inlet turbulence length scale, cm	0.127

To get the exit Mach number required a proper value of static pressure was specified at the outlet boundary located one axial chord downstream of the blade trailing edge. At the solid surfaces of the cascade the no-slip adiabatic-wall condition was imposed. Periodic boundary conditions were used in the pitchwise direction. For computational purposes, only half of the real span was considered, with the symmetry boundary condition at mid-span.

3. Computational Aspects

The 3D incompressible/compressible Navier-Stokes code SINF is based on the second-order finite-volume spatial discretization using the cell-centered variable arrangement and body-fitted block-structured grids. A detailed description of the solver for low-speed flows is given by Smirnov [4]. For transonic flow analysis, a high-order version of the Jameson's H-CUSP scheme [5] is implemented and tested [6] in combination with a regularization technique removing the difficulties of compressible flow computations in low-Mach-number regions.

The present numerical simulation of the transonic turbine cascade flow with the code SINF was performed using two multi-block grids, both of the H-O-H structure (see Figure 1). Details of the grids are given in Table 2. Due to a clustering, for both the grids the distance of a cell center adjacent to a solid wall, measured in wall units (y_p^+), was less than unity over the most part of the blade passage.

Table 2. Grid characteristics

	Coarse grid	Fine grid
Number of cells	129,936	364,308
Distance from any cell center adjacent to a wall	$0.2\times10^{-4}C_x$	$0.1\times10^{-4}C_x$
y_p^+, area-averaged	0.8	0.4

It has been established that the flow field and pressure loss data computed with the code SINF on the coarse grid and the fine one do not differ considerably. Consequently, results presented below for the fine grid are treated as practically grid independent.

The computations with the commercial CFD package FLUENT 5.5 were carried out on an unstructured grid with hexahedral cells identical to those of the fine SINF grid. In fact, the SINF grid had been transformed to the FLUENT unstructured- grid format using a specialized converter. The solver options assigned for second-order accuracy computations [7] were activated.

4. Results and discussion

4.1. Flow field

Figure 2 illustrates mid-span Mach number contours computed with two codes. The distributions look almost identical over the majority of the flow domain. Negligible distinctions are observed for the trailing-edge shock waves and the wake region only.

Static pressure distributions at various spanwise locations on the blade surface are given in Figure 3. Again one can see that results obtained with the two codes are practically identical. The distributions computed are in an excellent agreement with experimental data taken from the CD-ROM database [2]. Note that measurement data from the second half of the real span are also included in the figure, in compliance with the assumption of the flow symmetry. The strong affection of the spanwise location on the static pressure distribution over the suction surface is well reproduced in the CFD-analysis. At $x/C_x > 0.8$ both the codes predict an increase in pressure on the suction side that is due to the flow overexpansion and viscous-inviscid interaction phenomena in the trailing edge region (see Figure 2).

The sharply defined spanwise variations of the suction side static pressure, especially at 2.5% to 10% span positions, indicate a pronounced three-dimensionality of the flow under discussion.

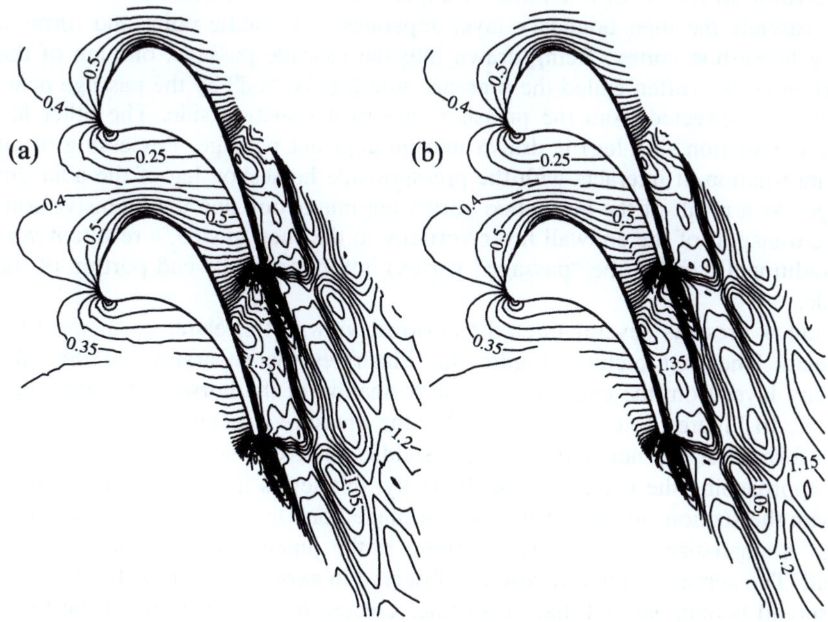

Fig. 2. Mid-span Mach number distributions computed with (a) SINF and (b) FLUENT

A detailed analysis of secondary flows developing in the cascade was performed using possibilities of a post-processor tool from the Computational and Informational System for CFD named SELIGER [8]. Data obtained with the code SINF were processed. In this analysis we were guided by generalizations of experimental findings as presented by Langston in a recent review [9].

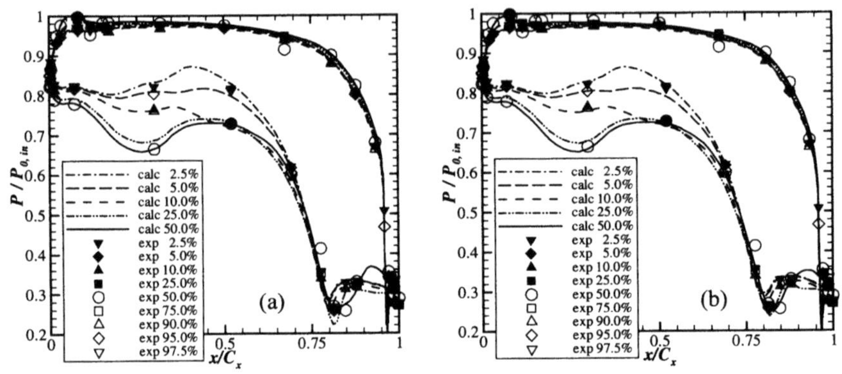

Fig. 3. Blade pressure loading distributions: results of (a) SINF and (b) FLUENT computations in comparison with experimental data

According to results of a number of experimental observations, at the endwall of the cascade the inlet boundary layer separates at a saddle point and forms a two-leg horseshoe vortex. Being drawn into the cascade passage, one leg of the horseshoe vortex (often called the pressure side leg) is "fed" by the passage near-endwall flow directed from the pressure side to the suction side. The other leg (called the suction side leg) is drawn into an adjacent passage. Possessing of an opposite rotation, it interacts with the pressure side horseshoe leg of the adjacent passage. As a result of the horseshoe vortex leg interaction and the intensive convective transport of the endwall layer vorticity to the suction side, a resultant vortex (traditionally called the "passage" vortex) is formed in the end portion of the cascade.

Figure 4 presents several three-dimensional views of streamlines in the flow computed. The first picture (Figure 4a) illustrates the separation of the inlet boundary layer from the endwall, and the formation of the horseshoe vortex with two legs. In Figures 4b,c one can see that the pressure side leg of the horseshoe vortex formed at the neighboring blade saddle point can split up for two parts when approaching the current blade. Partially, this leg is involved in the interaction with the suction side leg of the current blade horseshoe vortex. The other part "feeds" a small-size corner vortex forming at the juncture of the blade and the endwall. The corner vortex formation is observed in experiments as well [10].

It should be emphasized that in schemes created for explanations of the blade cascade secondary flow structure, the vortex motion is exaggerated using, for instance, spiral ribbon arrows [9, 10]. The actual rotation of the vortices is much less, and typically secondary flow streamlines do one or two rotations only. The latter is well illustrated by the present CFD-based visualization.

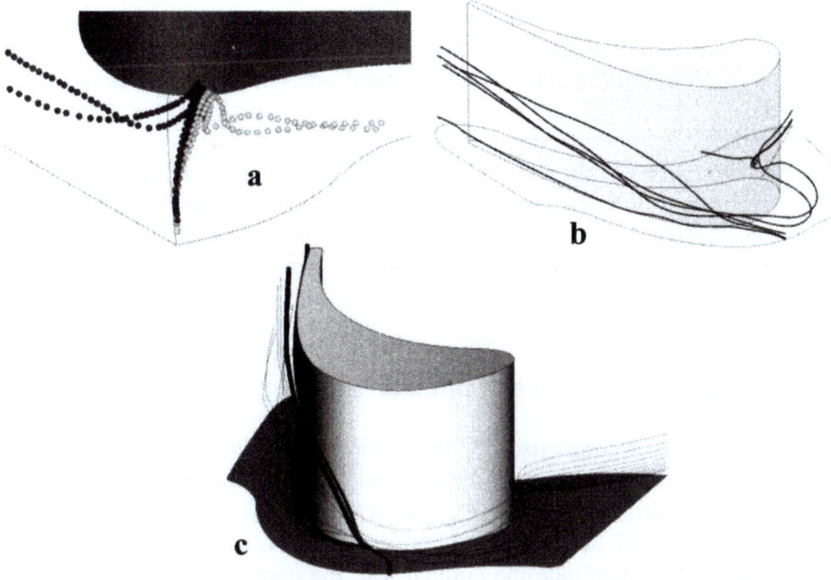

Fig. 4. Visualization of the secondary flows in the blade cascade: (a) inlet endwall boundary layer separation and horseshoe vortex formation, (b) evolution of the suction side leg of the horseshoe vortex and its interaction with the pressure side leg, (c) transformation of the pressure side leg of the horseshoe vortex into the corner vortex and the passage vortex

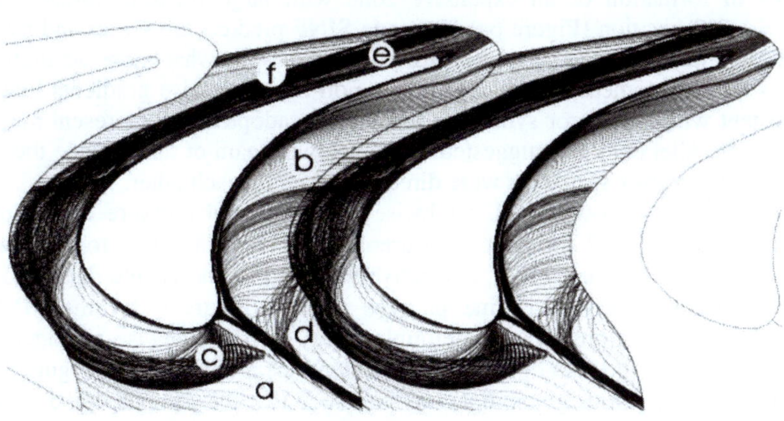

Fig. 5. Synthesized 2D view of streamline patterns: (a) inlet boundary layer, (b) endwall cross flow; (c) suction side leg of the horseshoe vortex system, (d) pressure side leg of the horseshoe vortex system, (e) resultant large-scale (passage) vortex; (f) boundary of the region with large pressure losses

In the cascade under consideration, a rapid drift of the interacting legs to the mid-span is observed in the second portion of the passage. Obviously, this drift is due to the passage pressure-to-suction endwall flow turning at the blade suction side. This flow together with the pressure side horseshoe vortex leg contributes to deviation of the suction-surface boundary layer fluid particles from the blade (see Figure 4c). As a result, a zone with large total pressure losses occurs near the suction side.

A synthesized view of computed streamline patterns is given in Figure 5. Despite the limitations of the 2D graphics, this figure gives a rather informative idea of the secondary flow specifics in the turbine blade cascade under consideration. A new interesting detail seen in the figure is the formation of a near-endwall bubble-type zone adjacent to the blade suction side near the leading edge. Streamlines in this zone are extremely sparse.

4.2. Pressure losses and exit pitch angles

Figure 6 presents a comparison of the computational and experimental data obtained for the total pressure field downstream of the cascade. Distributions of the total pressure coefficient, $C_{p,t} = (P'_{\text{in}} - P')/(P'_{\text{in}} - P_{\text{in}})$, are given for two measurement planes located at $x/C_x = 1.112$ (measurement Station 2) and 1.612 (Station 4). In order to facilitate the comparison, the computational data obtained for one pitch have been repeated three times with appropriate shifts in the pitchwise direction. Note that the left to right direction in the figure corresponds to the suction-to-pressure shift, i.e., to moving from the top to the bottom in Figures 2 and 5. The agreement of both the SINF and the FLUENT results with the measurement data is fairly good. The computations reproduce in detail the role of the passage vortex in formation of an expansive zone with large pressure losses. For the $x/C_x = 1.112$ position (Figure 6a), the code SINF predicts the vortex-induced core with high values of the total pressure coefficient somewhat more accurately. At $x/C_x = 1.612$, the measurement data show non-zero spanwise gradients that is inconsistent with the mirror symmetry formulation adopted at the present numerical simulation. Giel et al. [1] suggested that far downstream of the cascade the upper- and low-half vortex structures were directly impacting each other.

Data for the area-weighted, pitchwise averaged total pressure coefficient are compared in Figure 7 for three measurement planes (Stations 2 to 4) located at $x/C_x = 1.112$, 1.362, and 1.612, respectively. Note that the visible distinctions between the total pressure isoline patterns seen in Figure 6 for the SINF and FLUENT computations have a negligible effect on the pitchwise averaged distributions. So, only the data of the SINF computations are included in Figure 7. Analyzing the distributions given, one may conclude that the computed and the measured data on total pressure losses agree quite satisfactory.

Figure 8 compares computed (with the code SINF) and experimental distributions of the area-weighted, pitchwise averaged pitch angle. On the whole, there is a reasonable agreement between the distributions, especially if one takes into account the above mentioned deviations from the mirror symmetry in the actual flow.

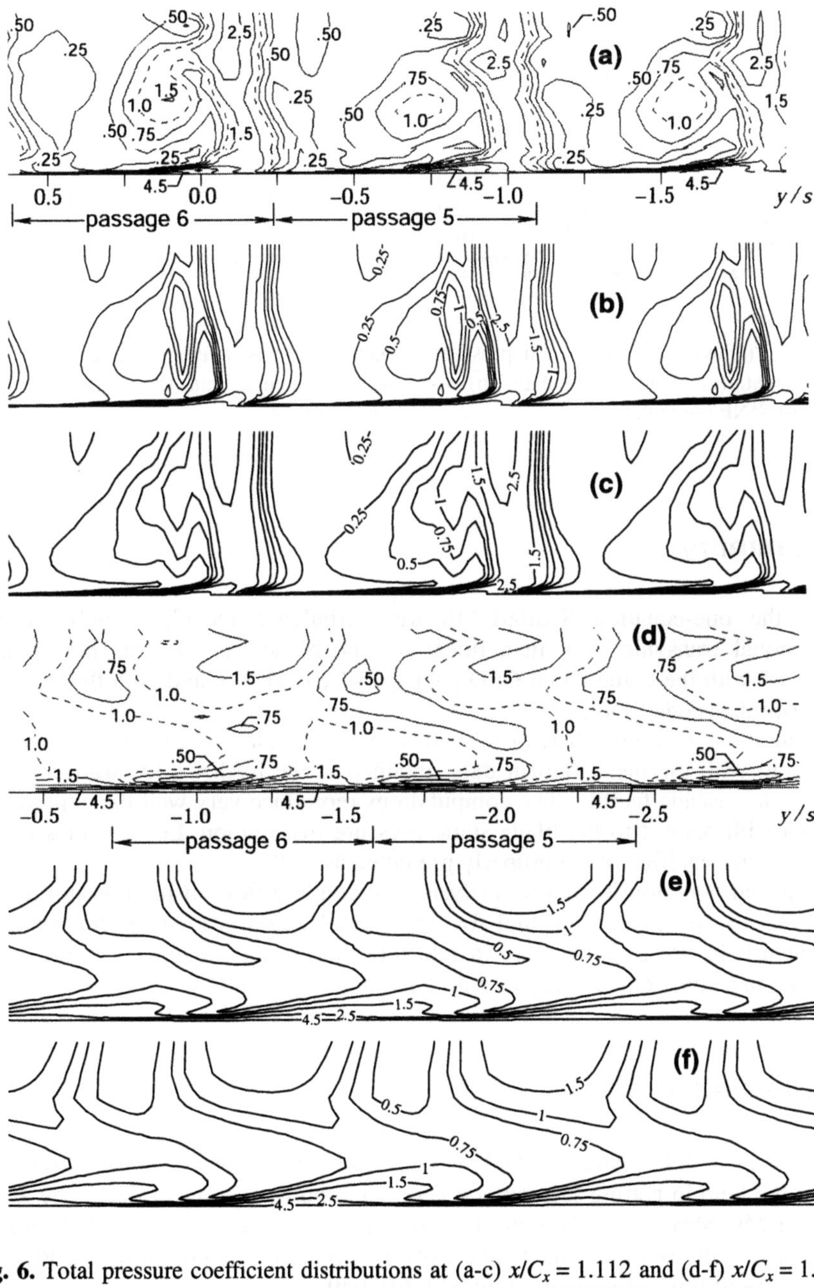

Fig. 6. Total pressure coefficient distributions at (a-c) $x/C_x = 1.112$ and (d-f) $x/C_x = 1.612$: (a, d) experimental data [1], (b, e) SINF computations, (c, f) FLUENT computations

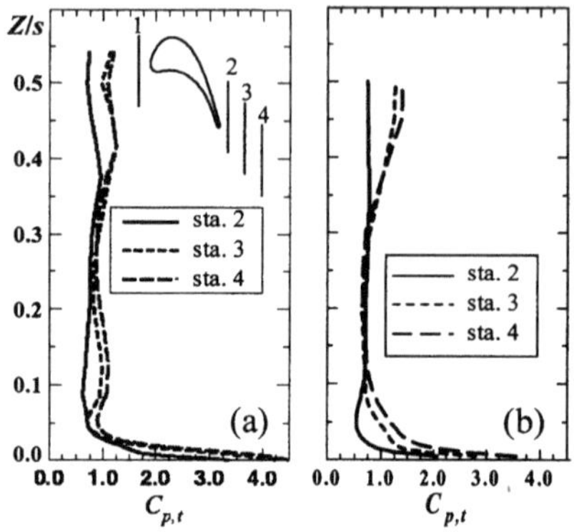

Fig. 7. Pitchwise averaged total pressure coefficient: (a) experimental data, (b) results of SINF computations

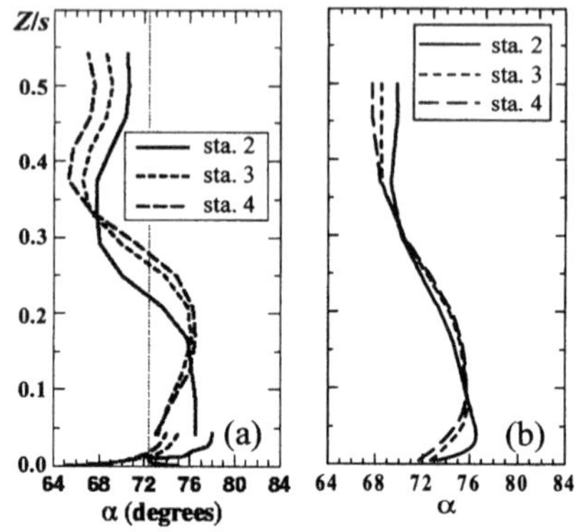

Fig. 8. Pitchwise averaged pitch angle: (a) experimental data, (b) results of SINF computations

5. Summary

Using the one-equation Spallart-Allmaras turbulence model, a highly three-dimensional transonic flow in a linear turbine blade cascade was numerically simulated with the commercial CFD package FLUENT 5.5, and with the in-house Navier-Stokes code SINF.

Being close to each other, flow field results generated by the two codes were compared with the measurement data obtained at the NASA GRC transonic turbine blade cascade facility. The computations reproduce very well the experimental data with respect to the blade static pressure distribution. Local values of the total pressure coefficient are properly predicted as well.

A special attention is paid to analysis of secondary flows using possibilities of an advanced post-processing tool for flow visualization. The streamline patterns presented help one to get a more insight into the features of secondary flows developing in turbine blade cascades with a large turning angle.

Acknowledgments

The authors wish to thank Dr. Paul W. Giel from the NASA Glenn Research Center for the help in getting the cascade geometry and measurement data. The assistance of Mr. Maxim Nikolaev in optimal using the H-CUSP scheme possibilities is also gratefully acknowledged. The work was partially supported by the Russian Foundation of Basic Research, grant 02-07-90049.

References

[1] Giel PW, Thurman DR, Lopez I, Boyle RJ, Van Fossen GJ, Jett TA, Camperchioli WP, La H (1996) Three-dimensional flow field measurements in a transonic turbine cascade. ASME Paper 96-GT-113.

[2] Giel PW, Gaugler RE (2001) NASA Blade 1. Endwall Heat Transfer Data. Version 1. NASA-Glenn Research Center, Turbine Branch, CD ROM.

[3] Spalart P, Allmaras S (1992) A one-equation turbulence model for aerodynamic flows. AIAA-Paper 92-0439.

[4] Smirnov EM (2000) Solving the full Navier-Stokes equations for very-long-duct flows using the artificial compressibility method. In: ECCOMAS-2000, Barcelona, Spain, CD-ROM proceedings, 17 p.

[5] Jameson A (1995) Positive schemes and shock modelling for compressible flows. Int. J. Num. Meth. Fluids 20: 743–776.

[6] Nikolaev MA (2001) Combination of the CUSP scheme with the compressibility scaling method for transonic flow calculations in domains of complex geometry (in Russian). In: Phys. Principles of Exp. and Math. Simulation of Heat and Mass Transfer and Gas Dynamics in Power Plants. Proc. 13th Int. Leontiev's School-Seminar, MPEI Publishers, Moscow, Russia, Vol.1, pp.108-111.

[7] FLUENT (1998) User's Guide.

[8] Goriatchev VD, Smirnov EM (1997) A computational and informational system for CFD: SELIGER. In: Proc. of the 15 IMACS World Congress on Scientific Computation, Modeling and Applied Mathematics, Berlin, pp. 47-53.

[9] Langston LS (2001) Secondary flows in axial turbines – a review. In: Heat Transfer in Gas Turbine Systems (Ed. by R.J. Goldstein), Annals of the New York Academy of Sciences, Vol. 934, pp. 11-26.

[10] Wang HP, Olson SJ, Goldstein RJ, Eckert ERG (1997) Flow visualization in a linear turbine cascade of high performance turbine blades. ASME J. of Turbomachinery 119: 1–8.

Numerical Optimisation of a High Pressure steam Turbine Stage

Piotr LAMPART
Institute of Fluid Flow Machinery, Polish Academy of Sciences,
ul. Fiszera 14, 80-952 Gdansk, Poland, e-mail: lampart@imp.gda.pl

Abstract Blading of a high-pressure (HP) steam turbine stage is optimised using an idea of direct constraint optimisation. The objective function to be minimised is the enthalpy loss of the stage. A simplex method of deformed polyhedron proposed by Nelder-Mead is used for optimisation. Current values of the objective function are found from 3D Reynolds-Averaged Navier Stokes (RANS) computations. To secure global flow conditions, there are constraints imposed on the mass flow rate, exit swirl angle, and reaction. The optimised parameters are here the stator and rotor blade numbers and stagger angles, rotor blade twist angle and parameters of stator blade compound lean at root and tip. Blade profiles are not changed. Optimisation gives a design with new 3D stacking lines of the blades and increased flow efficiency, compared to the original design.
Key Words 3D RANS solver, blade shape optimisation, HP axial-flow turbine stage

1. Introduction

Optimisation of 3D blading in turbomachinery is a relatively new field of research. It has become possible only with the development of 3D Navier-Stokes codes for turbomachinery applications capable of locating and quantifying loss generation processes in complex turbomachinery geometries, and also thanks to increasing capabilities of computing machines. A robust 3D solver in collaboration with a reliable optimisation technique can be a powerful design tool, also in the turbomachinery environment. At present two optimisation/design approaches become conspicuous with respect to turbine blading systems. One approach concentrates on development of 3D inverse design using Euler or Navier-Stokes codes where the shape of the blading changes during an iterative procedure until the target distribution of, for example, blade surface static pressure, or downstream velocity is reached, guaranteeing the required performance, e.g. [3, 14]. Another approach focuses on optimisation of global characteristics of the stage, and the final shape is obtained from minimising/maximising an objective function, for example the total energy loss or efficiency, total pressure loss of the stage etc. Values of the objective function for new geometries can be found directly from 3D viscous flow computations, e.g. [18, 7]. However, as the direct optimisation is time consuming and the main cost is time required for the Navier-Stokes solver, an interesting idea of approximate representation of the objective function using an artificial neural network trained over a data base of RANS solutions was put forward in [12]. In the process of optimisation new geometries can be selected using a number of optimisation methods, the most popular of which are the steepest gradient method, simplex methods and genetic algorithms.

Optimisation of 3D blading in low-pressure (LP) steam turbine stages, especially for exit stages, is shown in [7, 5] to bring considerable efficiency gains. There seems to be less room for efficiency improvements in low-load high-pressure (HP) steam turbine stages, [18, 6]. Efficiency of cylindrical (radially stacked and straight) blades is relatively high here unless a mistake was made while selecting basic parameters of the blading, like stator and rotor blade numbers and stagger angles. This, however, makes the optimisation task even more challenging.

A typical 3D design to raise the efficiency of HP stages is compound lean where some blade sections, usually at the root and tip, are linearly or non-linearly displaced in circumferential direction with respect to other sections. The effects of this design were investigated e.g. in [4, 13, 6]. It was found there that lean redistributes blade load, mass flow rate and loss span-wise, compared to the cylindrical blading, and can serve as a means of controlling secondary flow and tip leakage losses in HP turbines. Lean can change the state of boundary layers at the blade suction surface and at the end-walls, as well as can reduce span-wise variations of the exit swirl angle, which is likely to reduce downstream mixing losses and make stator/rotor matching easier. The quantitative effect of lean on the stage loss varies with turbine stage geometry and operation conditions. Due to a number of shape parameters involved in this design it is required that automatic optimisation techniques are used to find the optimum design.

The procedure for optimisation of an HP steam turbine stage is based on the concept of direct constrained optimisation, using the Nelder-Mead method of deformed polyhedron and a 3D RANS solver for turbomachinery applications. Direct optimisation is considered highly time consuming. The main cost is CPU time required for the Navier-Stokes solver that computes changing flow geometries. Therefore, 3D computational grids used during optimisation will be relatively coarse. However, some experience of the author shows that major tendencies in changing flow patterns with changing geometry of the turbine stage can already be discovered on coarse grids. Anyway, at least the original and final geometries will be checked on refined grids, and possible changes in flow patterns and efficiency gains will be implied based on the comparison of post-optimisation computations of the original and final geometries on refined grids.

2. Method of optimisation

Shape optimisation of turbine blading is carried out with the help of a code Optimus, [19]. Optimisation is understood here as an iterative procedure that seeks for an extremum of an objective function – a function of some shape parameters.

2.1. Optimised functions and parameters

In this investigation, the optimised (minimised) objective function is the enthalpy loss of the stator/rotor stage (the leaving energy not considered a loss and assumed to be used in the subsequent stage), defined as

$$\xi = (h_2 - h_{2s'})/(h_{0T} - h_{2s'}) \tag{1}$$

where $h_{0T}, h_2, h_{2s'}$ are the inlet total enthalpy, exit static enthalpy and isentropic enthalpy referred to the exit static pressure, respectively. Alternatively, one can also choose the total enthalpy loss of the stage including the leaving energy (as for the exit turbine stages), or stage power as an objective function. The following parameters of blade shape can be considered during the optimisation for each blade row: blade number, stagger angle, blade height, linear twist angle, linear lean angle, and linear sweep angle, 4 parameters of compound twist (2 at hub, 2 at tip), 4 parameters of compound lean (2 at hub, 2 at tip), and 4 parameters of compound sweep (2 at hub, 2 at tip). Each parameter is allowed to vary in a prescribed range of variation.

2.2. Flow constraints

To secure global flow conditions, there are also constraints imposed on the mass flow rate, exit angle, average reaction, reaction at tip and root. The exit angle and reactions are not allowed to assume values beyond the prescribed ranges, which is pronounced in the shape of the objective function f, that is $f = \zeta$, if the exit angle and reactions fall within the prescribed range, or $f = \infty$ (a very large number), otherwise, where ζ is a value of an optimised characteristic (enthalpy loss of the stage) obtained from the RANS solver. The penalty function is imposed on the mass flow rate if it falls beyond a required very narrow interval $[G_-, G_+]$: $f = \zeta$, if $G_- \leq G \leq G_+$, or $f = \zeta + min[(G - G_\pm)^2]/\varepsilon$, otherwise, where G - current mass flow rate, ε – penalty coefficient prescribed in a way that the objective function sharply rises to infinity with increasing distance from the limits of the assumed range of variation. As the pressure drop during the optimisation is kept constant and the mass flow rate is constrained in a very narrow range of variation, any change of power of the optimised stage can be attributed to the reduced flow losses only.

2.3. Method of deformed polyhedron

It is generally accepted that none of the available optimisation methods can be considered superior for all types of turbomachinery applications. The most popular seem to be the steepest gradient method, simplex-based methods and genetic algorithms. In this investigation, subsequent geometries in the course of optimisation are selected using a simplex method of deformed polyhedron proposed by Nelder & Mead [11]. In this method, optimisation of the objective function of n independent variables is performed using $n+1$ points being vertices of a polyhedron in the space of optimised parameters R^n. The initial polyhedron is usually generated in a random way from the prescribed range of variation of the optimised parameters, or the choice of initial vertices can also be supported by an educated guess, that is drawing on some engineering knowledge. First aim of each iteration during the optimisation procedure is to replace the least favourable vertex for which the objective function reaches the "worst" value with a new vertex, and to form a new polyhedron. This is done with the help of the following operations: symmetrical reflection with respect to a gravity centre, stretching, compression or reduction. These operations enable deformation of the polyhedron and its adapting to the topography of the objective function. An extremum of the objective function can be found even far away from the initial polyhedron. Of great importance for effective operation of the algorithm is proper selection of the reflection ratio α, compression ratio β, stretching ratio γ and reduction ratio δ. An adequately chosen reflection ratio is decisive for the rate of convergence of the algorithm and its ability to go beyond the vicinity of local extrema and to find a path towards a global extremum. Certainly there is no guarantee that an extremum to which the algorithm would converge is a global extremum. Anyway, there are at least three ways of further treatment if the designer feels that the algorithm holds on to a local extremum. First, to recognise the fact that a local minimum is a solution whose objective function is "better" than that of the original design. Second, to increase the reflection ratio, and third, to begin the optimisation process from another initial polyhedron.

Usually, Nelder-Mead's method of deformed polyhedron enables efficient optimisation of 5-10 geometrical parameters of 3D blading. Unlike for gradient-based methods, the efficiency of the deformed polyhedron method measured by the number of calculations of the objective function within one iteration, does not depend on the number of optimised parameters, and on average is limited to 2-3 RANS calculations per iteration. Yet it usually requires more iterations to converge, as compared to the steepest gradient method, however, fewer than in the case of genetic algorithms. Details of the used algorithm can be found in [18, 5].

2.4. 3D RANS solver

Values of the objective function are found from post-processing CFD computations performed with the help of a code FlowER - solver of viscous compressible flows through multi-stage turbomachinery [15]. The solver draws on the set of Reynolds-averaged Navier-Stokes equations for perfect gas. During the optimisation on coarse grids the effects of turbulence are taken into account with the help a modified algebraic model of Baldwin-Lomax [1], whereas verifying computations of original and final geometries on refined grids are made using the Menter SST turbulence model [10]. This is a two-equation eddy-viscosity model where the standard k-ω model is activated in the sublayer and logarithmic region, and then switched to the k-ε model in the wake region of the boundary layer, and where the eddy viscosity is redefined so as to guarantee the proportional relationship between the principal turbulent shear stress and the turbulent kinetic energy in the boundary layer. The governing equations are solved numerically based on the Godunov-type upwind differencing, high resolution ENO scheme and implicit operator δ of Beam & Warming, assuring second-order accuracy every-where in space and time. An H-type multi-grid refined at the endwalls, blade walls and trailing and leading edges is used. The computations are carried out in one blade-to-blade passage of the stator and rotor, and converge to a steady state, with the condition of spatial periodicity, and mixing plane approach assumed. The assumed inlet/exit boundary conditions impose the pressure drop and let the mass flow rate be resultant.

The code has recently been validated on a number of turbomachinery test cases, including Durham Low Speed Turbine Cascade, NASA Rotor37, NASA Low Speed Centrifugal Compressor and model air turbine of ITC Łódź. For details of the flow solver and its validation the reader is referred to [16, 17, 8, 9].

3. Optimisation of a high pressure steam turbine impulse stage

The optimised stage is an HP impulse stage of a 200 MW steam turbine with originally cylindrical blades and shrouded rotor blades. The stage power is about 5 MW, its average reaction 17%. The stage operates at a pressure drop p_2/p_0=0.9. Initial geometrical parameters of the stator and rotor are as follows: span/chord - 0.8 (stator) and 2 (rotor); stagger angles as defined between the profile chord and the normal to the cascade front – 46° (stator) and 72° (rotor), blade numbers – 50 (stator), 120 (rotor).

In this investigation, the 3D-shaped compound leaned stator blade is optimised together with stator and rotor blade numbers and stagger angles as well as rotor blade twist. There are nine geometrical parameters for the Nelder-Mead method - stator blade number and stagger angle, rotor blade number and stagger angle, rotor blade linear twist angle, and four parameters of stator blade compound lean – two compound lean displacements at each endwall as defined in Fig. 1. The blade sections (profiles) are assumed not to change during the optimisation. The objective function is the enthalpy loss of the stage not including the leaving energy. A penalty is imposed on the mass flow rate if it changes by more than ±0.5%, compared to the original geometry. The average reaction is assumed not to exceed the original value. To control changes of the exit energy the absolute exit swirl angle is not allowed to vary beyond the interval (-10o, 10o). Flow computations (3D RANS) are carried out for perfect gas assuming the specific heat ratio γ=1.3 and individual gas constant R=430J/kgK. Tip leakage flow over shrouded rotor blades is not evaluated here. Due to time restrictions, RANS computations in the course of optimisation are carried out on coarse grids of 100 000 cells (stator + rotor), also using the faster and less expensive turbulence model of Baldwin-Lomax. After optimisation, the original and optimised geometries were recalculated on refined grids – 800 000 cells (stator + rotor), using the Menter SST turbulence model.

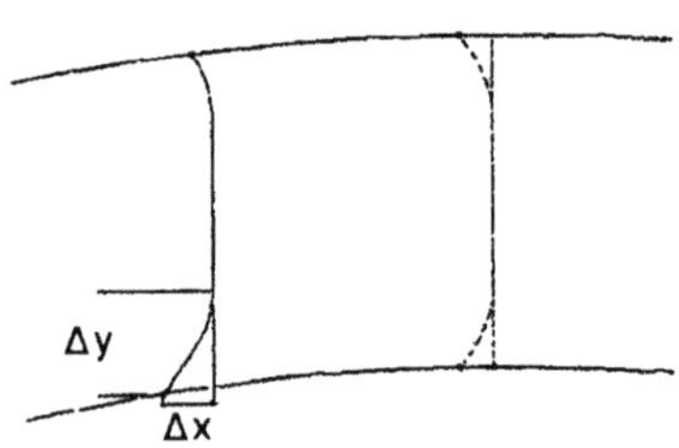

Fig. 1. Compound lean geometry of stator blade – leading and trailing edge in circumferential view

The optimisation process was completed after 65 iterations (with 120 geometries calculated). The objective function was decreased by 0.4% on a coarse grid. The previously cylindrical stator blades acquired compound lean shapes with the lean direction opposite to that of rotation of the rotor blades at both endwalls. At the trailing edge of the new stator blade the compound lean displacements at the tip are larger, compared to those at the root. Some rotor blade twist was introduced, closing throats towards the tip. Stator and rotor blade numbers and stagger angles were modified too. Changes of the optimised parameters are given in **Table 1**.

Table 1. Change of optimised parameters of HP turbine stage (l - blade height).

Optimised parameter	Its change
Stator blade number	3
Rotor blade number	-3
Stator stagger angle at hub[a] [°]	0.7
Rotor stagger angle at hub[a] [°]	-0.5
Rotor twist angle[b] [°]	-1.3
Stator compound lean displacement at tip $\Delta x/l$[c]	-0.08
Stator compound lean displacement at tip $\Delta y/l$	0.28
Stator compound lean displacement at hub $\Delta x/l$[c]	-0.05
Stator compound lean displacement at hub $\Delta y/l$	0.20

[a] Positive value of stagger angle increment opens throats, negative value closes throats.
[b] Positive value of twist angle opens throats towards the tip, negative value closes throats towards the tip.
[c] Positive value when the stator blade at the hub/tip is protruded with rotation of the moving blades.

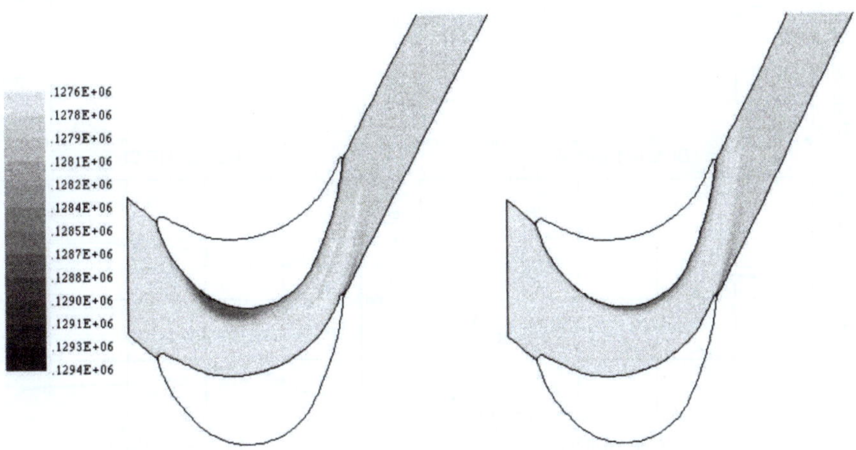

Fig. 2. Entropy function contours in rotor 10% blade span from root; original (left), optimised (right).

A comparison of the flowfield results for the original and optimised stage (obtained on refined grids using the Menter SST turbulence model) is presented in Figs. 2 to 6. The comparison exhibits significant differences in total pressure and entropy patterns in characteristic sections of the stator and rotor, as well as in span-wise distribution of pressures, velocities and losses. Changes of each optimised parameter have an effect on the stage performance. Two groups of parameters can be distinguished whose effects can clearly be separated - the first group consisting of stator and rotor blade numbers and stagger angles plus rotor blade twist (5 parameters), the other group consisting of four parameters of stator blade compound lean at the hub and tip. The effect of the first group is best observed in **Fig. 2** showing the entropy function contours in the rotor 10% of the blade span from the root. The contours exhibit the presence of a separation zone at the front part of the rotor blade suction surface at the root in the original design. The size of the separation zone at the root in the optimised design is consider-ably reduced. This is owing to changes of stator blade number and stagger angle which decrease the incidence on the rotor blade by about 2 degrees (in the rotating frame of reference). At the same time changes in the rotor blade number and stagger angle together with introduced rotor blade twist keep the well-streamlined flow patterns at the mid-span and tip, as well as the average reaction unchanged.

The effect of the second group of parameters is best observed in Figs. 3 to 5. The chosen direction of compound lean (opposite to that of rotation of the moving blades) increases pressure at the endwalls in the stator, unloads stator blades and reduces velocities there – **Fig. 3**.

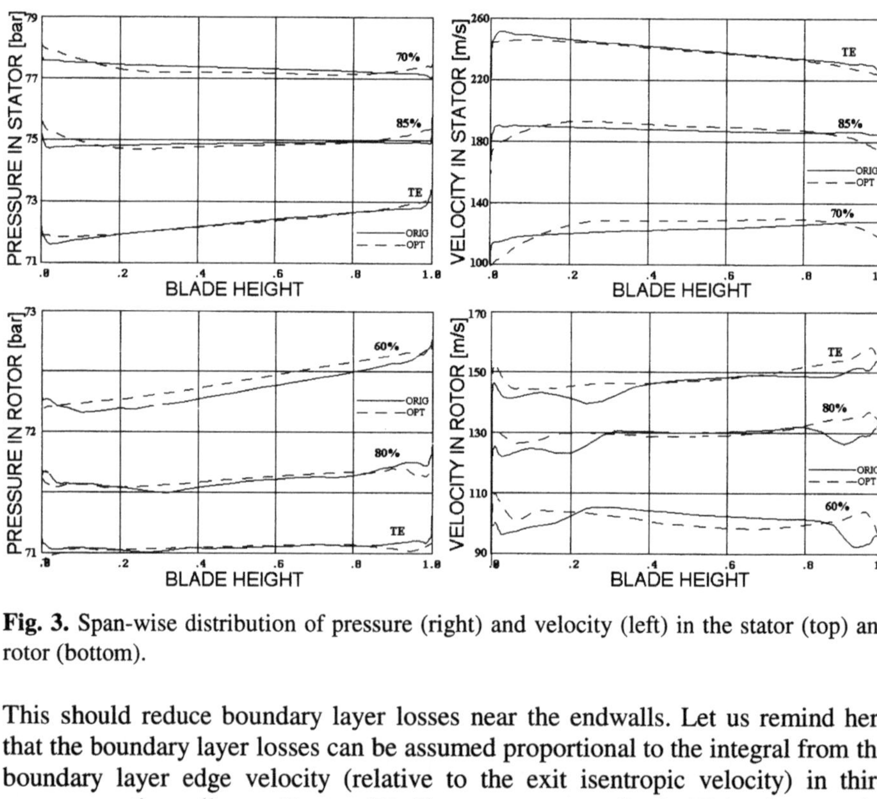

Fig. 3. Span-wise distribution of pressure (right) and velocity (left) in the stator (top) and rotor (bottom).

This should reduce boundary layer losses near the endwalls. Let us remind here that the boundary layer losses can be assumed proportional to the integral from the boundary layer edge velocity (relative to the exit isentropic velocity) in third power over the wall area, Denton [2]. However, as a result of additional span-wise pressure gradient there is increased convection of boundary layer fluid toward mid-span sections in the optimised design, which can be observed from total pressure contours downstream of the stator presented in **Fig. 4**. The secondary flow maxima downstream of the stator in the mixing plane are less intensive in the optimised design, however the wake seems to be slightly thicker at mid-span. The streamline curvature changes in the axial gap between the stator and rotor. In the optimised design, the pressure and velocity near the endwalls in the rotor exhibit opposite tendencies than in the stator. The pressure decreases and velocity increases at the endwalls in the rotor, as compared to the original design. More mass is passed through the endwall regions which is likely to increase endwall losses. However, the centres of loss due to secondary flows stay nearer to the endwalls, see total pressure contours at the rotor trailing edge presented in Fig. 4.

Fig. 5 shows a comparison of span-wise distribution of enthalpy losses in the stator, rotor and stage for the original and optimised design.

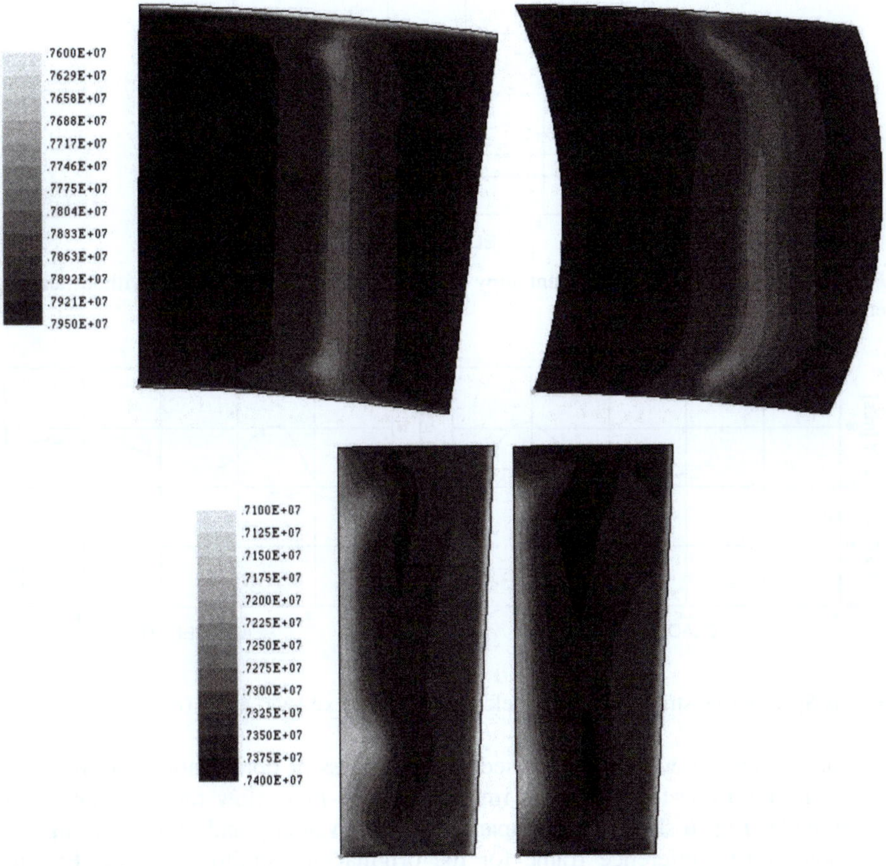

Fig. 4. Total pressure contours downstream of the stator (top) and at the rotor trailing edge; original (left), optimised (right).

In the optimised design, stator losses are considerably decreased near the endwall sections, and negligibly increased at the mid-span. Quantitatively, the stator losses (mass-averaged span-wise) are decreased by 0.3%. Rotor losses are increased at the endwalls. However, peaks due to secondary flows and separation at the root are considerably lower. Quantitatively, the rotor losses are decreased by 0.8% here. The combination of stator and rotor losses shows that stage losses without the exit energy are decreased near the endwalls, and 3D peaks are lower, than in the original design. Quantitatively, the stage losses are decreased by 0.4% (the level of loss decrease is the same as on a coarse grid). Note that the stator losses are determined at the mixing plane, whereas the rotor and stage losses at a section located 40% of the rotor axial chord downstream of the rotor trailing edge.

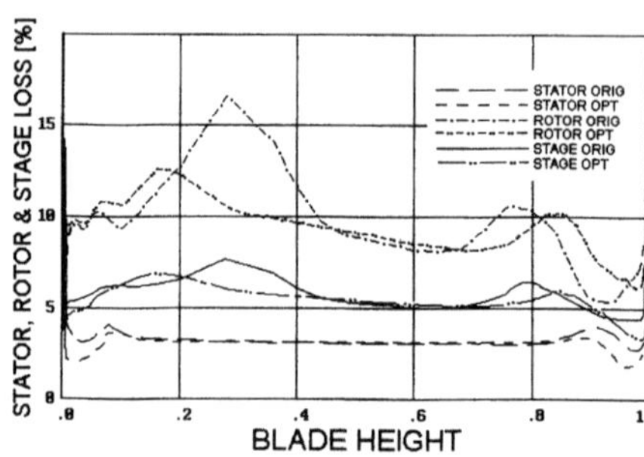

Fig. 5. Span-wise distribution of enthalpy losses in stator, rotor and stage (without leaving energy)

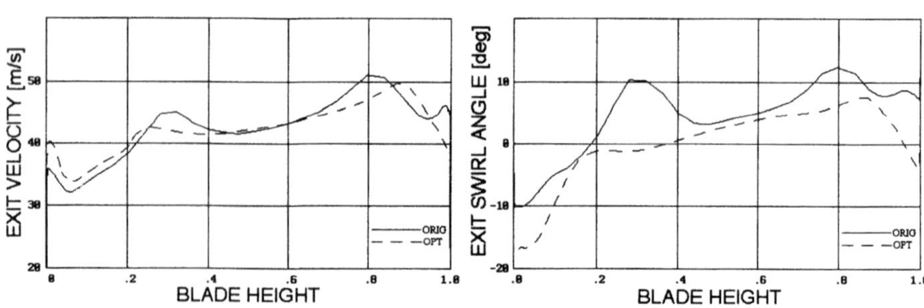

Fig. 6. Span-wise distribution of exit velocity (left) and exit swirl angle (right)

The exit energy was not optimised in the process of optimisation, but its value was controlled due to constraints imposed on the mass flow rate and mean exit swirl angle. **Fig. 6** shows the comparison of exit velocity and exit swirl angle in the non-rotating reference frame for the original and optimised stage. For approximately the same mass flow rate (it turned out to be decreased by 0.1% compared to the original design), the mean exit velocity is slightly decreased in the optimised design. This is largely due to reduction of 3D peaks in exit swirl angle. Apart from the root section, the redistribution of the exit swirl angle should reduce downstream mixing loss and provide more favourable inlet conditions to the subsequent stage.

4. Summary

Stator blade compound lean was optimised together with stator and rotor blade numbers and stagger angles, as well as rotor blade twist in a low-load HP steam turbine stage. Nine geometrical parameters were optimised to reduce the stage loss (without the leaving energy). The idea of direct constrained optimisation was presented. Geometries changing in the course of optimisation were found using Nelder-Mead's method of deformed polyhedron, whereas the corresponding values of the objective function were found from a 3D RANS solver. There were also constraints imposed on the mass flow rate, exit swirl angle, and reaction. As a result of optimisation, the stator of the tested stage acquired a new 3D stacking line, and the calculated stage loss without the leaving energy was decreased by 0.4%. The loss decrease can be attributed here to a reduced size of the separation zone at the rotor root, and decreased endwall and secondary flow losses. The obtained efficiency gains are not impressive here. However, further efficiency gains can still be expected due to a more favourable distribution of exit velocity and swirl angle, which is likely to reduce the downstream mixing loss and provide more favourable inlet conditions to the subsequent stage.

References

[1] Baldwin BS, Lomax H (1978) Thin layer approximation and algebraic model for separated turbulent flows. AIAA Paper 78-257

[2] Denton JD (1993) Loss mechanisms in turbomachines. Trans ASME J Turbomachinery 115: 621-656

[3] Demeulenaere A, Van Den Braembussche R (1998) Three-dimensional inverse method for turbomachinery blading design. Trans ASME J Turbomachinery 120: 247-254

[4] Harrison S (1992) The influence of blade lean on turbine losses. Trans ASME J Turbomachinery 114: 184-190

[5] Lampart P (2002) Numerical optimisation of stator blade sweep and lean in an LP turbine stage. ASME Paper IJPGC2002-26161

[6] Lampart P, Gardzilewicz A (1999) Numerical study of 3D blading in HP impulse turbines. Cieplne Maszyny Przepływowe (Turbomachinery) 115: 297-310

[7] Lampart P, Yershov S (2001) Direct constrained CFD-based optimisation of 3D blading for the exit stage of a large power steam turbine. Trans ASME J Engng Gas Turbines & Power 125: 385-390

[8] Lampart P, Świrydczuk J, Gardzilewicz A (2001) On the prediction of flow patterns and losses in HP axial turbine stages using 3D RANS solver and two turbulence models. TASK Quarterly 5: 191-206

[9] Lampart P, Yershov S, Rusanov A (2002) Validation of turbomachinery flow solver on turbomachinery test cases. Cieplne Maszyny Przepływowe (Turbomachinery) 122: 63-70

[10] Menter FR (1994) Two-equation eddy-viscosity turbulence models for engineering applications. AIAA J 32: 1598-1605

[11] Nelder JA, Mead R (1965) A simplex method for function minimisation., Computer J 7: 308-313
[12] Pierret S, Van Den Braembussche R (1998) Turbomachinery blade design using a Navier-Stokes solver and artificial neural network. ASME Paper 98-GT-4
[13] Singh G, Walker PJ, Haller BR (1995) Development of three-dimensional stage viscous time marching method for optimisation of short height stages. In: Proc. European Conference on Turbomachinery, Fluid Dynamics and Thermodynamic Aspects, Erlangen, Germany, March 1-3
[14] Tiow WT, Zangeneh M (2000) A three-dimensional viscous transonic inverse design method. ASME Paper 2000-GT-0525
[15] Yershov S, Rusanov A (1996) The application package FlowER for the calculation of 3D viscous flows through multi-stage turbo-machinery. Certificate of state registration of copyright. Ukrainian state agency of copyright and related rights. February 19 (in Russian)
[16] Yershov S, Rusanov A (1996) The high resolution method of Godunov's type for 3D viscous flow calculations. In: Proc. 3 Colloquium on Process Simulation, Espoo, Finland, June 13-16, pp 69-85
[17] Yershov S, Rusanov A, Gardzilewicz A, Lampart P, Świrydczuk J (1998), Numerical simulation of 3D flow in axial turbomachines. TASK Quarterly 2: 319-347
[18] Yershov S, Rusanov A, Shapochka A, Lampart P, Świrydczuk J, Gardzilewicz A (2002) Shape optimisation of two turbine stages using the deformed polyhedron method and a 3D RANS solver. Proc Inst Mech Engrs Part A J Power Energy 216: 203-213
[19] Yershov S, Shapochka A, Rusanov A (2000) 3D shaping of turbine blading based on 3D solutions of viscous compressible flow and optimisation. In: Proc. Conference on Improvements in Turbomachinery Using Methods of Mathematical and Physical Modelling, Kharkov-Zmiev, Ukraine, September 18-22, pp 171-178 (in Russian)

An Experimental Investigation of the Unsteady Flow in a Two-Stage Low-Pressure Research Turbine

Edward Canepa, Andrea Cattanei, Marina Ubaldi, Pietro Zunino

Dipartimento di Macchine, Sistemi Energetici e Trasporti, University of Genova

Abstract

Results of an experimental investigation of the unsteady flow in a two-stage low-pressure large-scale research turbine are presented. Velocity and turbulence measurements were performed upstream and downstream each blade row at midspan by means of a two-sensor hot-wire probe. Results show a complex unsteady flow with remarkable wake generated and potential flow interaction effects.

Keywords: axial flow turbine, hot-wire measurements, rotor-stator aerodynamic interaction.

Nomenclature

c	absolute velocity
g	stator pitch
r	radial coordinate
t	time
T	rotor blade passing period
w	relative velocity
x	axial coordinate
y	circumferential coordinate
α	absolute flow angle
β	relative flow angle
ω	rotational speed

Subscripts and superscripts

nom	nominal condition
x, *y*	in axial, tangential direction
'	fluctuating component
~	ensemble averaged
~'	ensemble averaged rms of fluctuating components

1 Introduction

Current design of axial turbines is carried out under the hypothesis of steady flow, regardless the flow in bladed components is unsteady and blade boundary layer development and associated losses are strongly influenced by unsteady effects.

Aerodynamic rotor-stator interaction constitutes the major cause of unsteadiness in turbomachines working at nominal conditions. Each cascade produces a non-uniform flow field which is due to potential flow effects as well as viscous effects acting through blade wakes and secondary flows. Therefore a blade row in relative motion with respect to the preceding one is affected by strong periodic variations of all the thermofluid-dynamic quantities and operates under strongly unsteady flow conditions.

In general the perturbation causes the stage loss to increase compared with the ideal condition of isolated blade row [1-2]. However the interaction between blade rows in relative motion has sometimes also positive effects, as it happens when the periodic perturbation promotes the boundary layer transition before laminar separation takes place [3-4]. This is the case of the low pressure stages of aero-engine gas turbines operating at low Reynolds numbers (50000-200000) during high altitude cruise. To avoid severe boundary layer laminar separations, characteristic of such low Re conditions, the designer is forced to limit the blade loading and to reduce the boundary layer laminar extension, with a consequent increase of losses at the high Reynolds number operating conditions [5]. Therefore, only a deep knowledge of the phenomenon can allow the designer to remove the aerodynamic loading restriction obtaining a more compact and efficient turbine design.

Results from fundamental experiments on wake-boundary layer interaction on flat plates and in cascades operating with inlet flow perturbed by the wakes are available in the technical literature (among the most recent [6-8]). They provide detailed insights into the physics of the rotor-stator interaction phenomena, contributing to increase the understanding of the physics and stimulating ideas applicable to wake-induced transition modelling to be used for CFD calculations.

In order to extend these unsteady flow principles to real machine environment and incorporate the appropriate physical models into the design procedures a further step is essential [9, 10], which consists in detailed experiments on realistic geometries representative of multi-stage turbine environment. For this reason a large-scale research turbine has been conceived, designed and constructed in the framework of a National Research Program (PRIN 1998) and it is in operation in the Aerodynamics and Turbomachinery Laboratory of the University of Genova [11]. This facility is suitable for both basic and applied experiments focused on the exploitation of blade row interaction effects for low-pressure turbine performance improvement.

The present paper describes the large-scale low-pressure research turbine and presents the results of a preliminary investigation aiming at identifying and quantifying the aerodynamic interaction mechanisms in the two-stage turbine model.

2 Test Facility and Experimental Procedure

2.1 Axial flow turbine

The test facility, shown in Fig. 1, is a large-scale low-speed two-stage axial flow research turbine representative of low-pressure gas turbine stages and its overall dimensions are 3766 mm x 2170 mm x 2269 mm (length x width x height). The working fluid is atmospheric air fed by a centrifugal fan mounted downstream of the turbine. The rig allows chord based Reynolds number variations from 50000 up to 500000. To obtain repetitive flow conditions on the two stages as well as to vary the operating point, the turbine is provided with a variable stagger centripetal distributor producing pre-swirled flow at the first stage inlet.

The turbine is braked by means of a 40 kW D.C. reversible electric motor and the fan is driven by a 60 kW electric motor, both of them equipped with a variable speed electronic control. The desired operating point is obtained by acting simultaneously on the rotational speeds of the electric motors and on the guide vane stagger angle.

The turbine model has been especially designed to study unsteady flow phenomena in LP turbines. All four rows consist of unshrouded blades mounted on discs, whose circumferential and axial positioning may be varied, thus allowing both rows clocking and gapping. The Plexiglas casing is designed to allow inde-

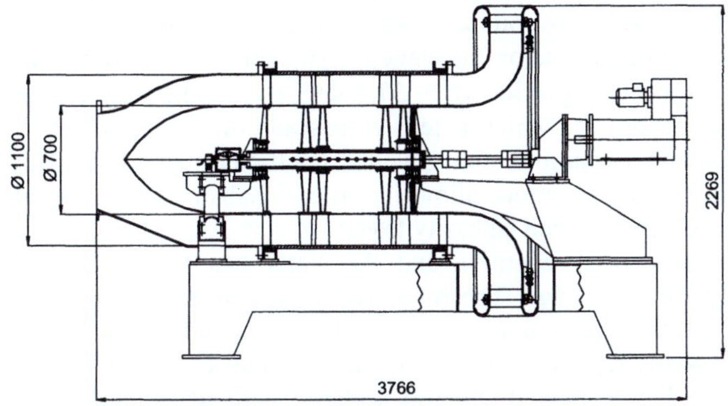

Fig. 1. Meridional section of the turbine model

Fig. 2. Two-stage low-pressure research turbine

pendent rotation. A long longitudinal opening, where modular caps or an optical glass window may be fitted, allows for stationary probes insertion between the rows and for LDV measurements. A photograph of the research turbine is shown in Fig. 2.

Axial and radial positioning of the probes is realised by means of a two-axis traversing system mounted on the casing, while angular positioning is obtained by means of the casing rotation system; all movements are computer controlled with minimum steps of 10 μm.

2.2 Flow conditions and turbine geometry

The meridional channel mean diameter is 900 mm and the blade height and chord are 200 mm and 120 mm, respectively; the bladings may be axially moved within a maximum length of 925 mm. At the design operating condition the rotational speed is 450 rpm and the flow rate 11.8 kg/s, with a total pressure drop per stage of about 1200 Pa and an overall ideal power of about 24 kW.

In the present configuration the stage is symmetric at midspan and is provided with MTU T106 blade profile (β_1 = 37.7 deg, β_2 = -63.2 deg, angles measured from axial direction) for which extensive cascade investigations exist (e.g. [12–13]); at nominal conditions the Zweifel coefficient is 1.055. However, the particular mechanical solutions allow for an easy substitution of the bladed disks and different geometries can be easily tested.

The two stages are repetitive and the stator has prismatic blades radially stacked on the trailing edge. The rotor has twisted blades, designed to obtain an outlet flow similar to the one generated by the centripetal distributor at the turbine inlet, so that nearly repetitive stages are obtained. Particular care has also been put in obtaining a highly loaded stage at midspan without extensive separations at both stator and rotor tips where solidity is lower.

2.3 Experimental procedure

Measurements of the unsteady velocity components were made using a two-component crossed hot-wire probe (Dantec P64) and a two-channel constant temperature anemometer (Dantec 55M10). Data were sampled by means of a Metrabyte DAS 58 Sample & Hold AD converter board and the instantaneous velocity components were determined by means of the directional calibration and non-iterative hot-wire analysis method of Schroeder [14].

In order to measure the velocity components on the blade-to-blade surfaces, the probe was traversed radially at 5 axial positions upstream and downstream each blade row. A schematic of the turbine blade rows with the measuring sections is shown in Fig. 3. In the following only the results obtained at midspan surfaces are reported. In the present experiment the axial gap between blade rows at midspan was 0.35 of the axial chord.

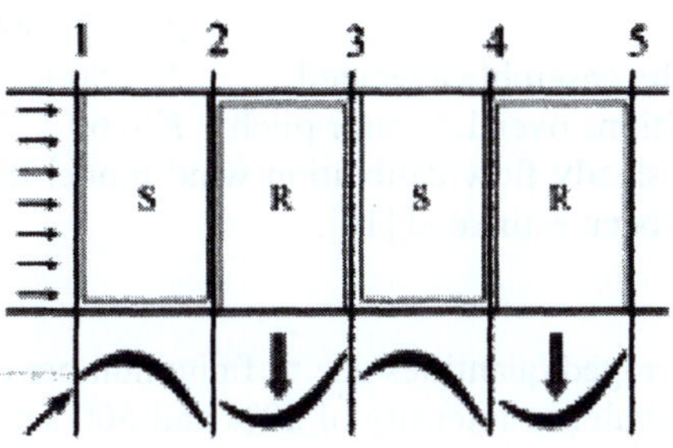

Fig. 3. Schematic of measuring sections

A phase-locked ensemble average technique [15], with the reference signal obtained from the shaft encoder, has been applied to separate periodic velocity variations associated with the blade passing frequency from random fluctuations.

In order to investigate rotor-stator interaction flow effects, measurements at each axial position are repeated for several different relative circumferential positions between probe and stator vanes [16].

The instantaneous velocity is a function of time (t_j) or rotor circumferential coordinate, of the data record (n), of the stationary probe circumferential position with respect to the stator vane (θ_k) and of the axial coordinate (x).

Omitting for simplicity the axial coordinate, the equations defining the ensemble average procedure are as follows:

- instantaneous velocity

$$c(t_j,\theta_k,n) = \tilde{c}(t_j,\theta_k) + c'(t_j,\theta_k,n)$$

- ensemble averaged velocity

$$\tilde{c}(t_j,\theta_k) = \frac{1}{N}\sum_{n=1}^{N} c(t_j,\theta_k,n)$$

- root mean square of the random unsteady fluctuations

$$\sqrt{\widetilde{c'^2}(t_j,\theta_k)} = \sqrt{\frac{1}{(N-1)}\sum_{n=1}^{N}\left[c(t_j,\theta_k,n) - \tilde{c}(t_j,\theta_k)\right]^2}$$

where:

$n = 1 \ldots N$ is the index of the sequence of records to be ensemble averaged or rotor revolutions considered,
$j = 1 \ldots J$ is the index of the time coordinate t or order of the sampled signal in the record,
$k = 1 \ldots K$ is the index of the circumferential position of the probe.

Typical data acquisition parameters for the present investigation are as follows:

Rotational speed	$\omega = 20.94$ rad/s
Sampling frequency	$f_s = 10$ kHz
Number of data for each rotor blade passage	$I = 100$
Number of measured blade passages	$z = 5$
Dimension of each sampled record	$J = z\,I = 500$
Number of rotor revolutions or records to be ensemble averaged	$N = 500$
Number of circumferential measuring positions over 1.5 stator pitch	$K = 60$

From verification tests performed in a steady flow calibration wind tunnel the following experimental uncertainties have been estimated [14]:
velocity magnitude = ± 1 %
yaw angle = ± 1 deg

Statistical uncertainties of ensemble averaged quantities due to finite number of samples with confidence level of 95%, turbulence intensity of 20% and 500 samples are as follows:
$\varepsilon_c = \pm 1.75$ %
$\varepsilon_{c'} = \pm 6.2$ %

3 Results and Discussion

The test Reynolds number is based on the stator vane outlet velocity and, as a consequence, depends on the turbine expansion ratio, which is determined by the fan rotational speed. In the present experiment a Reynolds number of 230000 was prescribed. Once the Reynolds number is prescribed, the nominal turbine operating condition can be searched for by imposing null incidence condition on both stators 1 and 2 at midspan. While the incidence angle on stator 1 is easily achieved by adjusting the inlet guide vane stagger angle, the null incidence on stator 2 has to be sought by changing the turbine rotational speed.

Figure 4 shows the ensemble averaged relative velocity and absolute flow angle distributions downstream of rotor 1 at midspan, with the rotational speed as varying parameter. For $\omega/\omega_{nom} = 1$, the mean yaw angle results 38 deg, i. e. the zero incidence condition on stator 2 is achieved. At this rotational speed the rotor wake velocity defect and the standard deviations of the random fluctuations (Fig. 5) indicate a suitable non-separated rotor flow with a moderate tangential extension of the wake.

The pressure side is on the right of the rotor wakes. As expected, the wake defect is larger on the suction side, where also the unresolved unsteadiness is larger. The increase of unresolved unsteadiness in the mid-pitch region of the rotor passage is a trace of the vane wake of stator 1. The chopped nozzle wake segments

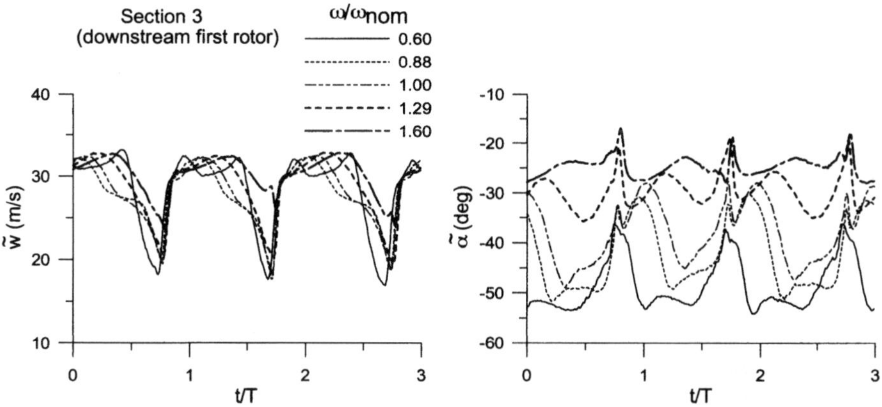

Fig. 4. Ensemble averaged relative velocity and absolute flow angle for different turbine rotational speeds

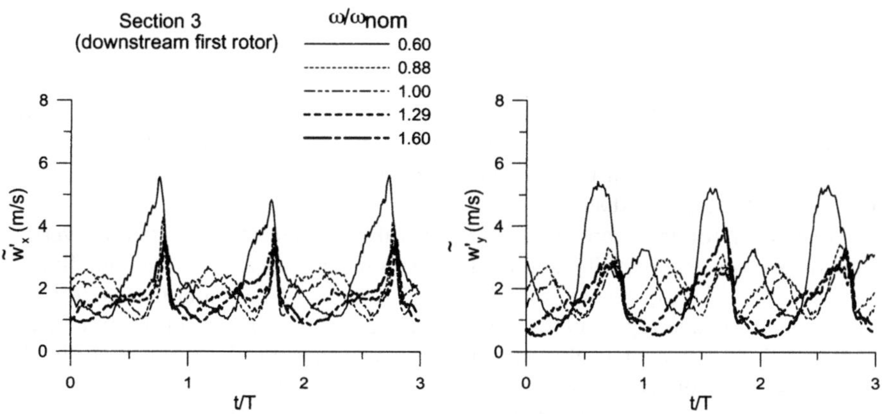

Fig. 5. Ensemble averaged velocity fluctuations for different turbine rotational speeds

are conveyed through the rotor passages and distorted by the velocity gradients in the rotor passage.

In order to identify flow interaction effects between blade rows, ensemble averaged velocity components and random fluctuations rms are represented in space-time diagrams (Figs. 6–11). Data for three blade passing periods are plotted on the vertical axis *t/T*. The normalised circumferential positions of the probe in the absolute frame of reference *y/g* are plotted on the horizontal axis. Data were measured over 1.5 times the stator pitch *g*. Stator vane wakes are represented in the plots by vertical lines (fixed absolute circumferential positions), while rotor blade wakes appear as inclined lines since disturbances reach the probe in the absolute frame of reference with the peripheral velocity *u*.

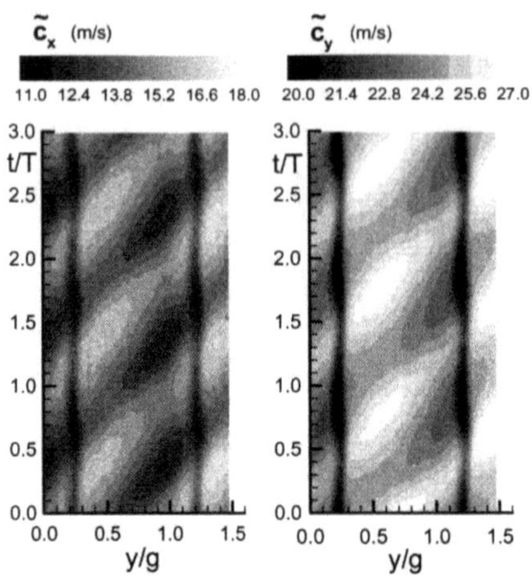

Fig. 6. Space-time plots of absolute velocity components: section 2

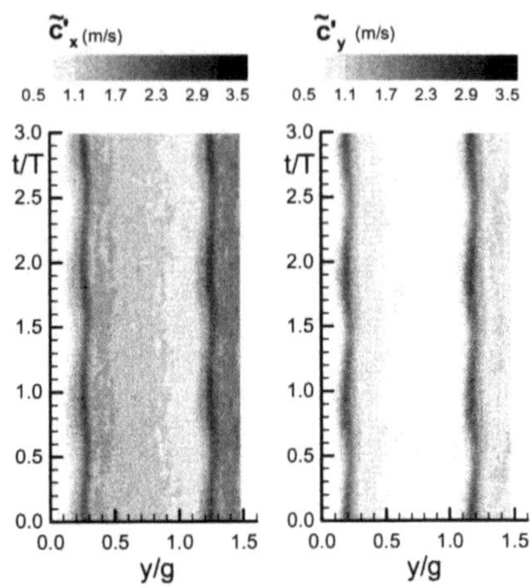

Fig. 7. Space-time plots of rms of velocity fluctuations: section 2

Probe is traversed circumferentially through the vane wake from the suction to the pressure side, therefore the vane suction side is on the left of the vane wakes. Rotor blade wakes hit the probe sensor with the suction side first and after leave the probe with the pressure side. Therefore the rotor blade suction side is on the lower side of the inclined strips.

At station 2 (downstream stator 1) vane wakes are clearly identified by vertical lines displaying the velocity defect (Fig. 6) and the increase of random fluctuations (Fig. 7). The turbulence level based on the rms of the random velocity fluctuations and the averaged velocity varies between 2–3 per cent in the free-stream flow and 8–10 per cent in the vane wakes. As expected, the turbulence level is only weakly influenced by the downstream passing rotor blades.

On the contrary the potential flow effect from the downstream rotor blades on the stator outflow is remarkable. The main effect of the passing rotor blades is to depress both components of the velocity on the suction side of the stator passage, inducing a strong periodic increase of the vane wake defect. The opposite happens on the vane pressure side where the outcoming flow is periodically accelerated.

At station 3 (between rotor 1 and stator 2) the flow pattern is dominated by the wakes shed from rotor blades, identified by the inclined lines showing the velocity defect and the random fluctuations increase (Figs. 8 and 9).

In Fig. 9 nuclei of relatively large values of rms of the velocity fluctuations are located at a fixed position around $y/g = 0.75$ in the absolute frame of reference. These high turbulence flow patches are the segments of the stator vane wakes chopped by the rotor blades, transported through the rotor and finally shed by each individual rotating channel. In the relative frame these vane wake segments are located between mid-pitch and blade pressure side. According to the analysis of Zaccaria and Lakshminarayana [17] in highly loaded rotor passages the nozzle

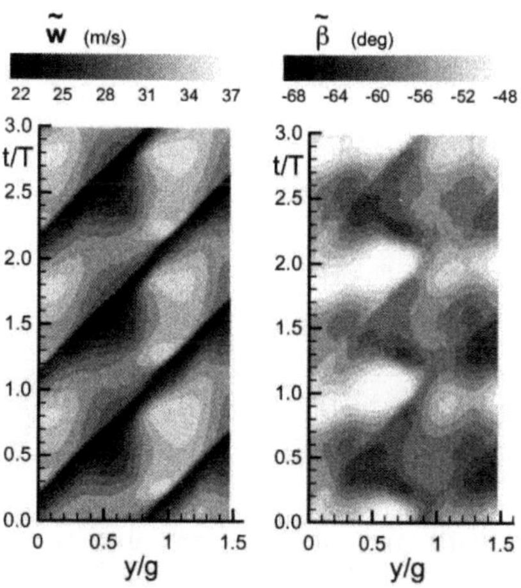

Fig. 8. Space-time plots of relative velocity and flow angle: section 3

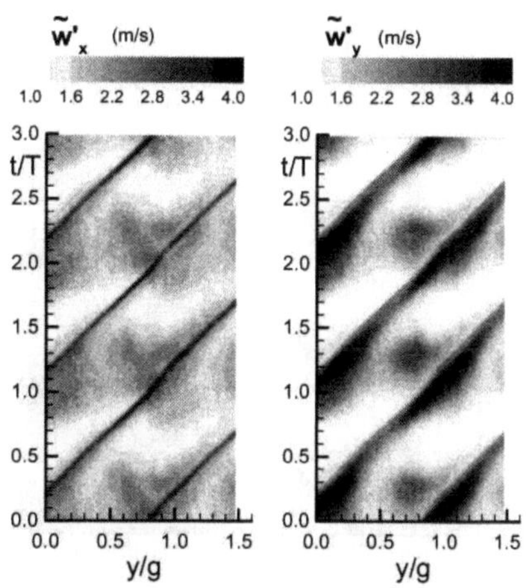

Fig. 9. Space-time plots of rms of velocity fluctuations: section 3

wake segments are re-oriented near the rotor leading edge and become nearly parallel to the rotor pressure side.

This type of wake redistribution is further confirmed by the results of recent DNS and LES numerical simulations of the wake periodically perturbed flow in a linear cascade equipped with T106 profiles [18]. The numerical results show that large levels of turbulence kinetic energy occur at the passage exit near the vane pressure side where the simulations predict that the bow apex of the chopped wake segment is located.

A second important spatial non-uniformity of the turbulence distributions is the remarkable increase of turbulence around $y/g = 1$ on the suction side of the wake. Presently it is not clear if it is due to a viscous effect (stator 1 vane wakes) or to the potential flow effect of stator 2. Looking at the relative velocity plots of Fig. 8, it comes out that these high turbulence wake regions correspond to zones with relatively weak velocity defect. The blade wake velocity defect, in fact, is non uniform in the absolute frame of reference, and a region with large velocity defect and relatively low turbulence alternates in y/g with a region of relatively low velocity defect and large turbulence.

Fig. 8 shows the high non-uniformity of the relative flow angle both in time (through the rotor passage) and in space. In general the flow changes from a more tangential direction to a more axial one from suction to pressure side through the blade wake. However the most striking feature of the flow angle distribution is the presence of nuclei of low absolute values (flow more axial) fixed in space at $y/g = 0.6$. The flow becomes more axial after the rotor blade pressure side has passed in front of the vane leading edge of stator 2 (located at $y/g = 0.3$).

A comparison between the distributions of relative yaw angle at station 3 (downstream rotor 1, Fig. 8) and at station 5 (downstream rotor 2, without any other blade row downstream, Fig. 10) confirms the significant upstream potential

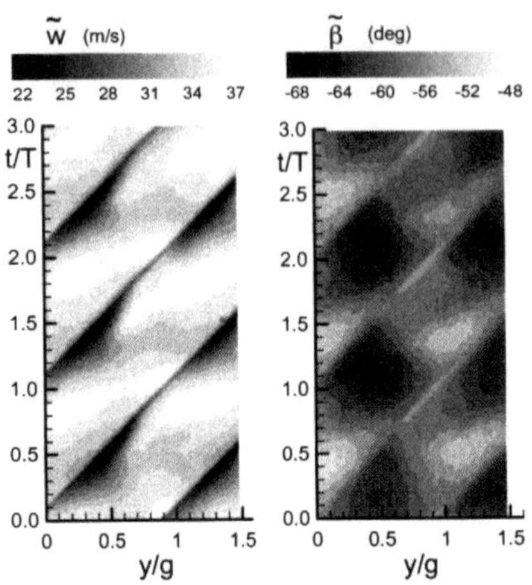

Fig. 10. Space-time plots of relative velocity and flow angle: section 5

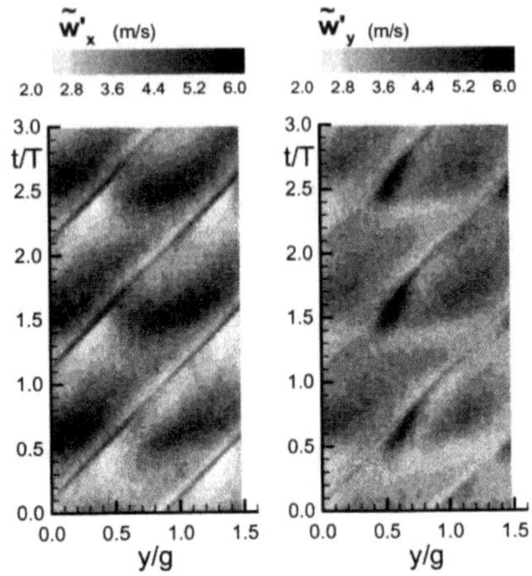

Fig. 11. Space-time plots of rms of velocity fluctuations: section 5

flow effects generated by the downstream vanes of stator 2 on the flow coming out from rotor 1. However, at station 5 it is still evident a certain circumferential non-uniformity resulting from the presence of the two stators upstream. Overall the flow leaving rotor 2 is more tangential than the flow leaving rotor 1. Since the geometry of the two rotors is identical, the downstream stator seems to influence the outflow angle and consequently also the rotor work extraction.

Comparing the relative velocity distributions of Fig. 8 and Fig. 10 it comes out that the kinematic fields downstream of the two rotors are very similar. The main difference is that at station 5 the relative velocity is moderately larger than at station 3. That is due to the combined effects of a larger endwall flow obstruction and larger absolute values of relative flow angle at station 5.

At station 5 the rms of random fluctuations (Fig. 11) are increased compared with the distributions at station 3, because turbulence is continuously produced through the stages. Turbulence of the second stage nozzle wake segments is added to the diffused background turbulence. The mixing of increased turbulence has smeared out the details of the remains of preceding blade row wakes and also the blade wakes of the second rotor are less distinguishable.

4 Conclusions

Results from an experimental investigation of the unsteady flow in a two-stage low-pressure large-scale research turbine were presented. The flow was surveyed upstream and downstream each blade row by means of hot-wire anemometry. In order to study rotor-stator aerodynamic interaction, the flow has been investigated at each measuring axial station moving the probe circumferentially over 1.5 pitches of the stators.

The potential flow unsteadiness generated by the downstream rows is remarkable and comparable to the wake generated unsteadiness. Rotor wakes present a significant spatial non-uniformity in term of turbulence level and relative velocity defect due to the combined viscous and potential flow effects.

Segments of the wakes of the preceding stator vanes can be easily identified downstream each rotor. Due to the high aerodynamic loading of the blade profiles, the segments are distorted and shifted to mid-pitch and toward the pressure side of the rotor wake streets.

Random fluctuations increase continuously through turbine blade rows. Downstream of rotor 2 the inter-blade turbulence associated to the remains of wake segments of preceding rows is comparable with rotor 2 blade wake turbulence.

In the near future investigations will be performed in order to separate the effects of wake generated and potential flow interaction mechanisms.

Acknowledgments

This work was supported by MURST (Italian Ministry of University and Scientific and Technological Research) through the PRIN 98 Research Programme.

References

1. Hodson HP (1984) Boundary layer and loss measurements on the rotor of an axial-flow turbine. ASME Journal of Engineering for Gas Turbines and Power 106: 391-399
2. Schulte V, Hodson HP (1998) Unsteady Wake-Induced Boundary Layer Transition in High Lift LP Turbines. ASME Journal of Turbomachinery 120: 28-35
3. Schroeder Th (1991) Investigations of blade row interaction and boundary layer transition phenomena in a multistage aero engine low-pressure turbine by measurements with hot-film probes and surface-mounted hot-film gauges. In: Boundary Layer in Turbomachines. Lecture Series 1991-06, VKI
4. Halstead DE, Wisler DC, Okiishi T, Walker GJ, Hodson HP, Shin HW (1997) Boundary Layer Development in Axial Compressor and Turbines. ASME Journal of Turbomachinery 119: 225-237
5. Hourmouziadis J (1989) Aerodynamic Design of Low Pressure Turbines. AGARD L. S. No. 167, pp. 8.1-40
6. Howell RJ, Ramesh ON, Hodson HP, Harvey NW, Schulte V (2000) High Lift and Aft Loaded Profiles for Low Pressure Turbines. ASME Paper 2000-GT-0261
7. Brunner S, Fottner L, Schiffer H-P (2000) Comparison of Two Highly Loaded Low Pressure Turbine Cascades under the Influence of Wake-Induced Transition. ASME Paper 2000-GT-268

8. Stieger RD, Hodson HP (2003) Unsteady Dissipation Measurements on a Flat Plate Subject to Wake Passing. Proc. 5th European Conference on Turbomachinery, Praha, Czech Republic, pp. 901-910
9 Solomon WJ (2000) Effects of Turbulence and Solidity on the Boundary Layer Development in a Low Pressure Turbine. ASME Paper No. 2000-GT-0273
10. Sieverding CH (2000) Trends in Experimental Aero-Thermal Turbine Research. Invited Lecture, 55° Congresso ATI, Bari, Italy
11. Arnone A, Cattanei A, Nurzia F, Pacciani R, Zunino P (2000) A two-stage low pressure gas-turbine model for rotor-stator aerodynamic interaction investigations. Proc. 15th Symposium on Measuring Techniques in Transonic and Supersonic Flows in Cascades and Turbomachines, Florence, Italy
12. Hoheisel H, Kiock R, Lichtfuss HJ and Fottner L (1987) Influence of Free-Stream Turbulence and Blade Pressure Gradient on Boundary Layer and Loss Behavior of Turbine Cascades. ASME Journal of Turbomachinery 109: 210-219
13. Acton P (1998) Untersuchung des Grenzschichtumschlages an einem hochbelasteten Turbinengitter unter inhomogenen und instationären Zuströmbedingungen. PhD thesis, Universität der Bundeswehr München
14. Schröder Th, Zunino P (2000) A New Method for the Calibration of the Directional Sensitivity of X-Hot-Wire Probes. Proc. XVth Bi-Annual Symposium on Measuring Techniques in Transonic and Supersonic Flows in Cascades and Turbomachines, Florence, Italy
15. Lakshminarayana B (1981) Techniques for Aerodynamic and Turbulence Measurements in Turbomachinery Rotors. ASME Journal of Engineering for Power 103: 374-392
16. Ubaldi M, Zunino P, Barigozzi G, Cattanei A (1996) An Experimental Investigation of Stator Induced Unsteadiness on Centrifugal Impeller Outflow. ASME Journal of Turbomachinery 118: 41-54
17. Zaccaria MA, Lakshminarayana B (1995) Unsteady Flow Field Due to Nozzle Wake Interaction with the Rotor in an Axial Flow Turbine: Part I - Rotor Passage Flow Field. ASME Paper 95-GT-295
18. Michelassi V, Wissink JG, Rodi W (2003) DNS, LES and URANS of Periodic Unsteady Flow in a LP Turbine Cascade: a Comparison. Proc. 5th European Conference on Turbomachinery Fluid Dynamics and Thermodynamics, Praha, Czech Republic, pp. 1185-1195

About the Onset of Partload Instability of Swept and Unswept Axial Pump Blades

Martin FORSTNER**, Helmut JABERG*
Institute for Hydraulic Fluid Machinery, Graz University of Technology
A-8010 Graz

*Corresponding author: Helmut Jaberg, o.Univ.-Prof. Dr.-Ing., Institute for Hydraulic Fluid Machinery, Kopernikusgasse 24, A – 8010 Graz, Tel. +43 316 873 7570, Fax. +43 316 873 7577, Email: Helmut.Jaberg@hfm.tugraz.at

**Present affiliation: Martin Forstner, Dipl.-Ing. Dr.techn., BIOS BIOENERGIESYSTEME GmbH, Sandgasse 47, A-8010 Graz

Abstract The partload flow field of forward, backward and unswept pump impellers was investigated in air using the hot-wire anemometry. The pump impellers were equipped on their suction side with three hot-wire anemometers at three circumferential positions with angles of sixty degrees between them allowing measurements close to the hub, the tip and in mid-channel. It was shown that each of the three cases investigated „tip stalls“ and how the unstable head curve “saddle” of forward swept blades appears at a much lower flow rate then the phenomenon for unswept or backward swept blades.
Key Words: Axial flow turbomachinery, blade sweep, rotating stall

1. Introduction

LDA measurements in the partload region of forward, unswept and backward swept blades give reason to the assumption that the stability of the flow mainly depends on conditions along the hub [1]. Forward swept blades show a very stable flow field along the hub together with a secondary flow in the rear part of the blading which is directed towards the hub. Even a local separation near the leading edge at these low flow rates seems to have no negative influence. The opposite happens on backward swept blades: A strong separation area at the hub and the leading edge is observed for extremely low flow rates. Presently our considerations indicate that the different separation behaviour of the different blade sweeps depends on the shift of the blade loading to the aft-running profiles and the assumption that the collapse of the flow field begins at the blade tip (tip-stalling). Forward swept blades have lower loading at the tip whereas backward sweep increases the blade tip loading.

This investigation clarifies whether the collapse of the flow begins at the blade tip or the blade hub and will give insight into the criteria which cause the blade stall.

2. Earlier Investigations

Partload behaviour of compressors has been in the focus of experimental (and numerical) investigations since long and so the literature is very numerous out of which only a rough overview will be given in this contribution. The different publications cited will give further insight without being outlined here in detail.

Day [2] investigates the onset of partload instability of aircraft compressors. He relates his investigations to an earlier work of Emmons et al. [3].Both authors explain the onset of blade stall with so-called “stall cells” (figure 1a) which appear in one single channel and is thus of very short circumferential extension. This is the “classical” rotating stall propagating against the direction of rotation. Stall cells are “irreversible” as short after their first appearance the lead to complete pressure break down within only a few revolutions, they always start to develop at the blade tip. Stall cells can appear in one single channel or sometimes extend also to neighbouring blade channels, where the cells with smaller circumferential extension rotate much faster than the bigger ones. Aircraft compressor as the ones investigated by Day usually have a very large number of blades as compared to our runners with just five so that we have to concentrate in our investigation on stall cells with a larger circumferential extension.

If larger circumferential flow disturbances are to be investigated the so-called “pre stall waves” or “modes” appear which have smaller variations of axial speed in the circumferential direction (figure 1b). Modes appear short before stall i.e. at a slightly larger flow rate and can be “reversible” as they can either be increased or even also disappear when the flow is throttled.

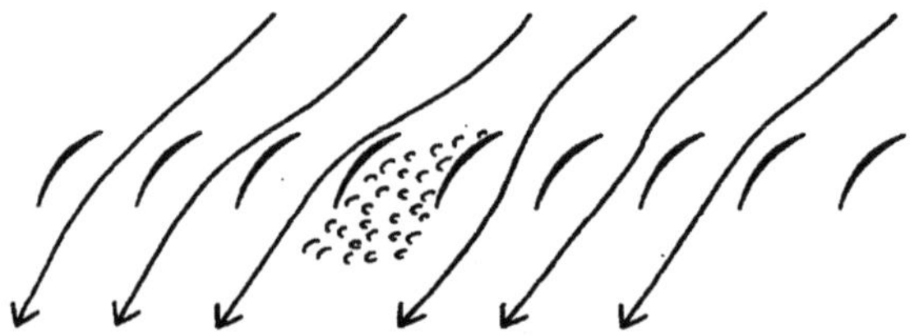

Fig. 1.a. “stall cells”

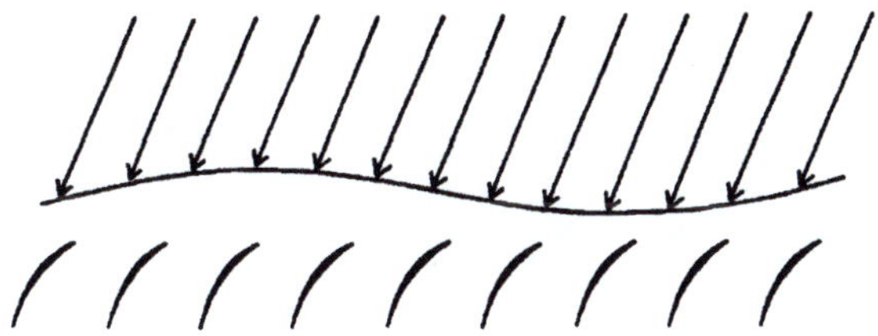

Fig. 1.b. "modes"

Garnier [4] observed modes which appeared 200 revolutions before the flow stalled. When modes appear and the flow is further decreased the intensity of the modes increases until a stall cell appears which will lead to the collapse of the flow.

Although both phenomena are basically different they appear both of them close to the stability limit of turbomachinery. But the collapse itself will always be initiated by a stall cell.

Camp and Day [5] call stall cells with very short circumferential extension „spikes“ with a rotational speed of 60 to 80 % of the runner speed, the smaller the faster. Modes on the contrary rotate with less than half the runner speed. At the highest point of the head curve characteristic all blades work close to the stability limit thus each velocity deficit which is created by a mode can lead to flow separation. It is typical that a flow disturbance close to the hub will cause a mode with a large stall cell, whereas spikes initiate at the tip.

Soundranayagam and Elder [6] investigate a blower with a lower hub to tip-ratio similar to the one of our work. It is also a free – vortex design but equipped with 27 blades. They tried to force the flow to collapse at the hub, but even very large separation areas up to 25 % of the blade height did not stall. The stability of the total flow regime obviously depends on the flow close to the tip - which is additionally influenced by the tip leakage and the outward flow of boundary layer material with low energy. Further investigations of boundary layer influence – in this case in front of the runner - is investigated by Saathoff [7] and Dobat [8] who call this phenomenon „spill forward“.

3. Experimental Set-up

The geometry of the three runners investigated is shown in figure 2 and the details of the geometry are given in table 1. The measurements were performed in air and the test bed is given in figure 3. The hub-to-tip ratio is 0.508 and the hub radius is 150 mm, the tip radius 295. The specific speed of all three blades is $n_q = 210\ \text{min}^{-1}$.

backward sweep   no sweep   forward sweep

Fig. 2. impeller geometries

Table 1. Blade geometry

specific speed	n_q	210 min^{-1}
head rise coefficient	ψ	0.298
flow coefficient	φ	0.288
hup to tip ratio	r_H/r_T	0.508
number of blades	z	5
efficiency	η_u	88%
rel tip clearance	s/D	0.071%
Rational speed	n	1440 min^{-1}
Pump head	H	30 m
Flow rate	Q	3.5 m^3/s
Radius of hub	r_H	150 mm
Radius of casing	r_T	295 mm
Reynolds number	Re	160000

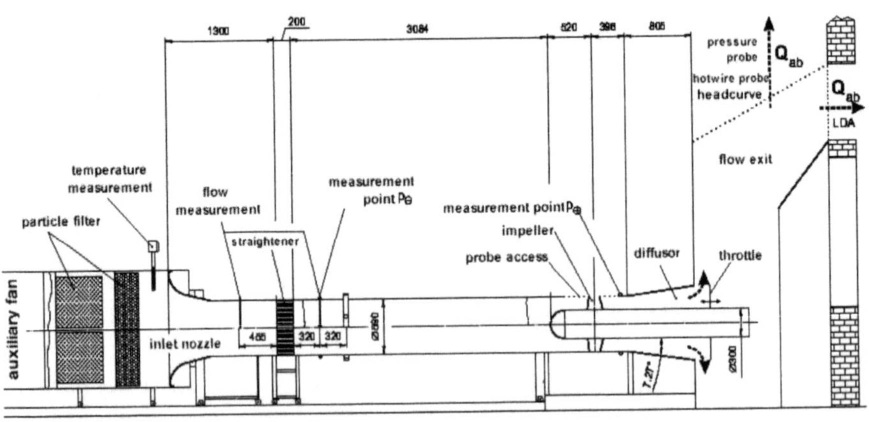

Fig. 3. Test bed

The arrangement of the hot-wire anemometers is given in detail in figure 4a and 4b which shows the meridian section, the hot wire measurements were made at position 1. (fig. 4a). The circumferential section is given in figure 4b.

In all the following figures of the measurement results the hot wire output is shown and not the velocity as usually as the case. The reason is that due to the unstationary character of the flow field close to and at instability the flow direction varies very much and is not orthogonal to the hot wire as usually is required. Furthermore the position (fig. 4a) of the hot wire is differently far away from the blade leading edge for each of the three geometries investigated. This is why we deliberately did not evaluate the velocity itself. We believe that this phenomenonlogical description of the flow field will yield the appropriate results described below.

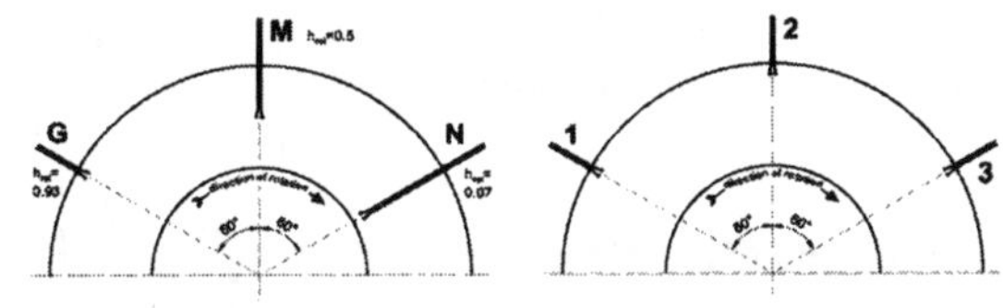

Fig. 4.a. Angular position of hot wire probes

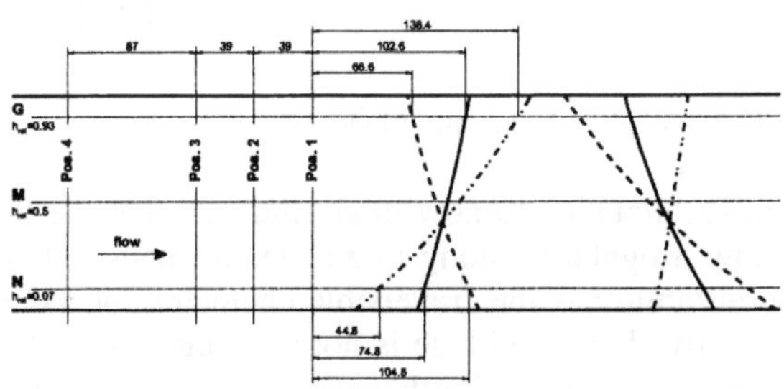

Fig. 4.b. : Hot wire probes and blade access section

4. Results

In figure 5 the results for the unswept blade is exhibited for a flow rate of 51 percent of the best efficiency point where the flow still is just stable and is than slightly throttled to provoke instability. Figure 5 shows the findings for the unswept blade. The upper line (N) gives the hot wire output at the hub, the middle line along blade mid and the bottom line the hot wire output for the blade tip.

The abscissa gives the time scale and each single rotor revolution is marked. In the beginning of the recording we see a very even course of the velocity for all three positions, although this is very distinct at mid blade and looks a bit unquiet along hub and tip, but this unquietness does not alter its shape. From revolution 7 on the flow along mid blade starts to oscillate slightly, the oscillations growing until at revolution 18 something extra ordinary may be observed at the tip where two channels show a very remarkable deterioration much stronger than any deterioration along mid blade or at the hub until this time. Three revolutions later (rev. 21) the flow collapses completely from tip over mid channel to hub. It is interesting to observe that the oscillations showing first along mid channel rotate at roughly 80 % of the rotational speed, which is found from the distance between each of the dash-dotted lines being slightly bigger than one revolution indicating that it rotates slower, thus at about 80 % of the rotor speed. The oscillations along hub and tip on the contrary only begin once the said disturbance at the tip is observed and than they directly synchronise together with the mid blade oscillations.

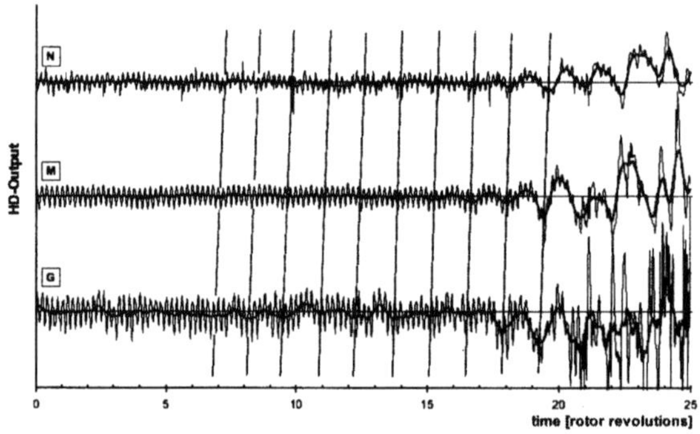

Fig. 5. Forward swept blade N: hub, M: mid blade, G: tip

Although the observations up to now might indicate that the flow instability is caused by some flow irregularity along midblade, we believe that the tip disturbance initiates the instability as the „reversible character“ of the modes to be observed before instability along midblade is not so dangerous as the obvious „stall cell“ at the tip at the revolution 18. Further support to this arguing is given by the finding for the forward and backward swept blade – besides arguments that can be found in the above cited literature.

For the backward swept blade (fig. 6) we find a comparatively unquiet flow alone the tip with different distortions with different velocity, three of them are indicated by the first three doted lines in fig. 6. These disturbances seem to level out without causing any inconvenience.

The first spike is observed in revolution 22 at the tip. But even prior to this appearance the uneven flow along the tip has formed into regular oscillations from revolution 16 on, oscillations which can also be observed along blade mid and from revolution 18 on also along the blade hub. These oscillations rotate with roughly 50 % of the rotor speed until the spike appears at revolution 22.

Looking more closely to the flow measurements (fig. 6 below) we can nicely see that the disturbance of the first spike propagates into midchannel about one revolution later and than needs a bit longer to extend to the hub.

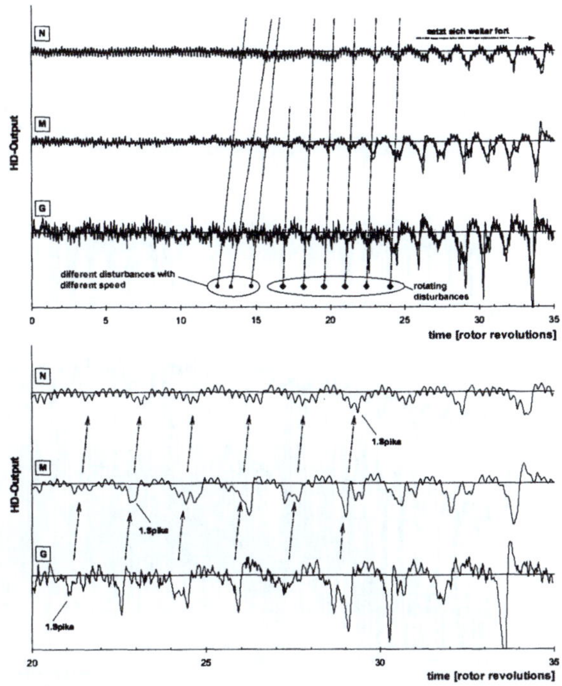

Fig. 6. Backward swept blade N: hub, M: mid blade, G: tip

The forward swept blade is investigated at a flow rate of 44% of best efficiency flow rate where the lower value was chosen because it is the onset of head curve instability. The flow field (fig. 7) is characterised by oscillations throughout the impeller from hub to tip. That they are stronger than the ones observed before for the unswept or backward swept impeller is certainly due to the fact that at this lower flow rate the difference between optimum and actual angle of attack is even bigger than in the other two runners with different points of operation.

Obviously at revolution 36 a distinctive spike is again observed at the tip of the blade. During the revolutions before this event we see along the blade tip the formation of modes and also at blade mid and at the hub. But only after appearance of the spike at the blade tip (rev. 36) the head curve instability appears also. In the lower part of figure 7 the result of the revolution 36 event is shown. We see (as also in the furthest right part of fig. 7 above) that the spike-like decrease of the meridional velocity at the tip increases the meridional velocity at blade mid and hub and in the further history of the flow in this impeller also at the blade tip. This increase of the meridional velocity as the flow rate decreases is a consequence of the insensitivity of the hot wire anemometer to changes of the flow direction. From the shape of fig. 7 below it can be concluded that the separation area extends from tip to blade mid but not down to the blade hub, where the flow shape is naturally seriously affected but by far not to the same extend as at the other two locations.

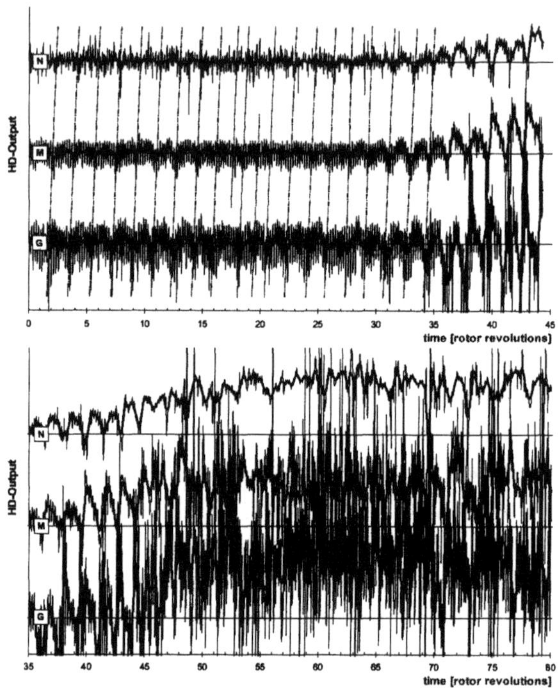

Fig. 7. Forward swept blade N: hub, M: mid blade, G: tip

5. The rotating of the stall

As all three blades can be regarded „tip stalling" the extension of the rotating stall in circumferential direction can be investigated by concentrating on the tip side flow region. Therefore the three hot wire anemometers were all positioned at 93% of the blade height and at an angular position of 60° away from each other (fig. 4b).

Looking at the unswept blade first (fig. 9) a very small disturbance (rev. 12) increases as well as intensity or size is concerned. Within two revolutions (rev. 14) the total pressure as produced by the runner decreases with 21 Pa/rev [9]. 54 further revolutions onwards these disturbances produce a total collapse of the flow (rev. 20) where no patterns what so ever can be observed. Modes could neither be observed during the hole development of this stall. It should be noted that the first disturbance that is stretched over more than one anemometer (and thus over more than one blade passage) may be observed at revolution 11. All disturbances before – if they are any – are limited only to one blade passage and are only very weak. It must also be born in mind that all these observations are made close to the tip at 93% blade height.

The backward swept blade exhibits some thirty revolutions before break down (fig. 8) very small disturbances which rotate with 20 to 70 % of the rotor speed and extend over small angular regions as they can only be detected either at anemometer 1 or 2 or 3. These observations can be seen on the left part of figure 8 and can only exceptionally be connected by a single line which is a very vivid token that they are very limited in the angular direction. Even the ones which are connected fade out. The disturbance that finally leads to the stall (rev. 26) can be observed 9 to 10 revolutions before and during this time span, their strength seems to vary but the frequency remains the same. Periodic oscillations of the whole system can be excluded as a source for these disturbances [9]; modes in the sense of reversible oscillations which might appear up to 200 revolutions before stall can not be detected as the before mentioned disturbances (left portion of fig. 8) are to much limited in circumferential direction. The oscillations before the first spike (fig. 6) appear so directly before stall (0.2 seconds) that they are regarded a stall effect. From revolution 27 on the pump head curve drops with a gradient of 5 Pa/rev and the stall extends circumferentially over two or three blade passages.

The forward swept blade (fig. 10) shows pronounced rotating oscillations long before stall, much more than 200 revolutions, where every 20 revolutions "irregularities" appear as shown in fig. 7 (rev. 17 – 20). The oscillations which finally lead to the collapse of the flow can be observed some 50 revolutions long without any interruption. These oscillations are „reversible" as they appear when the throttle is further wide open as in the moment of the collapse. By further opening the throttle also operating points without any oscillations can be adjusted. Thus we conclude that these oscillations are in fact the „modes" as described by Camp and Day [2,5] and sketched in paragraph 3 of this work. The rotational speed of all these modes is roughly 58% of the rotor speed. The amplitude of the modes increases and forms a stall cell which first hits only one single channel and leads to the collapse after one further revolution. Again this collapse happens all of a sudden and without any remarkable rotating patterns. In fig. 10 this increasing amplitude can first be detected at revolution 42 at anemometer two (middle line) than at anemometer 3 (top line) and than anemometer 1 (bottom line). The drop of the pressure head curve which goes in parallel to these observations in the runner is measured with a gradient of just 1.6 Pa/rev.

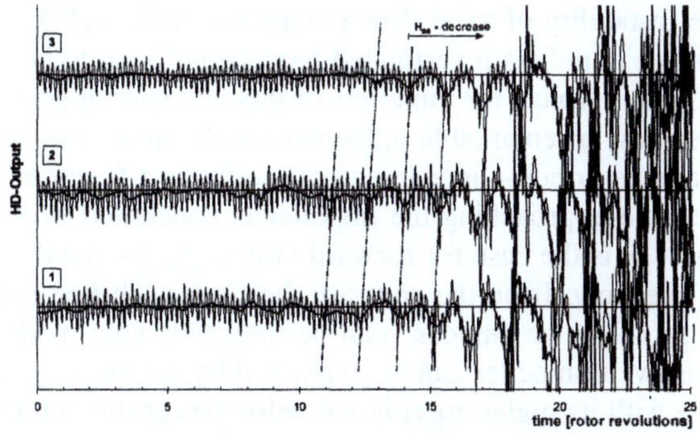

Fig. 8. Unswept blade: Angular position of stall cells

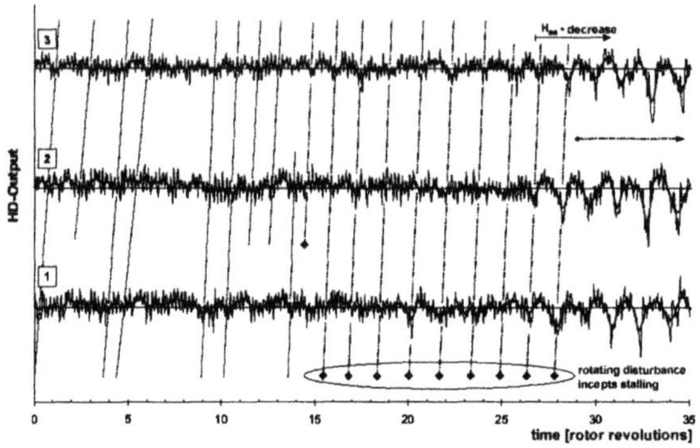

Fig. 9. Backward swept blade: Angular position of stall cells

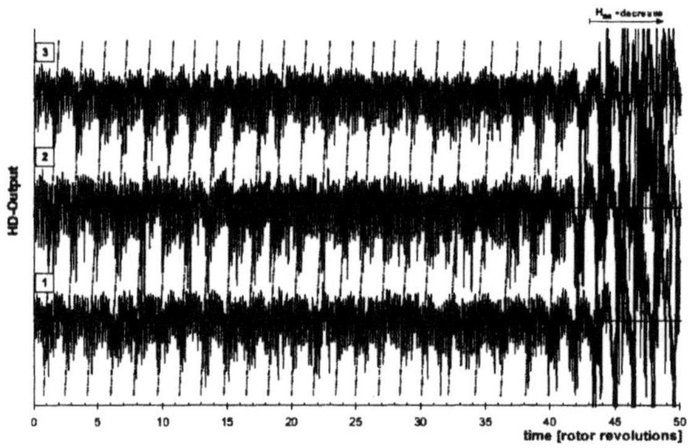

Fig. 10. Forward swept blade: Angular position of stall cells

6. Conclusions

The head curve instability of axial flow pumps is – as is well known – created by stall cells. All three impellers investigated have been found to be "tip stalling" and we believe that this is typical for impellers of this and similar geometries.

Our findings show a remarkable agreement with earlier measurements [2,5] as far as the appearance or not – appearance of modes as a function of the radial distribution of the fluid approaching the impeller is concerned. If the flow is shifted closer to the tip – as is the case for forward sweep [1,9] - „modes" appear prior to the collapse of the flow. If on the contrary the flow is shifted to the hub – as for backward sweep [1,9] – no „modes" can be detected. That modes can not be observed for the unswept impeller can be explained by the velocity profile in front of the leading edge with its higher meridional velocities at the hub than the ones presumed as design conditions with cm (r) = constant [9]. The short oscillations of roughly half a second prior to the collapse were not observed by Camp and Day

[5] for which the reason is most probably the very different number of blades. In our case the blockage of one channel is equivalent to 20% of the circumference, for the higher blade numbers of Camp and Day it was just 2%.

The loss of the head curve proceeds with different gradients for the impellers investigated. This steepest gradient (21pa/rev.) appears for the unswept blade for which the flow collapses at the highest point of the head curve (fig. 11). The flow in the backward swept impeller stalls already when the head curve still increases, but the drop is not so steep (5pa/rev.). The comparatively stable flow along the hub has most probably a stabilising influence. The still increasing head curve with stalling flow at the tip of a backwards swept impeller means that most of the work is done by the flow closer to the hub.

For the forward swept blade the collapse of the flow appears right after the head curve has started to decrease. The total pressure loss which results in this said drop of the head curve is obviously caused by the reversible oscillations that appear long before collapse of the flow inside the impeller. Although the forward swept blade exhibits these strong but „reversible“ oscillations the head curve keeps rising until much lower flow rates (fig. 11) than the forward swept or unswept impeller. So in spite of these oscillations the forward swept blade is the best geometry which should have potential for further optimisation by a better leading edge design which should lower these pre-stall oscillation and their consequence.

Conclusively it may be stated that our measurements confirm well the model by Camp and Day [5]. The shift of the velocity profile and thus a change of the incidence angle at the blade tip – both as a consequence of blade sweep – are responsible for the different behaviour and the partload instability of these impellers.

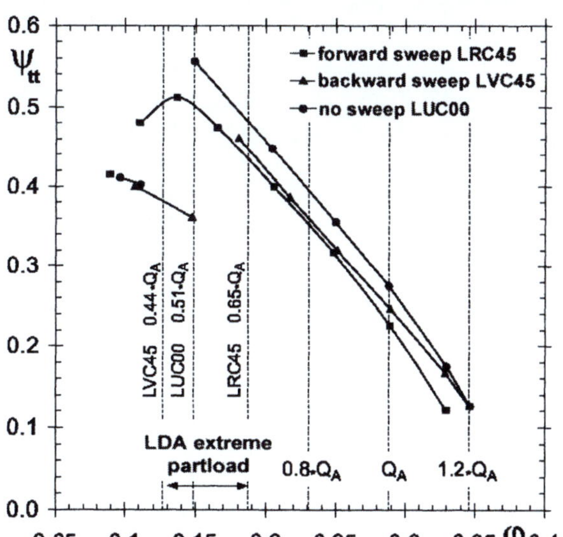

Fig. 11. Head curve characteristic for forward, backward and unswept impeller

References

[1] Forstner, M., Kuhn, K., Glas, W. and Jaberg, H., 2001, The flow field of pump impellers with forward and backward sweep. Proc. 4th European Conference on Turbomachinery - Fluid Dynamics and Thermodynamics, Florence, Italy, pp.. 577 –588.

[2] Day, I. J., 1993, Stall inception in axial flow compressors, ASME Journal of Turbomachinery 115, pp. 1-9.

[3] Emmons, H. W., Pearson, C. F. and Grant, H. P., 1955, Compressor surge and stall propagation, Transactions of the ASME 79.

[4] Garnier, V. H, Epstein, A. H. and Greitzer, E. M., 1991, Rotating waves as a stall inception indication in axial compressors, ASME Journal of Turbomachinery 113, pp. 290 – 302.

[5] Camp, T. R. and Day I. J., 1998, A Study of spike and modal stall phenomena in a low – speed axial compressor, ASME Journal of Turbomachinery 120, pp. 393 – 401.

[6] Soundranayagam, M. and Elder, R. L., 1993, A study of stall in a low hub-tip ratio fan, ASME Journal of Turbomachinery 115, pp. 10-18.

[7] Saathoff, H. and Stark, U., 2001, Tip clearance flow on a low-speed compressor cascade. Proc. 4th European Conference on Turbomachinery - Fluid Dynamics and Thermoynamics, Florence, Italy, pp. 81-91.

[8] Dobat, A., Saathoff, H. and Wulf, D., 2001, Experimentelle Untersuchung zur Entstehung von Rotating Stall in Axialventilatoren, in: Ventilatorenentwicklung – Planung – Betrieb, Braunschweig, pp. 345- 360, VDI-Verlag, Düsseldorf.

[9] Forstner, M., 2002, Experimentelle Untersuchungen an vor- und rückwärtsgepfeilten Axialpumpenschaufeln, Dissertation, Technische Universität Graz.

Mixing interface algorithm for 3D turbulent flow analysis of the GAMM Francis turbine

Sebastian MUNTEAN
Center of Advanced Research Engineering Sciences, Romanian Academy-Timisoara Branch
Bv Mihai Viteazu 24, 300223 Timisoara, Romania
Tel : (+40) 256 403692
Fax : (+40) 256 403700
E-mail : seby@mh.mec.utt.ro

Romeo F. SUSAN-RESIGA, Ioan ANTON
Department of Hydraulic Machinery "Politehnica" University of Timisoara

Abstract The paper presents a methodology for computing the 3D turbulent flow in both distributor and runner of Francis turbines. The 3D computational domains correspond to interblade channels for the GAMM Francis turbine distributor and runner, respectively. In order to couple the steady absolute distributor flow field with the runner steady relative flow, a mixing interface technique is developed and employed on the conical distributor-runner interface. Our mixing algorithm removes the circumferential variation of velocity components, pressure and turbulence quantities using a piecewise polynomial least squares algorithm. Comparison of numerical results with available experimental data validates the methodology and assesses its accuracy.
Key Words: Francis turbine, mixing interface method, turbulent flow

Nomenclature

$c_r = V_r \big/ \sqrt{2E_{ref}}$	[-]	radial velocity coefficient
$c_u = V_u \big/ \sqrt{2E_{ref}}$	[-]	tangential velocity coefficient
$c_z = V_z \big/ \sqrt{2E_{ref}}$	[-]	axial velocity coefficient
$c_m = \sqrt{c_r^2 + c_z^2}$	[-]	meridian velocity coefficient
$c_p = (p - p_{ref})/(\rho E_{ref})$	[-]	pressure coefficient

g	[m/s^2]	gravity
k	[m^2/s^2]	turbulent kinetic energy
p	[Pa]	static pressure
l	[m]	turbulence length scale
$s = length/R_{ref}$	[-]	curvilinear abscissa
$y+$	[-]	turbulent boundary layer non-dimensional parameter
V	[m/s]	absolute velocity
W	[m/s]	relative velocity
$E_{ref} = gH_{ref}$	[J/kg]	specific hydraulic energy
H	[m]	net head
Q	[m^3/s]	discharge
R	[m]	radius
I	[%]	turbulence intensity
ρ	[kg/m^3]	water density
ε	[m^2/s^3]	turbulent dissipation rate
ω	[rad/s]	angular velocity
$\varphi = Q/\pi\omega R^3_{ref}$	[-]	discharge coefficient
$\psi_{ref} = 2E/\omega^2 R^2_{ref}$	[-]	energy coefficient
$\nu = \varphi^{0.5}/\psi^{0.75}$	[-]	specific speed
μ	[Pa.s]	dynamic viscosity
μ_t	[Pa.s]	turbulent dynamic viscosity
λ	[-]	relaxation parameter

Subscripts and Superscripts

r	radial direction
u	tangential direction
z	axial direction
ref	reference section (draft tube inlet section)
in, *out*	inlet section, outlet section

1. Introduction

Developments in computer software and hardware made possible the computation of three-dimensional flows in turbomachines, [10]. However, computing the real flow (viscous and turbulent) through the whole hydraulic turbine requires large computer memory and CPU time even for the present day computers. As a result, a simplified simulation technique must be employed to obtain useful results for turbine analysis, using currently available computing resources.

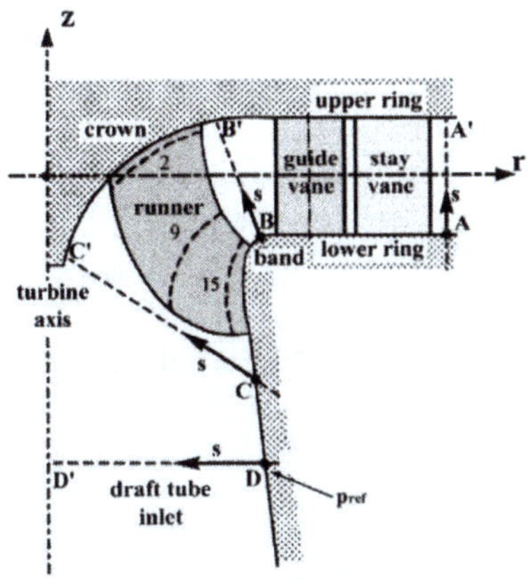

Fig. 1. Meridional cross section through the GAMM Francis turbine model with the flow survey axes [1, 9]

The turbomachinery flow is essentially unsteady due to the rotor-stator interaction. On the other hand, rigorously speaking, the geometrical periodicity of the stator/rotor blade rows cannot be used since there are differences in flow from one inter-blade channel to another. However, with carefully chosen and experimentally validated assumptions, one can devise a methodology for computing the turbine flow, such that very good and engineering useful results are obtained [5, 6, 7, 8].

This paper presents such a methodology, and validates the numerical results with experimental data for the GAMM Francis turbine model, [1, 9]. The model corresponds to a medium/high specific speed Francis turbine with specific speed $\nu = 0.5$. The distributor consists of 24 stay vanes and 24 guide vanes whilst the runner has 13 blades with its external diameter 0.4 m, so the reference radius is $R_{ref} = 0.2$ m.

The distributor (stay vanes and guide vanes) computational domain corresponds to an inter-blade channel, bounded upstream by a cylindrical patch generated by AA' and downstream by a conical patch BB', Fig 2. The distributor inlet section corresponds to the spiral casing outlet section, while the outlet section is conventionally considered to be the distributor-runner interface.

Fig. 1 shows the GAMM Francis turbine meridional view with the survey axes used to investigate the velocity and pressure fields.

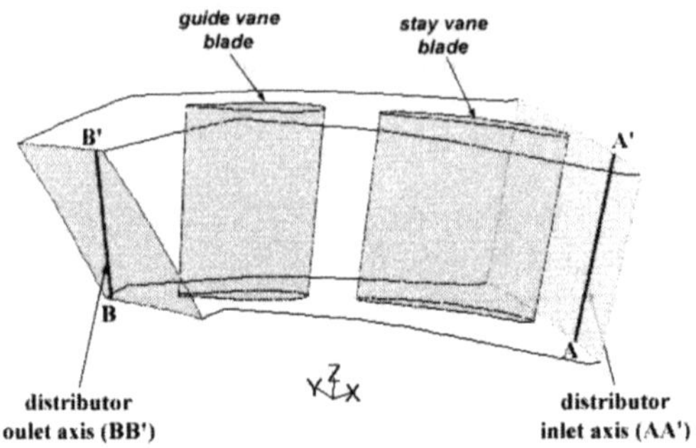

Fig. 2. Three-dimensional view of the GAMM Francis turbine model distributor domain

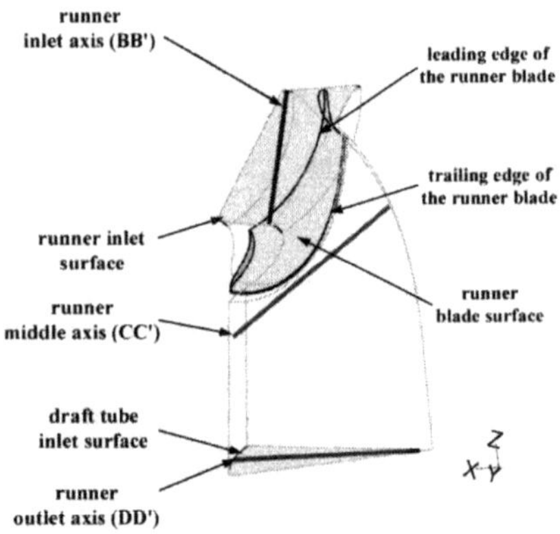

Fig. 3. Three-dimensional view of the GAMM Francis turbine model runner domain

The runner computational domain also corresponds to an inter-blade channel bounded upstream by a conical patch (wrapped on the same conical surface as the distributor outlet) and extended downstream up to the draft tube inlet of radius DD', Fig 3.

The boundary conditions for both distributor and runner computational domains are presented in section 2.

In this paper, we introduce a mixing interface technique for coupling the distributor and runner velocity and pressure fields as well as the turbulence quantities. This approach performs a circumferential averaging of the velocity components and pressure while the turbulence quantities (turbulence intensity and turbulence length scale) on the distributor outlet are averaged on the whole section. The iterative process employed for achieving continuity for both absolute velocity and pressure across the distributor-runner interface is presented in section 3.

This circumferential averaging technique, outlined in section 4, was carefully chosen in order to preserve the velocity and pressure profile shapes on the mixing interface. Two additional requirements are satisfied by our averaging (mixing)

technique: the sharp velocity gradients near solid surfaces are correctly represented and the global discharge is conserved during the iteration process.

Our numerical results and available experimental data are compared in section 5. First, we examine the velocity and pressure distributions on the mixing interface. Next, the pressure distribution on the runner blade is investigated in three representative sections where data are available.

The paper conclusions are summarized in section 6.

2. Absolute and relative flow equations. Boundary conditions

The governing equations for the incompressible and viscous flow are:

$$\nabla \cdot \boldsymbol{V} = 0, \tag{1a}$$

$$d(\rho \boldsymbol{V})/dt = \rho \boldsymbol{g} - \nabla p + \mu \Delta \boldsymbol{V} . \tag{1b}$$

The left hand side in the momentum equation has the following expression in the inertial frame of reference:

$$\partial(\rho \boldsymbol{V})/\partial t + \nabla \cdot (\rho \boldsymbol{V}\boldsymbol{V}). \tag{2}$$

When the equations of motion are solved in a rotating frame of reference, the relative velocity is introduced,

$$\boldsymbol{W} = \boldsymbol{V} - \boldsymbol{\omega} \times \boldsymbol{r} , \tag{3}$$

and the left-hand side of the momentum equation can be written as:

$$\partial(\rho \boldsymbol{W})/\partial t + \nabla \cdot (\rho \boldsymbol{W}\boldsymbol{W}) + 2\boldsymbol{\omega} \times \boldsymbol{W} + \boldsymbol{\omega} \times (\boldsymbol{\omega} \times \boldsymbol{r}) + \rho\, \partial \boldsymbol{\omega}/\partial t \times \boldsymbol{r} , \tag{4}$$

where Coriolis and centripetal terms are included.

In this case, we are using the FLUENT code [3] to solve the Reynolds-averaged Navier-Stokes (RANS) equations

$$\partial(\rho U_i)/\partial x_i = 0 \tag{5a}$$

$$\begin{aligned} \partial(\rho U_i)/\partial t + \partial(\rho U_i U_j)/\partial x_j = \rho g_i - \partial p/\partial x_i + \partial\left[(\mu + \mu_t)\partial U_i/\partial x_j\right]/\partial x_j \\ - \rho \in_{ijk} \in_{klm} \omega_j \omega_l x_m - 2\rho \in_{ijk} \omega_j U_k \end{aligned} \tag{5b}$$

where U is the generic symbol for absolute or relative velocity, $-\rho \in_{ijk} \in_{klm} \omega_j \omega_l x_m$ is the centripetal term and $-\rho \in_{ijk} \omega_j U_k$ is the Coriolis term, due to the rotating coordinate system.

The turbulent viscosity, μ_t, is evaluated using the standard k-ε model. The transport k-ε model equations for the turbulent kinetic energy, k, and the dissipation rate, ε, read

$$\partial(\rho k)/\partial t + \partial(\rho k U_i)/\partial x_i = P - \rho\varepsilon + \partial\left[(\mu + \mu_t/\sigma_k)\partial k/\partial x_j\right]/\partial x_j , \tag{6a}$$

$$\partial(\rho\varepsilon)/\partial t + \partial(\rho\varepsilon U_i)/\partial x_i = (C_{1\varepsilon}P - C_{2\varepsilon}\rho\varepsilon)\varepsilon/k + \partial[(\mu + \mu_t/\sigma_\varepsilon)\partial\varepsilon/\partial x_j]/\partial x_j, \tag{6b}$$

where P is kinetic energy production and the turbulent viscosity, μ_t, is computed by combining k and ε as follows:

$$\mu_t = \rho C_\mu k^2/\varepsilon \tag{7}$$

where the constants are $C_{1\varepsilon}=1.44$, $C_{2\varepsilon}=1.92$, $C_\mu=0.09$, $\sigma_k=1.0$ and $\sigma_\varepsilon=1.3$.

The distributor computational domain is discretized with a structured mesh (250250 hexahedral cells) while the runner domain, due to its three-dimensional complex shape, is discretized using unstructured meshes (282325 tetrahedral cells), [4]. The grid resolves the boundary layer to have $30 \leq y^+ \leq 100$ and the non-equilibrium wall function is used.

The method developed in this paper is designed to solve only for steady flows (absolute or relative), and therefore the partial time derivatives vanish.

The absolute flow equations are the natural choice for the distributor, but for the runner one may choose to solve either the absolute or the relative flow. Since our goal is to develop an iterative technique to couple the distributor and runner hydrodynamic fields, it is convenient to use absolute velocity conditions at the runner inlet section.

The boundary conditions for computing the flow within a solution domain are as follows:

- velocity field is prescribed on the inflow section for both distributor and runner domains. The velocity profile prescribed on the distributor inlet section is the measured profile adjusted to satisfy the specified discharge, [2]. The mixed velocity profile on the distributor outlet section is imposed on the runner inlet section. The mixing algorithm is presented in section 4;
- turbulence quantities (turbulence intensity and turbulence length scale) are prescribed on the inflow section for both distributor and runner domains. For the computations made in this paper, a turbulence intensity of 6% and a turbulence length scale of 0.14 of the inlet distributor height are assumed on the distributor inlet section. Also, at each iteration the turbulence quantities (turbulence intensity and turbulence length scale) are imposed on the runner inlet section. This values are computed as follows:

$$I = 100\sqrt{2k/3V^2}\ [\%] \tag{8}$$

and the turbulence length scale

$$l = C_\mu^{3/4} k^{3/2}/\varepsilon \tag{9}$$

the I and l are averaged over the distributor outlet section;
- pressure distribution is imposed on the distributor outlet section. At the first iteration, an arbitrary (e.g. constant, or experimental when available) outlet pressure distribution is used to initialize the iterative process;
- on the runner outlet (draft tube inlet) pressure conditions should be specified. However, both numerical and engineering considerations lead to the conclusion

that this is not the best choice. Numerical experiments have shown that enforcing a measured pressure distribution at the runner outlet produces spurious recirculation after the flow leaves the blades, in disagreement with the measured velocity field. On the other hand, from the engineering viewpoint, neither velocity nor pressure distributions are usually known in industrial practice. As a result, one should employ a condition that is a velocity pressure relationship. Such condition can be devised by assuming that in this section there is no radial flow, i.e. $V_r = 0$ or negligible. The radial projection of the momentum equation becomes,

$$\partial p / \partial r = \rho V_u^2 / r \,. \tag{10}$$

This is the so-called *radial equilibrium outlet condition.* However, this condition defines the pressure up to an additive constant. Therefore, a pressure value should be specified at $r = 0$. In consequence, radial equilibrium condition Eq. (10) is prescribed on the runner outlet section. This assumption was used with excellent results for inviscid flow, [6, 7];

- periodic conditions are imposed on the periodic boundary;
- wall conditions are imposed on the stay, guide, and runner blades, as well as on the distributor upper/lower rings and runner band and crown, respectively.

The computational domains are discretized with 3D (hexahedral and tetrahedral cells) [4], and a segregated solver is used within the FLUENT code, [3].

3. Distributor-runner coupling algorithm

Since the flow computation is performed separately for the distributor and runner domains, a coupling technique is required in order to obtain a continuous velocity and pressure fields across the conical surface generated by BB'. In this paper, an iterative coupling method is employed, with the algorithm outlined below, see Fig 4.

First, the distributor flow is computed, using the inlet velocity profile [2], and an arbitrary (e.g. constant, or experimental when available) outlet pressure distribution. Next, iterations are performed as follows:

- compute the runner flow, using the inlet velocity distribution obtained at the distributor outlet, and the radial equilibrium outlet condition for outlet pressure;
- compute the distributor flow, keeping the inlet velocity constant and using the outlet pressure distribution obtained at the runner inlet.

The stopping criterion is that the pressure distribution on the distributor-runner interface is practically unchanged from one iteration to another.

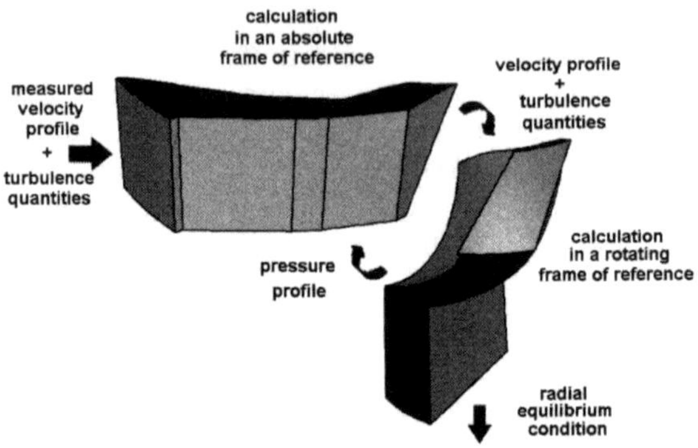

Fig. 4. Mixing interface algorithm

Under-relaxation is used for both velocity and pressure, i.e,

$$\left(p_{out}^{i}\right)_{distributor} = \lambda\left(p_{in}^{i-1}\right)_{runner} + (1-\lambda)\left(p_{out}^{i-1}\right)_{distributor} \tag{11a}$$

$$\left(V_{in}^{i}\right)_{runner} = \lambda\left(V_{out}^{i-1}\right)_{distributor} + (1-\lambda)\left(V_{in}^{i-1}\right)_{runner} \tag{11b}$$

The relaxation parameter value was set as $\lambda=0.5$, but further investigations can be carried out in order to speed up the convergence. Numerical investigations have shown that 7...9 iterations lead to an acceptable convergence from an engineering viewpoint, provided that a good pressure distribution is used for the first distributor computation.

The chief difficulty in using the above algorithm lies in transferring data from distributor to runner and vice versa. When looking at the Figs 2 and 3, one can easily observe that the distributor outlet surface do not match the runner inlet surface, although both lie on the same cone. Moreover, the distributor has 24 blades, while the runner has only 13. The problem is even more complicated by the use of an unstructured mesh.

As a result, assuming a steady relative runner flow requires that there is no circumferential variation in both velocity and pressure. This assumption actually neglects the unsteady interaction between guide vane wakes and runner blades.

On the other hand, the FLUENT code requires that a so-called *radial profile* should be provided in cylindrical coordinate system. This is a set of ordered points (e.g. (r_i, p_i), $i=1...n$) with radius values starting at $r_{B'}$ and ending at r_B for the BB' axis.

The above considerations led to the following procedure for processing the interface data:

- first, the numerical values (for velocity components or pressure) are plotted against the radius (measured from the turbine axis) for all mesh points on the distributor outlet or runner inlet surface;

- second, a piecewise polynomial least square regression is performed, such that a best average value is obtained at each radius. The piecewise polynomial least square technique is outlined in next section;
- third, for the velocity components a correction is performed such that the prescribed discharge is conserved, without altering the flow angles.

Since this approach performs a circumferential averaging, it is equivalent to the full mixing of the wakes (or any other circumferential non-uniformities). As a result, it is known as the "*mixing interface method*".

In order to implement the above algorithm, we have developed a FORTRAN code, using the IMSL library. The procedure is called at each iteration step, performing the "mixing", discharge correction, and relaxation.

4. The mixing algorithm

Let us assume that all data on the distributor-runner interface (either velocity components or pressure) are given as sets $(x_i, f_i), i = 1,...,N$, with N the number of nodes on the interface. These sets are not necessarily ordered, and may have several f values at the same x. The mixing problem is to find a function $f(x)$, which preserves the corresponding velocity or pressure profile shape, and is preferably smooth. Since no particular shape can be apriori prescribed for $f(x)$, it seems that a polynomial regression might be a reasonable choice. However, it has been found during our preliminary numerical experiments that a good data approximation requires high order polynomials (degree five or higher), which in turn displays spurious oscillations. As a result, a least squares piecewise polynomial has been found to be the best choice.

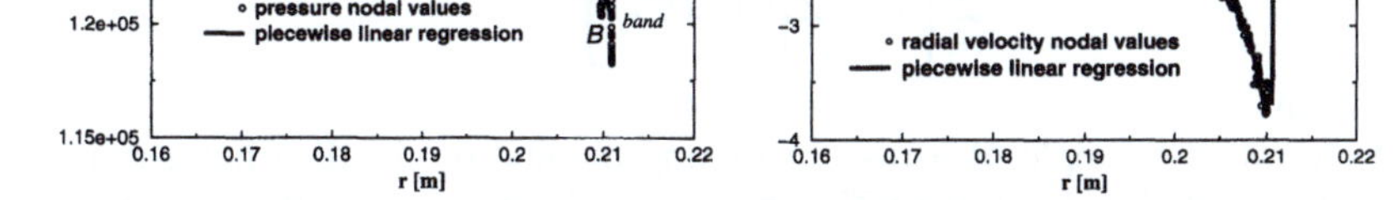

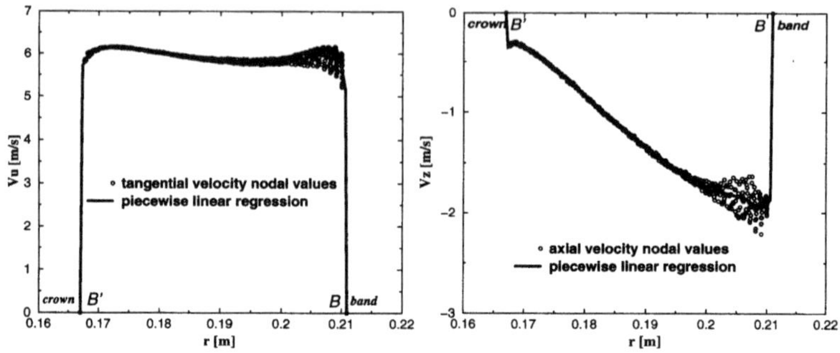

Fig. 5. The pressure and velocity components on the mixing interface. Solide lines are piecewise linear least squares fits

B-splines provide a particularly convenient and suitable basis for a given class of smooth piecewise polynomial functions. Such a class is specified by giving its breakpoint sequence, its order *k*, and the required smoothness across each of the interior breakpoints. The corresponding B-spline basis is specified by giving its knot sequence $t \in R^M$. Assuming that x_1, and x_N are the endpoints of the interval of interest, one chooses the first k knots equal to x_1 and the last *k* knots equal to x_N. This can be done since the B-splines are defined to be right continuous near x_1 and left continuous near x_N.

When the above construction is completed, we will have generated a knot sequence *t* of length *M*, and there will be $m=M-k$ B-splines of order k, say B_1, ..., B_m. That is, each piecewise polynomial function has a unique representation

$$f(x) = a_1 B_1(x) + \ldots + a_m B_m(x)$$

as a linear combination of B-splines. B_i is a nonnegative function that is nonzero only on the interval $[t_i, t_{i+k}]$.

The IMSL routine BSLSQ computes a weighted discrete approximation from a spline subspace to a given data set $(x_i, f_i), i = 1, \ldots, N$. In other words, it finds the B-spline coefficients a_1, ..., a_m, such that

$$\sum_{i=1}^{N} \left(f_i - \sum_{j=1}^{m} a_j B_j(x_i) \right)^2$$

is a minimum.

Fig 5 shows an example of using our mixing method on the distributor-runner interface. The first plot shows the 897 nodal values for pressure on the runner inlet section. The solid curve corresponds to the piecewise linear regression computed with the above algorithm using 25 break points. This is further used as distributor outflow boundary condition for the next iteration.

The next three plots display the velocity components V_r, V_u and V_z, each of them having 564 nodal values on the distributor outflow section. The solid curve corresponds to the piecewise linear regression computed with the above algorithm. These curves are further used at the next iteration step as velocity conditions on the runner inflow section. Special care should be taken to preserve the correct flow rate value, Q, prescribed for the operating point under investigation. As a result, after performing the mixing step, we compute the flow rate,

$$\tilde{Q} = 2\pi \int_{R_{min}}^{R_{max}} (V_r ctg\alpha + V_z) r\, dr ,$$

then we multiply the "mixed" velocity components by $Q/\tilde{Q}$. Here α denotes the semi-angle of the cone corresponding to the mixing section, and R_{min}, R_{max} are the radii where the cone intersects the turbine crown and bend, respectively.

Our numerical experiments showed that $\tilde{Q}$ is no more than 0.5% off the prescribed value Q.

5. Numerical results

In order to validate our methodology and to assess the accuracy of the numerical results, we have performed extensive comparisons with available experimental data. Fig 6 shows the velocity components and pressure coefficients on the distributor/runner interface, after the iterative procedure has converged. One can see an excellent agreement between numerical velocity components and the measured values. This indicates a correct velocity field at runner inlet. The computed pressure distribution agrees well with experimental data. We obtained the turbulence intensity of 2 % and turbulence length scale of 0.5 mm on the mixing interface. Since the present study is performed at the best efficiency point, one can see that only a slight improvement is achieved by considering a turbulent flow in comparison with our former inviscid calculations [6, 7].

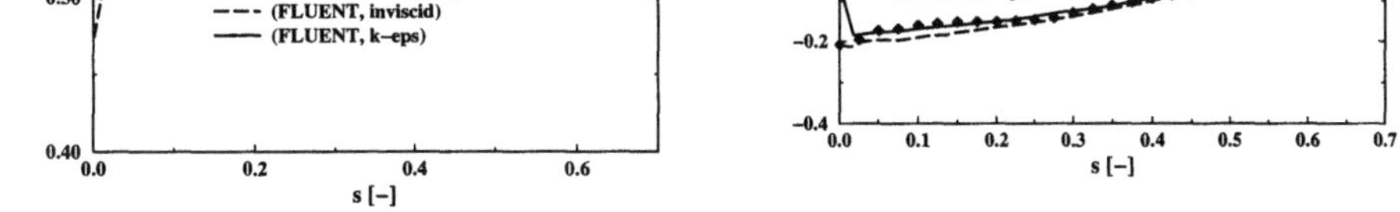

Fig. 6. The pressure and velocity coefficients on the distributor outlet/runner inlet (mixing interface) at best efficiency operating point

Next, we investigate the pressure distribution on the runner blade, Fig 7. Near the crown, S2 section, one can see that both turbulent and inviscid computations give reasonable results on the pressure side, while failing to predict the minimum pressure on the suction side, near the trailing edge.

A good agreement between our computations and the experimental data is obtained on the mid section S9. A more significant departure of numerical results from the measured values can be observed on the blade suction side on the section S15 in the band neighborhood. This is an interesting result, since the flow angle on the runner inflow section, near the band, is correctly predicted, Fig 6. Good results for suction side of S15 can be obtained if the flow angle at the runner inlet is slightly altered, as shown by Nilsson [8].

However, we consider that our methodology offers the correct runner inflow velocity field by computing the distributor and runner flows.

6. Conclusions

The paper presents a methodology for computing the 3D incompressible turbulent flow in a Francis turbine. The computational domain corresponds to the distributor and runner interblade channels.

As a result, steady absolute flow is assumed in the turbine distributor, and a steady relative flow is considered for the runner. Only velocity conditions are required at the distributor inlet (constant in the circumferential direction).

There are two key ingredients in our methodology. (1) The distributor-runner coupling is achieved through a conical mixing interface. The circumferential averaging (mixing) is performed using a least squares B-splines procedure, which practically preserves the flow rate value (within 0.5% variation). However, the mixed velocity profile is further corrected to insure the exact prescribed discharge. Good convergence of the iterative mixing interface algorithm is achieved, with no more than 7-9 iterations. (2) The radial equilibrium condition is used on the draft tube inflow section. As a result, the draft tube is not included into the computational domain. Moreover, this boundary condition does not require apriori knowledge of either velocity or pressure distributions. It has been found that the radial equilibrium condition leads to very good results for variable operating points, and is able to deal with recirculating flows as well.

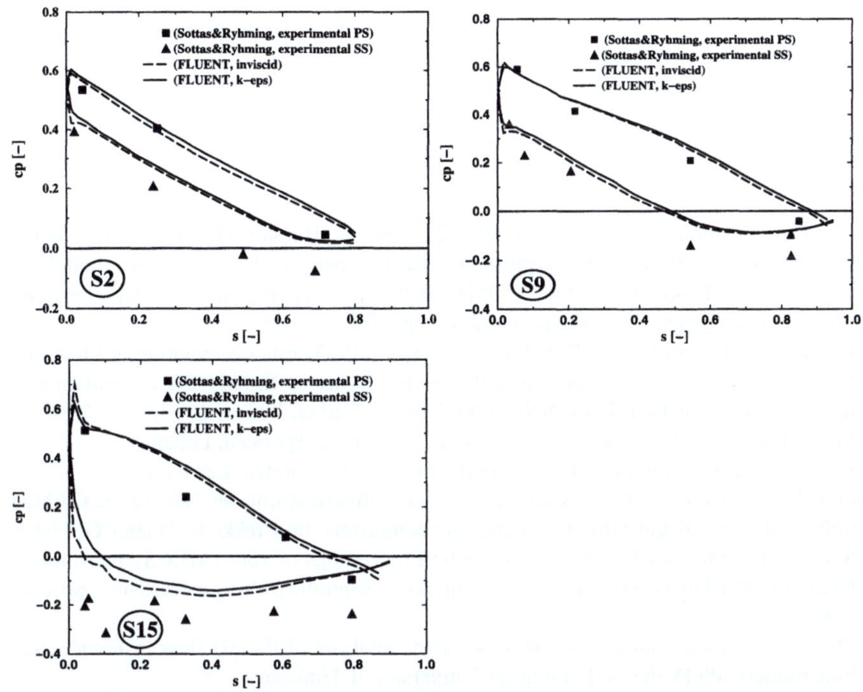

Fig. 7. The pressure coefficient on the S2 (near crown), S9 (middle) and S15 (near band) sections of the runner blade at best efficiency operating point

We apply our methodology to compute the velocity and pressure fields as well as the turbulence quantities for the GAMM Francis turbine. Numerical results are presented for the best efficiency point. A very good agreement with experimental data is obtained on the mixing interface. Particularly, one can conclude that the velocity field on the runner inlet is the correct one.

The pressure distribution on the runner blade is well predicted near the crown and on a mid section. However, the minimum pressure on the suction side near the band is over estimated.

In conclusion, the mixing interface methodology we have presented in this paper is robust and suitable for current 3D turbulent analysis in both distributor and runner of Francis turbines.

Acknowledgments

The present work has been supported from the National University Research Council Grant (CNCSIS) 109/2002, 29/2003 and 220/2003. Numerical computations have been performed at the Numerical Simulation and Parallel Computing

Laboratory from the "Politehnica" University of Timisoara, National Center for Engineering of Systems with Complex Fluids.

References

[1] Avellan F, Dupont P, Farhat M, Gindroz B, Henry P, Hussain M, Parkinson E, Santal O (1990) Flow survey and blade pressure measurements in a Francis turbine model. In: Pejovic S (ed) Proceedings of the 15th IAHR Symposium on Hydraulic Machinery and Cavitation, Belgrade, Yugoslavia, vol 2, I5, pp 1-14

[2] Bottaro A, Drotz A, Gamba P, Sottas G, Neury C (1993) Euler Simulation of Flow in a Francis Distributor and Runner. In: Sottas G and Ryhming IL (eds) 3D-computation of incompressible internal flows, NNFM 39, Vieweg Verlag, Braunschweig, pp 77-84

[3] Fluent Inc (2001) FLUENT 6. User's Guide, Fluent Incorporated, Lebanon

[4] Fluent Inc (2001) Gambit 2. User's Guide, Fluent Incorporated, Lebanon

[5] Gros L, Avellan F, Bellet L, Kueny J-L (1998) Numerical flow analysis of the GAMM turbine at nominal and off-design operating conditions. In: Brekke H, Duan CG, Fisher RK, Schilling R, Tan SK, Winoto SH (eds) Proceedings of 18th IAHR Symposium on Hydraulic Machinery and Cavitation. Singapore, Republic of Singapore, vol 1, pp 121-128

[6] Muntean S (2002) Numerical methods for the analysis of the 3D flow in Francis turbine runners. Ph.D. thesis, Politehnica University of Timisoara

[7] Muntean S, Susan-Resiga S, Anton I (2002) 3D Flow Analysis of the GAMM Francis Turbine for Variable Discharge. In: Avellan F, Ciocan G, Kvicinsky S (eds) Proceedings of the 21st IAHR Symposium on Hydraulic Machinery and Systems. Lausanne, Switzerland, vol 1, pp 139-146

[8] Nilsson H (2002) Numerical investigations of turbulent flow in water turbines. Ph.D. thesis, Chalmers University of Technology

[9] Parkinson E (1995) Test Case 8: Francis Turbine, Turbomachinery Workshop ERCOFTAC II

[10] Sottas G, Ryhming IL (eds) (1993) 3D - computation of incompressible internal flows, Proceedings of the GAMM Workshop, Notes Numerical Fluid Mechanics (NNFM) 39, Vieweg Verlag, Braunschweig

Study of Blade to Blade flows and Circumferential Stall Propagation in Radial Diffusers and Radial Fans by Vortex Cloud Analysis

R. Ivan LEWIS, Emeritus Professor
Newcastle University, UK
2-16 Bruce Building, Newcastle University, Newcastle upon Tyne, NE1 7RU, UK
Tel.: +44 191 2226201, Fax: +44 191 2226201, Email: R.I.Lewis@ncl.ac.uk

Abstract The objective of this paper is extension of vortex cloud simulation to the study of deep rotating stall cell propagation in radial turbomachines. Previous studies [1][2] provided the basic analysis, summarised in part here, for radial and mixed-flow blade rows but with identical blade-to-blade flow. Lifting of this restriction here permits the natural development of circumferential flow variations, revealing the growth of major upstream rotating stall cells for a radial diffuser with ten log-spiral blades but with high angle of attack (deep stall). For the same blade row run as a radial fan however stall cell formation and propagation is found to be inhibited. Additional studies are included for a cambered blade geometry typical of axial compressors. Used as an eight bladed radial diffuser, classical rotating stall is predicted. When operated as a rotor, rotating is again inhibited and modified.
Key Words: radial turbomachines, rotating stall, vortex dynamics

Nomenclature

c	absolute velocity
c_θ	absolute swirl velocity
$k(s_m, s_n)$	coupling coefficient (analytical)
l	blade chord
r, θ	polar coordinates
s	distance along body surface
t	blade pitch, time
w	Relative velocity

w_θ		Relative swirl velocity
x,y	$[m]$	Cartesian coordinates in z-plane
x_m, y_m		pivotal points
$K(s_m, s_n)$		coupling coefficient (numerical)
M		number of surface elements
N		number of blades in group
T		blade group pitch
U_∞, V_∞	$[m/s]$	components of W_∞
W_∞	$[m/s]$	vector mean velocity
$Z = re^{i\theta}$		radial diffuser coordinates in the Z plane
α_∞	[rad]	vector mean flow angle
α_1		absolute inlet swirl angle
β_1		relative inlet swirl angle
β_m	[rad]	profile slope
γ		surface vorticity strength
λ		stagger angle
θ		camber angle
$\zeta=\xi+i\eta$		blade coordinates in the cascade ζ plane
$\Delta\Gamma_j$		vortex element strength
Ω	[rad/s]	rotor angular velocity

1. Introduction

A full statement and development of the underlying equations for vortex cloud simulation of mixed-flow and radial turbomachines has been given in previous presentations [1][2] and a brief summary only of these techniques will be given here. A detailed exposition of the underlying methodolgy of vortex cloud analysis has been given in ref. [3] and a more recent review of numerical developments and applications in ref. [4]. Extensions to cascades in relative motion were presented in detail by Chunjun and Lewis in refs [5] and [6] including rotating stall simulations. A good review of the mechanisms of rotating stall were given by Horlock, 1978, ref. [7] with a more recent update by Cumpsty, ref. [8], following the work of MacDougall, 1988, ref. [9]. However, there is no known parallel work completed on radial cascades and rotors as reported here.

First, the basic equations for surface vorticity flow modelling will be given in Sections 2 & 3 including extension to vortex cloud modelling of turbomachine linear cascades. Second, the transformation techniques for applying these equations to radial cascades will be summarised in Section 4. Investigations of the stalling flow of a ten bladed radial diffuser will also be studied in Section 4, followed by studies of an identical blade row run as a fan rotor, Section 5. A second radial blade row will be considered in Section 6 based on typical axial compressor geometry with a cambered blade profile. A summary of conclusions will follow in Section 7.

2. Vorticity Modelling of Cascades

The fundamental basis of vortex element representation of the Navier-Stokes equations has been given by the author [3], beginning with the boundary integral equation for potential flow past an arbitrary body, Eq. (1), with reference to **Figure 1**,

$$\tfrac{1}{2}\gamma(s_m)+\oint k(s_m,s_n)\gamma(s_n)ds_n + W_\infty(\cos\alpha_\infty\cos\beta_m+\sin\alpha_\infty\sin\beta_m)=0 \tag{1}$$

where, for the boundary condition of zero internal velocity inside the body, the surface velocity v_{sn} is equal to the surface vorticity $\gamma(s_n)$.

$$v_{sn}=\gamma(s_n) \tag{2}$$

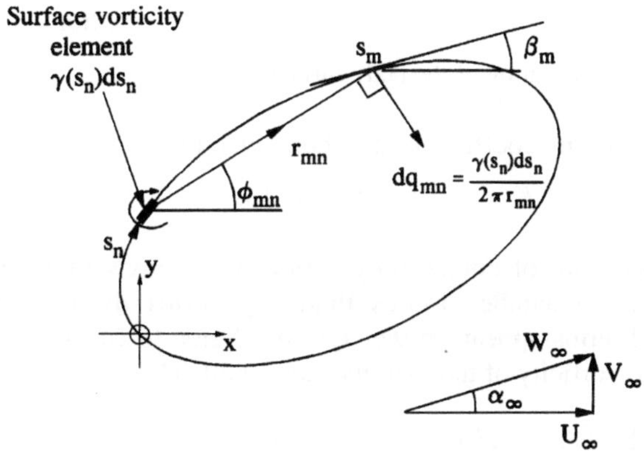

Fig. 1. Surface vorticity model for flow past a body in a uniform stream $W\infty$.

The coupling coefficient $k(s_m,s_n)$ linking points m & n on the body surface is given by

$$k(s_m,s_n)=\frac{1}{2\pi}\left\{\frac{(y_m-y_n)\cos\beta_m-(x_m-x_n)\sin\beta_m}{(x_m-x_n)^2+(y_m-y_n)^2}\right\} \tag{3}$$

For a turbomachine cascade blade periodic in the y direction with pitch t, this becomes

$$k(s_m,s_n)=\frac{1}{2t}\left\{\frac{\sin\frac{2\pi}{t}(y_m-y_n)\cos\beta_m-\sinh\frac{2\pi}{t}(x_m-x_n)\sin\beta_m}{\cosh\frac{2\pi}{t}(x_m-x_n)-\cos\frac{2\pi}{t}(y_m-y_n)}\right\} \tag{4}$$

The standard numerical strategy is to represent the body surface by M discrete elements as shown in **Figure 2**, whereupon Eq.(1) transforms to the set of M linear equations

$$\sum_{1}^{M} K(s_m, s_n)\gamma(s_n) = -U_\infty \cos\beta_m - V_\infty \sin\beta_m \tag{5}$$

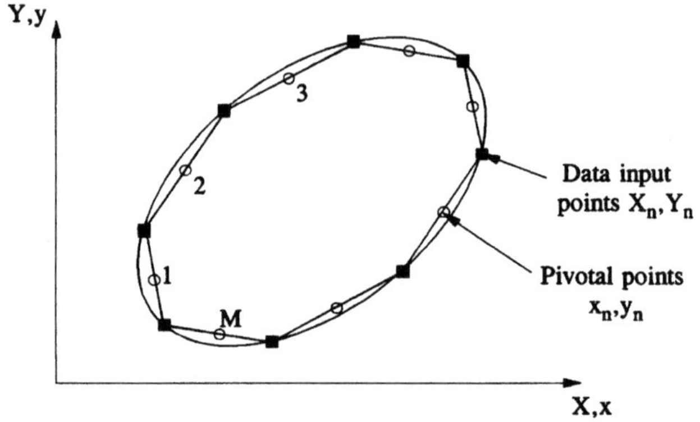

Fig. 2. Numerical discretisation of the body surface.

The revised coupling coefficients are then given by

$$K(s_m, s_n) = k(s_m, s_n)\Delta s_n \tag{6}$$

These equations are of course only applicable to inviscid/potential flows, but may be extended to handle viscous fluids by vortex dynamics modelling, for which additional terms appear on the right hand side to account for the influence of the distributed vorticity of the external flow field. Thus we have

$$\sum_{1}^{M} K(s_m, s_n)\gamma(s_n) = -U_\infty \cos\beta_m - V_\infty \sin\beta_m$$
$$-\sum_{j=1}^{Z} \Delta\Gamma_j (U_{mj} \cos\beta_m + V_{mj} \sin\beta_m) \tag{7}$$

where the spacially distributed vorticity has been discretised into a large number Z of small discrete vorticies $\Delta\Gamma_j$. The unit velocities induced by $\Delta\Gamma_j$ are given by

$$\left.\begin{aligned} U_{mn} &= \frac{1}{2t}\frac{\sin\frac{2\pi}{t}(y_m - y_n)}{\cosh\frac{2\pi}{t}(x_m - x_n) - \cos\frac{2\pi}{t}(y_m - y_n)} \\ V_{mn} &= -\frac{1}{2t}\frac{\sinh\frac{2\pi}{t}(x_m - x_n)}{\cosh\frac{2\pi}{t}(x_m - x_n) - \cos\frac{2\pi}{t}(y_m - y_n)} \end{aligned}\right\} \tag{8}$$

The numerical analysis is conducted over a series of small discrete time steps Δt for each of which a discrete vortex element $\Delta\Gamma_j$ is shed from each body surface element which, for a typical element j , will be of strength

$$\Delta\Gamma_j = \gamma(s_j)ds_j \tag{9}$$

The right hand side of the basic governing equation Eq. (7) thus includes the influence of the superimposed uniform stream W_∞ plus the vorticity already shed into the mainstream flow. In the author's previous publications [1] to [3], for cascades the assumption was made that each blade flow was indentical, a not unreasonable assumption for unstalled turbine or fan blade rows. However for off-design angles of attack leading to stall it is well known that there can be significant blade-to-blade variations particularly in fan and compressor cascades and the aim here is to extend the analysis to deal with such situations. We deal with this next.

3. Revised Cascade Modelling

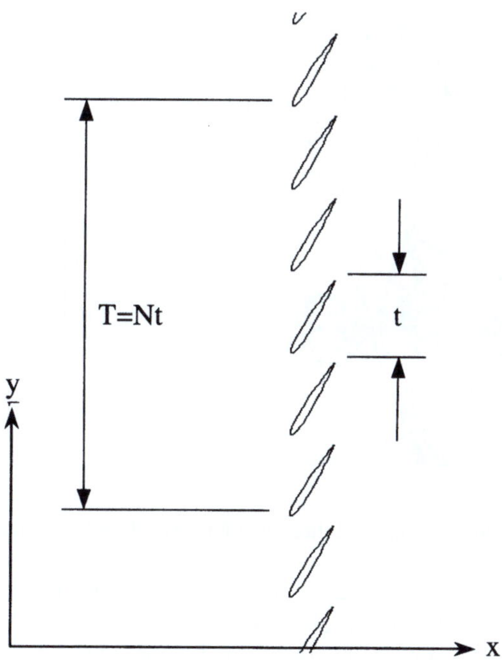

Fig. 3. Revised model for the cascade with N independent blade profiles and group pitch T.

Adaption of the previous vortex cloud model to allow simulation of at least some blade-to-blade variations is in fact relatively simple and is as illustrated above in **Figure 3**. The approximation here is that there are N independent blades, (in this example 5) and that the flow pattern will thus repeat at the revised cascade pitch T=Nt. The group of N=5 blades selected here may then be regarded as a single body shape when making up the $K(s_m,s_n)$ matrix of Eqs. (7). The blade pitch t in Eqns. (4) and (8) is then replaced by the group pitch T. In effect we are now analysing the flow through a group of N separate bodies, but cascaded as a group at pitch T in the y direction. Choice of the group size N is then a matter of computational limitations including available memory, time of execution and numerical accuracy available for inversion of the large $K(s_m,s_n)$ matrix. The outcome is best illustrated at this point by a typical solution as shown below in **Figure 4**.

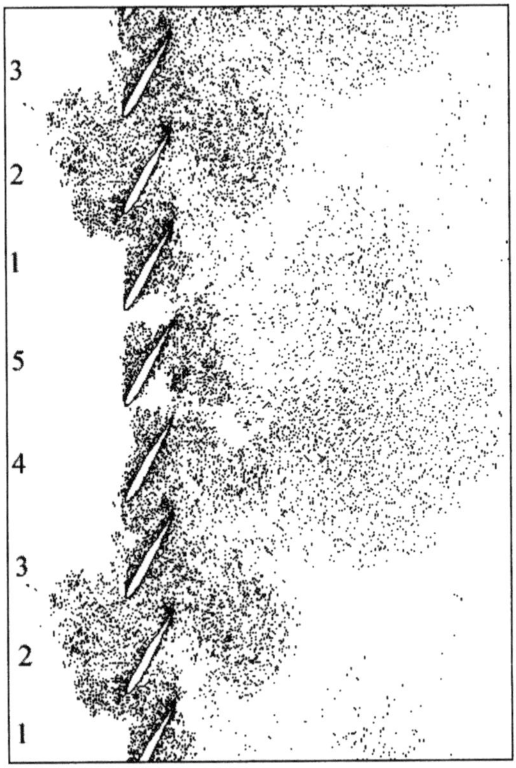

Fig. 4. Simulation of deep propagating stall through a fan cascade at high angle of attack.

The fan cascade shown here consists of five C4 profiled blades with zero camber θ, a stagger angle of $\lambda=45°$ and a pitch/chord ratio t/l=1.0. We note that the predicted pattern is repeated after every five normal blade pitches t as indicated by the repeated blade numbering from 1 to 5.

The inflow angle β_1 of 85° for this example corresponds to 20° angle of attack and is of course excessive, ensuring a state of deep stall. The outcome of this is well known but none the less remarkable. After a period of time as the motion proceeds, we observe the development of a large stall cell actually upstream of the blade row which propagates in the vertical direction. The reason for this behaviour can be deduced from the vortex dynamics simulation demonstrating the power of this CFD technique for prediction and diagnosis. Thus we may observe that the major blockage of the blade passage at position 2 due to the stall cell is causing increased flow through the passages adjacent to blade 5 due to the consequent decrease in angle of attack of the blades behind the moving stall cell. Shortly blade No.1 will also begin to unstall as the cell proceeds upward. Thus blades 4,5 and 1 are delivering a pressure rise locally across the blade row whereas the adjacent blades 2 and 3 are totally stalled. The outcome of this is that the stagnant fluid in the blade passages adjacent to blade 2 is blown upstream to augment the developing upstream stall cell causing a periodic fluctuation and thus likely strong aerodynamic excitation of the blades.

In this example there were only five independent blade profiles resulting in one large stall cell. In practice there could be several cells developing and propagating and one simulation with ten independent blades of a compressor cascade undertaken by the author revealed two cell regimes, i.e. one every 5 blade pitches. This situation is illustrated in the next example to be investigated in the next section, namely the case of a radial diffuser.

4. Extension to Radial Diffusers

Because the distributed vorticity of the mainstream flow has been replaced in the numerical model by a cloud of discrete vorticies, it is valid to apply conformal transformations to the linear cascade and its flowfield provided the vortex element strengths $\Delta\Gamma$ remain unchanged. Thus the previous analysis may be applied directly to radial stators by means of the following transformation [3][4] between a radial diffuser in the Z plane and a rectilinear cascade parallel to the η axis in the ζ plane,

$$\zeta = \ln Z \tag{10}$$

where $Z=re^{i\theta}$ and $\zeta=\xi+i\eta$. The coordinate and velocity transformations are then

$$\zeta = \ln r,\ \eta = \theta \qquad \text{(11)a}$$

$$q_\zeta = q_z r \qquad \text{(11)b}$$

where r,θ are polar coordinates of the actual radial blade row.

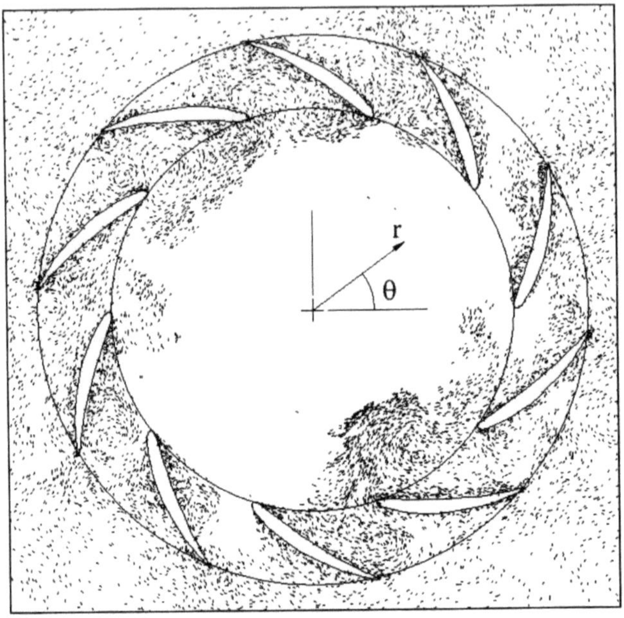

Fig. 5. Simulation of deep stall in a radial outflow diffuser with ten blades.

Figure 5 shows the outcome of applying this transformation to analysis of a ten bladed radial outflow diffuser. The blade geometry in the cascade ζ plane is identical to that used in section 3, namely a C4 base profile with $\theta=0°$, $\lambda=45°$, $t/l=1.0$ and with a pre-whirl inlet angle $\beta_1=85°$ (anti-clockwise outflowing swirl). For this computation the number of blades was set at N=10 to achieve full simulation of flow through the radial diffuser.

The main observation to note here is the development of two major rotating stall cells on the upstream (inner) side of the diffuser. These are not identical but very similar in size and extent and are found to precess around the inner region in the well known manner of rotating stall in ***axial*** compressors and fans. At circumferential locations between the two stall cells the blades are less stalled (and sometimes temporarily unstalled) due to their increased local distribution of the mass flow and hence reduced local angles of attack. In these regions pressure recovery through the diffuser is achieved resulting in higher pressure surrounding the exit regions. This higher pressure forces the stagnant fluid in the stall cell regions to be pushed back inwards with a consequent build up of the stall cells and a perpetuation of the large scale disturbances. In the case of a centrifugal fan it can be expected that these stall cells would cause serious fluctuating interference with the rotor and be a serious source of unwanted vibrations.

5. Extension to Radial Fan Rotors

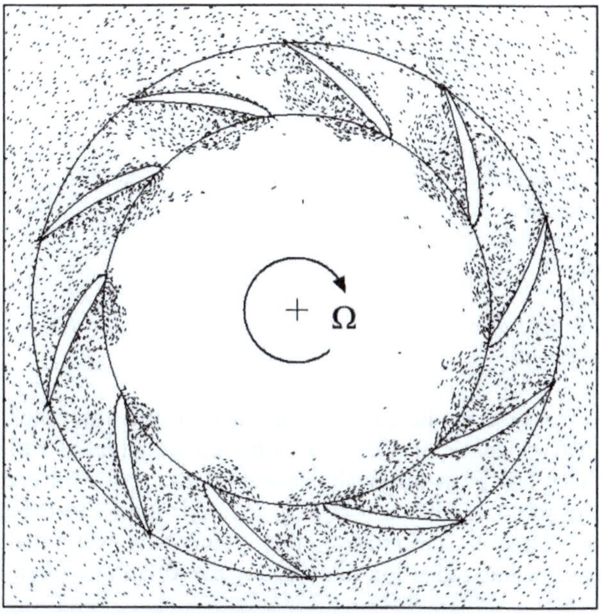

Fig. 6. Flow through a backward sloping radial fan with relative inlet angle β1=85°.

It is of considerable interest to extend these studies from radial *stators* to radial ***rotors***, **Figure 6**, since in the latter we would expect the additional influence of work input including those due to relative eddy and Coriolis forces. The question raised then is whether these additional effects would enhance stall cell development or inhibit it. Fortunately all the relevant equations to handle this have already been derived [3] and applied to potential flow through radial and mixed-flow turbomachines. To summarise, the governing Eq. (7) develops into the extended form

$$\left.\begin{aligned}\sum_{n=1}^{M} K(s_m, s_n)\gamma(s_n) = &-U_\infty \cos\beta_m \\ &-\left[V_\infty + \Omega\left\{r^2 - \tfrac{1}{2}\left(r_1^2 + r_2^2\right)\right\}\right]\sin\beta_m \\ &-\sum_{j=1}^{Z}\Delta\Gamma_j\left(U_{mj}\cos\beta_m + V_{mj}\sin\beta_m\right)\end{aligned}\right\} \quad (12)$$

where the new middle line includes the influence of blade row rotation with angular speed Ω. From the appearance of this term $\Omega\left\{r^2 - \tfrac{1}{2}\left(r_1^2 + r_2^2\right)\right\}$ one can immediately detect that it represents the additional influence upon the flow due to Coriolis accelerations of fluid particles as they pass through the rotor and one might anticipate some considerable influence upon the consequent flow. This is indeed the case as we shall see from the following example.

In this example the previous blade row as shown in Fig. (5) was rotated clockwise as a radial fan rotor with zero prewhirl but with sufficient angular velocity Ω to deliver the same relative inflow angle of β_1=85° and thus 20° angle of attack. A summary of the data is as follows:

Table 1. Design data for radial fan.

Inlet radius r_1	1.0m
Outlet radius r_2	1.3691m
Meridional velocity at r_1	1.0m/s
Angular velocity	109.15 revs/min
Prewhirl angle α_1	0°
Relative inlet angle β_1 at r_1	85°
Camber θ	0°
Stagger λ	60°
Pitch/chord ratio t/l	1.0
Time step Δt	0.005sec

As can be seen from the predicted flow pattern after 300 time steps, Fig 6, although there is clear evidence of stall with some circumferential variation from blade to blade, there has been no establishment of the upstream eddy regimes characteristic of the stationary blade row, Fig. (5). It is quite clear that the presence of Coriolis accelerations and relative eddy effects have had a major stabilising effect in the case of this radial fan rotor as compared with its stator equivalent Fig. 5.

6. Axial compressor type blade rows

In the foregoing fan *uncambered* blade row studies it was difficult to detect classical cascade stall propagation or regular rotating stall in the radial diffuser at smaller angles of attack. Although some stall propagation was present below 15° angle of attack it was largely intermittent. It was decided therefore to consider also a typical axial compressor blade row with the following specification.

Table 2. Axial compressor cascade design data.

Chord length	1.0m
Velocity normal to cascade	1.0m/s
Camber θ	44.49°
Stagger λ	41.11°
Pitch/chord ratio t/l	1.162
Design inlet angle β_1	54.59°
Predicted design outlet angle β_2	30.69°

Adopting here the cascade geometry studied in refs. [5]&[6], we note the introduction of typical camber θ=44.49° and stagger λ=41.11°. The design inlet angle given above is that for shock-free inflow for which the inlet stagnation point is precisely on the leading edge. Vortex cloud analysis was undertaken with a time step δt=0.015s for angles of attack of 0, 5, 10, 15 and 20° above this for a cascade with eight independent blades. Only for the last of these studies, with an inlet angle β_1=74.59°, was regular stall propagation actually found to develop in this cascade and a sample of the results is shown in **Figure 7** below.

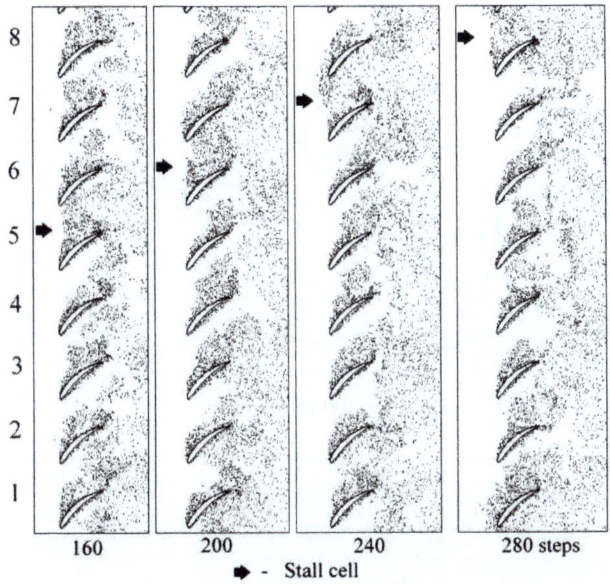

Fig. 7. Stall propagation in a compressor cascade (Table 2 data with β1=74.59° & δt=0.015).

As may be observed, after 160 time steps a stall cell has built up in blade passage No. 5. After successive intervals of 40 time steps this stall cell has propagated upwards to blade passages 6, 7 and 8. It should be noted that this pattern behaviour was found to occur only with 20° angle of attack. At lower inlet angles there was evidence of developing stall progagation but with less firm regularity.

To conclude, let us consider adaption of the above cascade geometry for design of an equivalent radial diffuser. Application of the conformal transformation Eqs. (10) and (11) results in the circular radial outflow cascade shown in **Figure 8**. Adopting an inlet prewhirl angle of β_1=74.59°, with time steps of magnitude δt=0.015 and a meridional velocity at the inlet radius r_1 of value 1.0, the predicted flow pattern after 300 and 335 time steps is shown in Figs. 8(a) and 8(b) respectively.

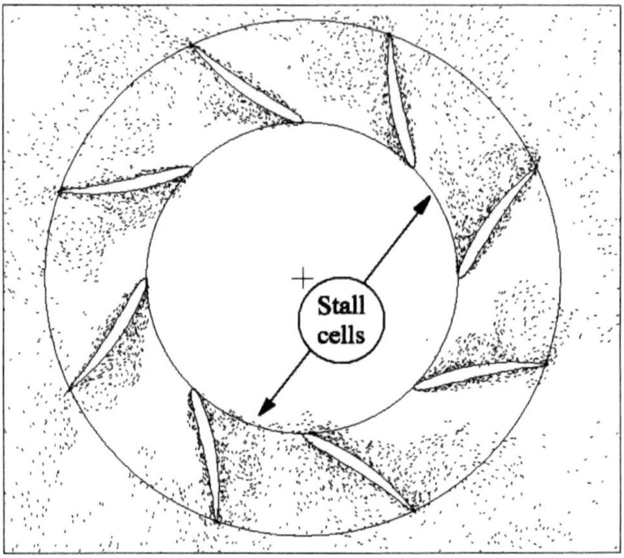

(a) After 300 time steps.

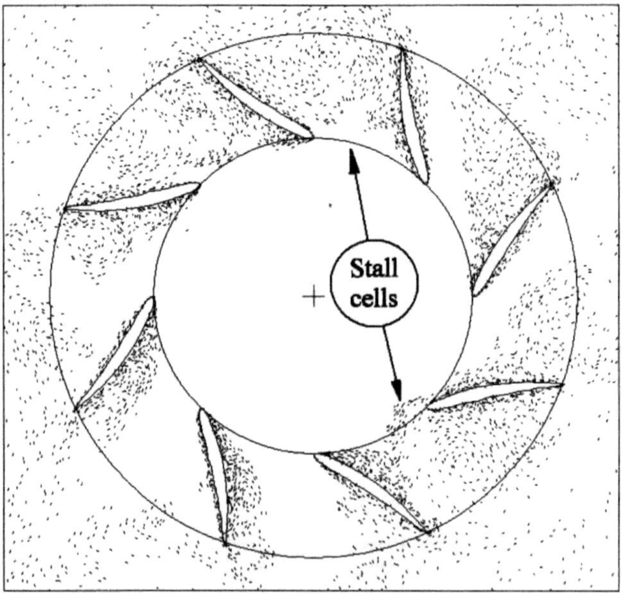

(b) After 335 time steps.

Fig. 8. Development of rotating stall cells in an eight bladed "axial compressor type" radial diffuser.

We observe from Fig. 8 the presence of a single passage stall cell at the bottom region which has moved anticlockwise into the next blade passage after 35 time steps. We also note the presence of a second but less concentrated cell located at the top of the diffuser also rotating anticlockwise at the same speed. Reference back to Fig. 7 in fact reveals the presence of a similar second stall cell region around passage 1 to begin with, moving upwards to passage 4 finally. The main difference fluid dynamically between the two axial and radial cascades lies in the influence of viscous diffusion. For infinite Reynolds number viscous effects would have little influence. For the cascade Reynold's number of 1×10^5 selected here on the other hand, viscous diffusion will differ considerably due to the additional mainstream radial diffusion in the case of the radial diffuser. Despite this stall cell propagation is remarkably similar for both cases considered comparing Figs. 7 and 8.

To conclude these studies, the flow through the previous cambered radial blade row was investigated when used as a radial fan rotor with two speeds of rotation, **Figure 9**. Fig. 9(b) shows the typical long term flow pattern for Ω=34.644revs/s, at which angular velocity zero prewhirl α_1 provides the same relative inlet angle as the stator just considered in Fig. 8, β_1=74.59°. Fig. 9(a) in the other hand shows the predicted flow pattern for half this speed of rotation, Ω=17.322 revs/s, but with sufficient prewhirl α_1=61.15° to give the same relative inlet angle β_1=74.59°. Three important observations may be made.

1. Although some rotating stall still occurs, there is decreasing evidence of major stall cells as the rotor angular velocity increases.
2. Major separation seems to be limited largely to the leading edge region with evidence of what seems to be an approach towards reattachment at mid-chord on the trailing surface ***t***.
3. From mid-chord towards the trailing edge region there is evidence of migration of diffused vorticity from the trailing surface ***t*** to the leading surface ***l***, indicated by ***M*** in Fig.9(b).

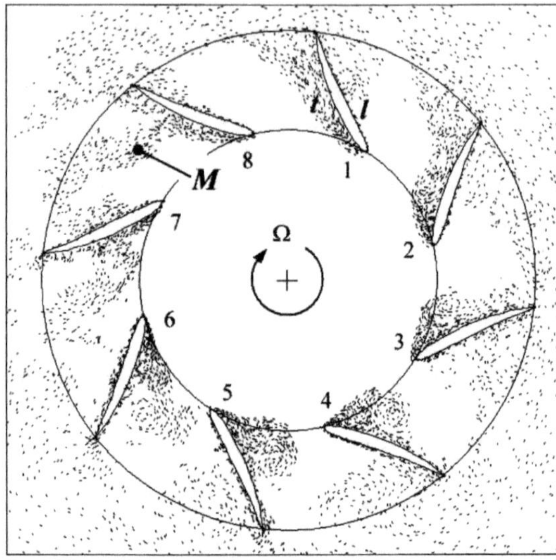

(a) Ω=17.322 revs/s, Prewhirl angle α_1=61.15° 280 iterations with δt=0.015.

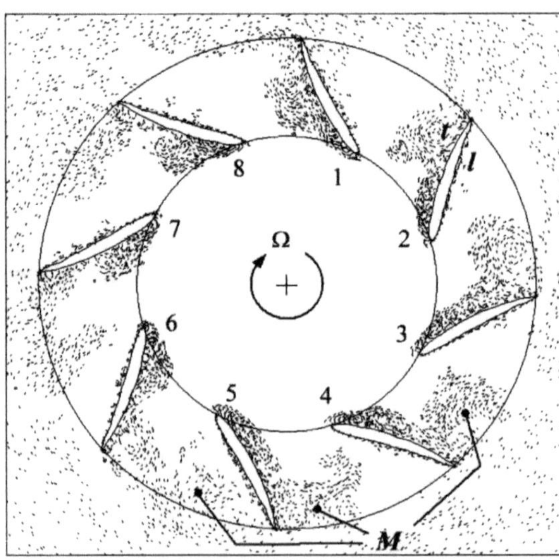

(b) Ω=34.644 revs/s, Prewhirl angle α_1=0.0°, 360 iterations with δt=0.01.
l - leading surface. ***t*** - trailing surface.
M - Circumferential vorticity migration.

Fig. 9. Flow through the radial diffusing blade row operating as a radial fan rotor.

Undoubtedly the explanation for the above is the influence of Coriolis accelerations upon the relative flow through these radial rotors. Specific work input due to associated Coriolis forces is dependent only upon the radial shift of the fluid from inlet r_1 to outlet r_2 and the rotor angular velocity Ω as shown by the Euler Pump Equation, namely

$$\left.\begin{aligned} \tfrac{1}{\rho}(p_{02} - p_{01}) &= U_2 c_{\theta 2} - U_1 c_{\theta 1} = \Omega(r_2 c_{\theta 2} - r_1 c_{\theta 1}) \\ &= \underbrace{\Omega(r_2 w_{\theta 2} - r_1 w_{\theta 1})}_{\substack{\textit{Aerodynamic} \\ \textit{specific work input}}} + \underbrace{\Omega^2(r_2^2 - r_1^2)}_{\substack{\textit{Coriolis specific} \\ \textit{work input}}} \end{aligned}\right\} \tag{13}$$

where w_θ is the swirl velocity measured relative to the rotor, and is related to the absolute swirl velocity c_θ through

$$c_\theta = w_\theta + r\Omega \tag{14}$$

It is clear that although there is leading edge flow separation due to the sharp angle of attack, the specific work input imposed by Coriolis effects has a stabilising effect as the fluid proceeds further through the blade passages. However, partly due to the anticlockwise "relative eddy rotation" or "slip flow" in the passage exit region, free vorticity tends to migrate from where it has been created on the unstable trailing surface ***t*** towards the leading surface ***l***.

7. Conclusions

The following summary of conclusions may be drawn from these fluid dynamic developments and studies:

1. The basic linear cascade vortex dynamics code developed here has successfully predicted the presence of rotating stall in fan and compressor cascades in agreement with earlier work.
2. At high incidence angles in deep stall, substantial stall cells intrude back upstream of the blade row.
3. A conformal transformation technique has succeeded in extending these flow simulations to deal with radial blade rows used as either stators or rotors.
4. At large relative inlet angles a couple of large rotating deep-stall cells were found to develop on the inlet side of a radial outflow diffuser stator comprising a backward swept zero camber fan cascade at high stagger.
5. For the cambered compressor type radial diffuser stator no deep-stall cells of this kind developed for the same leading edge angle of attack of 20°, although rotating stall was present.
6. For the same blade row employed as a radial fan rotor, work input due to Coriolis forces leads to general stabilisation of the initial leading edge stall regime. The same "relative eddy" effects result in vorticity migration at the outlet radii from the trailing (highly loaded) blade surface towards the (generally low velocity) leading edge surface.

References

[1] Lewis, R.I., 2001, "Development of Vortex Dynamics for Simulation of Turbomachine Cascades and Blade Rows." Journal of Computational and Applied Mechanics, Vol. 2, No. 1, pp. 73-85.

[2] Lewis, R.I. Sept 2001, "Extension of Vortex Methods to the Flow Simulation of Mixed-flow Turbomachines." Proc. of the Second International Conference on Vortex Methods, Istanbul, Turkey.

[3] Lewis, R.I. 1991 "Vortex Element Methods for Fluid Dynamic Anal,ysis of Engineering Systems." Cambridge University Press.

[4] Lewis, R.I. 1999, "Vortex Element Methods, the most natural approach to flow simulation - a review of methodology with applications." Proceedings of the First International Conference on Vortex Methods, Kobe, Japan, pp.1-15.

[5] Chunjun Ji 1997 "Vortex Cloud Method and its Applications to Turbomachine Cascades." Ph.D. thesis, Newcastle University.

[6] Lewis, R.I., Chunjun Ji, 1998. "Extension of Vortex Cloud Modelling to Cylinder Arrays and Cascades in Relative Motion.", Proceedings of the Fourth European Computational Fluid Dynamics Conference, Athens, p. 642-647, J. Wiley & Sons.

[7] Horlock, J.H. 1973, "Axial Flow Compressors." R.E.Kreiger Publishing Co., INC., New York.

[8] Cumpsty, N.A. 1989, "Compressor Aerodynamics.", Longman Scientific & Technical co-published with J.Wiley

[9] MacDougall, N.M. 1988, "Stall inception in axial flow compressors", PhD Thesis, University of Cambridge.

New Approach to Radial Compressor Return Channel Design

Árpád VERESS
Department of Aircraft and Ships
Budapest University of Technology and Economics
Sztoczek u. 6 J. ép. 426, H-1111 Budapest
Hungary
Tel.: (+36 1) 463 1992, Fax: (+36 1) 463 3080
Email: veress@rht.bme.hu

René VAN den BRAEMBUSSCHE
Turbomachinery and Propulsion Department
von Kármán Institute for Fluid Dynamics
Waterloose steenweg 72, B-1640 Sint-Genesius-Rode
Belgium
Tel.: (+32 2) 359 9609, Fax: (+32 2) 359 9600
Email: vdb@vki.ac.be

Abstract The design and optimisation of a vaneless diffuser, cross-over and return channel of a multistage radial compressor is presented. Special attention is given to the extension of the vanes upstream of the cross-over.

These unconventional vanes are designed by a new 3D procedure that is based on the equations for irrotational flow and accounts for meridional curvature and difference in contour length at hub and shroud. The flow analysis by means of a 3D Navier Stokes solver indicates a substantial improvement of the performance by eliminating part of the flow separation inside the cross-over.

The geometry is further optimised by means of a 3D inverse design method to obtain a smooth Mach number distribution along the vanes at hub and shroud. It resulted in a performance improvement by suppressing also the separation on the return vanes.

3D Navier Stokes analysis also shows a direct impact of lean on secondary flows and a performance increase by introducing negative lean.
Keywords: CFD, inverse design, return channel, radial compressor

Nomenclature

b	[m]	vane height
c	[m]	reference length
C_p	[-]	pressure recovery coefficient
m	[m]	meridional coordinate
$\dot{m}$	[kg/s]	mass flow
P	[Pa]	pressure
R	[m]	radius
s	[m]	curvilinear coordinate
W	[m/s]	velocity
z	[pcs]	number of vanes
β	[rad]	blade angle (measured from radial)
δ_{th}	[m]	blade thickness
θ	[rad]	angular coordinate
ρ	[kg/m^3]	density
ω	[-]	total pressure loss coefficient

Subscripts

bl	blade
fl	flow
LE	leading edge
m	meridional component
TE	trailing edge
th	thickness
ps	pressure side
ss	suction side
2	impeller exit – diffuser inlet
3	return channel exit
is	isentropic

Superscripts

–	average value
o	total conditions
s	static conditions

1. Introduction

Impellers of multistage compressors are linked by a diffuser, cross-over and return channel (Fig. 1). The main task of the vaneless or vaned diffuser is to transform as much as possible kinetic energy, available at the impeller exit, into static pressure rise. The purpose of the return channel, containing highly curved vanes, is to take out the swirl of the flow while guiding the flow to the inlet of the next impeller. Both components are connected by a vaneless 180.° cross-over bend.

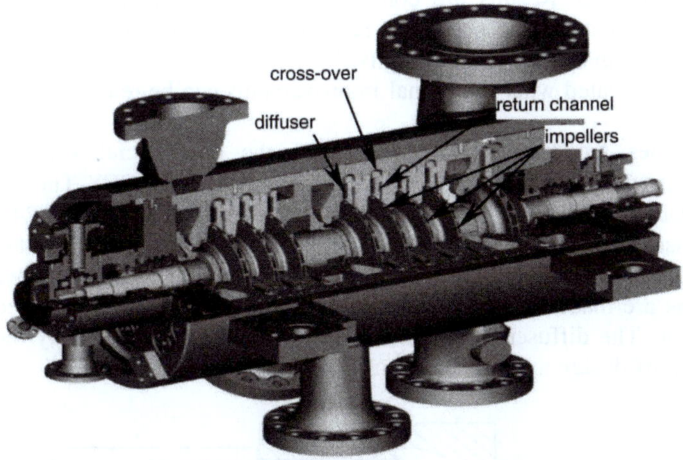

Fig. 1. Dresser-Rand DATUM multi-stage centrifugal compressor

Compactness of these non rotating parts is very important because a rapid deceleration of the fluid not only reduces the friction losses, proportional to the streamline length and square of the velocity, but also limits the cost. The last argument can be very important in case of compressors operating at high pressure (Fig. 1) where the cost of the barrel increases with the diameter to the power 3.

There are not many ways to increase the efficiency or pressure recover of the diffuser and return channel. The vaneless diffuser pressure rise depends on the inlet to outlet radius ratio and an increase of pressure recovery requires an increase of the overall dimensions. The use of vaned diffusers allows a more compact geometry but unfavourably affects the operating range.

Big losses may occur in the cross-over because of separation resulting from the strong meridional curvature and because of long flow paths resulting from the swirl velocity. It is generally not attempted to do any pressure rise in this part.

Making pressure rise in the return channel is anyway very difficult because of the inlet flow distortion resulting from the cross-over and the large meridional and circumferential turning that is required to bring the fluid back to axial at the inlet of next stage.

Thygesen [7] has shown that higher efficiencies are possible by extending the deswirl vanes upstream of the return bend. Similar experience is reported in [4]. When increasing the efficiency and pressure recovery by extending the vanes, less diffusion may be required in the vaneless diffuser, and the overall compressor size may be reduced without penalising the performance.

The traditional design approach is not suited to generate efficient vanes with controlled diffusion for a complex 3D flow with high inlet swirl in a strongly curved meridional contour.

Present paper describes the design of a vaned cross-over duct in combination with the return channel by means of an analytical method and by means of a 3D inverse design method. Results are compared to the ones of a classical return channel geometry that will be used as a reference to quantify the improvements that are achieved with the advanced designs.

It is further evaluated what additional improvements can be realized by leaning the blades.

All geometries are analyzed by means of the commercial solver CFX-TASCflow using the k - ε turbulence model with wall function. CFX-TurboGrid is used to generate a 3D grid with 128 points in the streamwise direction, 24 points in the pitchwise and 26 points in the spanwise direction. Care has been taken to have a y+ value of less than 1.0 for turbulent flows at the first point off the wall.

The designs are made for a typical compressor stage with specific speed 0.52 operating in air. The diffuser inlet conditions are the ones predicted by the VKI 1D design and off-design analysis program CCOD and listed in Table 1.

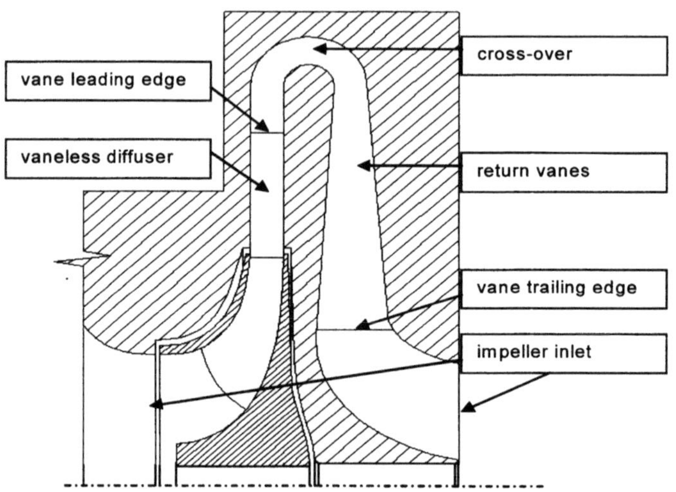

Fig. 2. Meridional cross section of the compressor stage

The meridional cross section, shown on Fig. 2, remains unchanged during the design process. The inlet section is defined by the impeller exit width and the outlet section is defined by the next stage inlet section.

2. Reference Geometry

Reference geometry is a vaneless diffuser and cross-over followed by a vaned return channel. The vanes are defined by the method described by Rothstein [5,6]. This method starts from a central streamline that is tangent to the incoming flow at the leading edge and radial/axial at the outlet. The blade suction and pressure side are defined by specifying a channel of constant or slightly diverging width around this central streamline (Fig. 3). The purpose is to turn the flow to axial while controlling the average velocity. The intersection between the different flow channels results in a sharp leading- and trailing edge. Only the leading edge is rounded off.

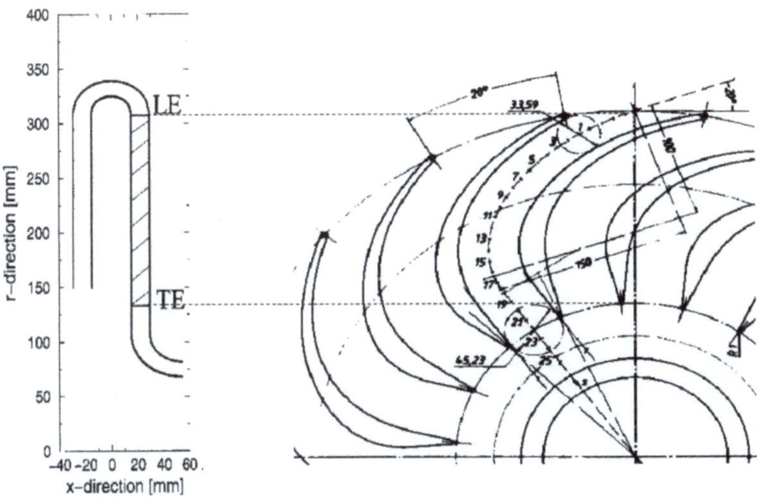

Fig. 3. Return vane design by the method of Rothstein

The shortcomings of this geometry are best illustrated by the velocity vectors, calculated by the 3D Navier Stokes solver, on a grid surface near the vane suction side (Fig. 4). The meridional velocity, plotted in the vaneless diffuser and first part of the cross-over bend, shows a return flow on the shroud side near the inlet of the cross-over. It is the mainly due to the large deceleration of the meridional velocity as a consequence of the strong meridional curvature. The total velocity vector is plotted downstream of the cross-over bend. It shows a second zone of separated flow on the hub contour near the vane leading edge. This one is caused by the sudden deceleration of the meridional velocity at the end of the cross-over in combination with a non-optimum incidence resulting from the large flow variation from hub to shroud.

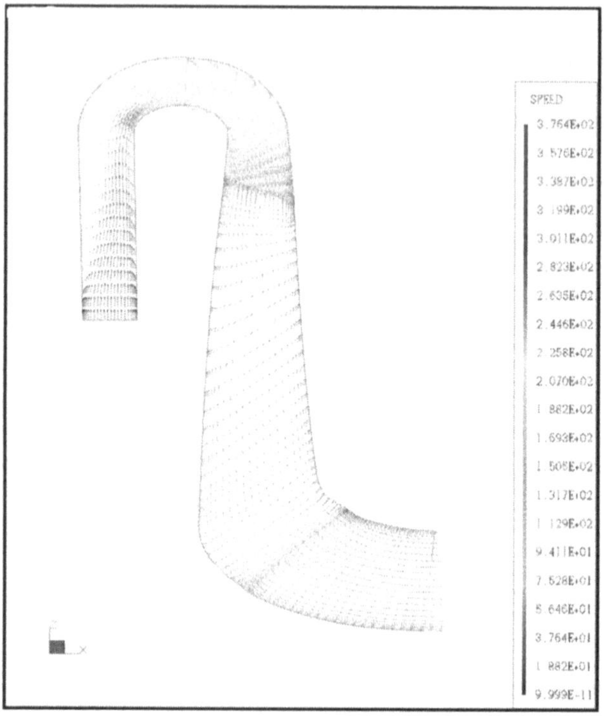

Fig. 4. Velocity vectors on a grid surface near the suction side of the return vanes

3. 3D Vane Design

In order to avoid the massive flow separation observed in the reference return channel, it has been decided to extend the vanes upstream of the cross-over duct up to the vaneless diffuser exit (Fig. 2). The vane leading edge is located sufficiently away from the impeller exit not to cause any reduction in operating range. The trailing edge radial position and the number of vanes are kept unchanged at 18.

It is not obvious how a good 3D vane should look like in terms of blade camber and thickness distribution. This is especially complex in radial geometries where the blade loading depends as well on camber as on radius change and meridional curvature. An analytical design procedure, called BLADECONTOUR, in which the blade loading can be prescribed, has been developed for this purpose.

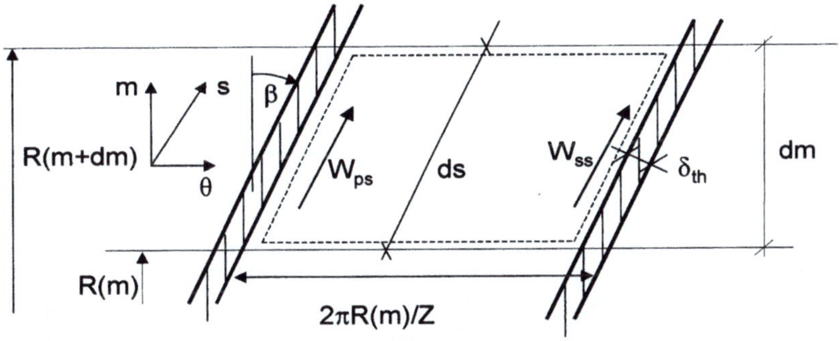

Fig. 5. Control surface for analytical vane design method

The design requires the definition of the optimum blade circumferential position (θ) and thickness (δ_{th}) at hub and shroud. BLADECONTOUR starts by defining the mid-span β_{bl} distribution corresponding to an irrotational flow with controlled blade loading from leading edge to trailing edge.

$$\nabla \times \overline{V} = 0 \tag{1}$$

Hence, there must be zero circulation on the closed contours defined by part of the suction and pressure side of two adjacent vanes and two lines at constant radius (Fig. 5). This provides following relation between vane loading and the change in flow angle β [8]:

$$W_{ss} - W_{ps} = \cos\beta_{bl}\, d/dm\left((2\pi R/z - \delta_{th}/\cos\beta_{bl})W_m tg\beta_{fl}\right) \tag{2}$$

Replacing the local meridional velocity by the pitchwise averaged value

$$\overline{W}_m = \dot{m}/(\rho z b(2\pi R/z - \delta_{th}/\cos\beta_{bl}))$$

and assuming that $\beta_{fl} = \beta_{bl}$, one obtains following relation between blade angle and loading:

$$W_{ss} - W_{ps} = \dot{m}/(\rho z b \cos\beta_{bl})d\beta_{bl}/dm \tag{3}$$

Assuming that the blade camber is tangent to the mean flow direction at midspan is a good approximation except for a small zone near the leading edge, in case of non-zero incidence, and near the trailing edge where the deviation may increase with vane loading and eventual flow separation.

Imposing $(W_{ss}\text{-}W_{ps})\rho\ zb/\dot{m} = Ct$ corresponds to constant blade loading from LE to TE if ρ zb/$\dot{m}$ is constant. The corresponding blade camber angle β_{bl} at any meridional position is obtained by integrating (4) from leading edge to that position.

$$\beta_{bl} = C^t \int_{LE} \cos\beta_{bl}\, dm \tag{4}$$

The leading edge blade angle $\beta_{bl\text{-}LE}$ is set equal to the local flow angle, which corresponds to zero incidence. The C^t is adjusted until the exit flow is axial.

The blade angle distribution at hub and shroud requires some modification of the calculated average distribution, indicated on Fig. 6, to account for the difference in curvilinear length and loading at hub and shroud.

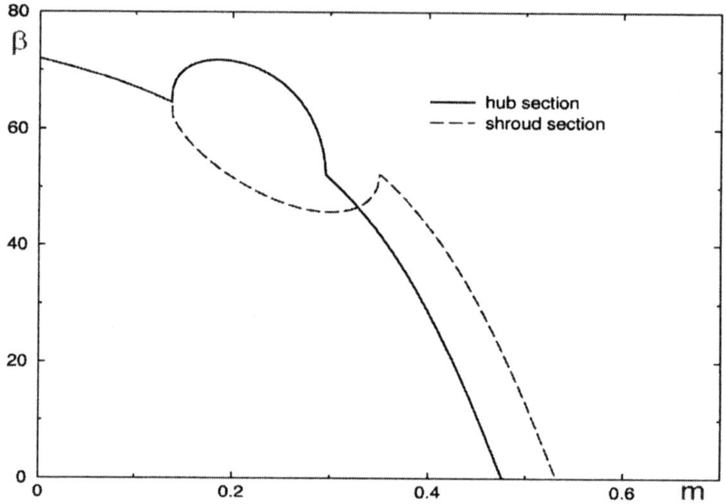

Fig. 6. Flow angle distribution imposed at hub and shroud

Although meridional velocity decrement causes flow turning more tangential at the outlet of cross-over near the hub section (Fig 8. at point C), which means more loading – remains to be improved –, in this case, a simple elliptic variation of the blade angle is introduced in the cross-over area to decrease the average blade loading at the hub section and increase the average blade loading at the shroud section.

The blade camber line θ distribution is then calculated by integrating the vane angle along the meridional length:

$$\theta(m, \beta_{bl}) = \int_{LE}^{TE} \tan \beta_{bl} / R \; dm \tag{5}$$

$\theta_{hub}=\theta_{shroud}$ at leading edge. The amplitude of the elliptic angle variation at the cross-over is adjusted until the value of $(\theta_{hub}-\theta_{shroud})_{TE}$ is within a prescribed limit corresponding to the maximum acceptable lean at trailing edge.

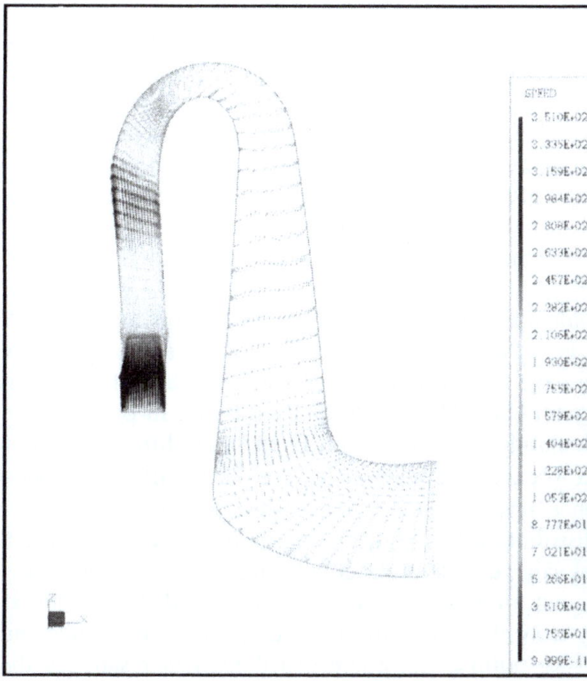

Fig. 7. Velocity vectors on a grid surface near the suction surface of the CBL vanes

The blade design is completed by adding a thickness distribution to the camberline. The thickness is defined by an ellipse in the first 25% of the blade length. The short axis is 10% of the longitudinal one, which results in a 2.5% maximum blade thickness relative to the total vane length. A linear variation is imposed in the second part between the point of maximum thickness and the prescribed trailing edge thickness.

The result of a 3D analysis by means of the Navier Stokes solver is shown on Fig. 7. The meridional contour seems somewhat distorted on this figure because one has tried to get the best view on the distorted 3D grid surface. The separation at the inlet of the cross-over has disappeared but it is still present at he exit of the cross-over. As will be explained in next section this is mainly due to a local variation in blade loading resulting from the meridional curvature.

4. Inverse Design of Vanes

In order to get a better control of the flow in this complex geometry it has been decided to redesign it by means of the inverse design method developed by Demeulenaere et al. [1,2]. It is an iterative procedure in which an initial geometry is modified until the desired/prescribed velocity or pressure distribution on the blade contour is obtained. It makes use of a 3D Euler solver with modified boundary conditions to calculate the normal velocity corresponding to the required velocity on the present blade contour. This V_n distribution is then input for a transpiration technique to define the new blade geometry corresponding to the streamlines of the Euler solution. It is a full 3D method, in which the designer decides in how many sections along the blade span he wants to specify the Mach number distribution on suction and pressure side. In present design the Mach number distribution is specified only at hub and shroud. The final distribution imposed at every grid point on the blade is then obtained by a linear interpolation between these two sections.

The main problem when making an inverse design is to define a Mach number distribution that is not only optimum in terms of losses and pressure rise but that corresponds to a vane that satisfies the constraints related to vane thickness, flow angle and maximum lean. This is especially difficult in present case because the vanes are very long and of short height. Even small differences between hub and shroud may lead to excessive lean angles (i.e. large differences between θ_{hub} and θ_{shroud}). Experience has shown that the easiest way to assure that the imposed Mach number distribution is a realistic one, is by deriving it from the one calculated on an initial vane, taking into account some basic flow relations.

Although it is not a requirement it is recommended that the initial geometry is not too far away from the optimum to speed up the convergence and to facilitate the definition of an optimum pressure distribution that satisfies the mechanical constraints.

Present inverse design starts from the previously defined 3D geometry and the imposed Mach number distribution is derived from the one obtained by a 3D Euler analysis and indicated by +++ on Fig. 8.

Two critical areas where improvements are possible are easily distinguished in these initial Mach number distributions. The first one is at the start of the cross-over bend. The meridional velocity increases near the hub side because of the convex curvature. As the tangential velocity remains unchanged, the flow becomes more axial (lower β_{fl}) and the blade loading locally reduces to almost zero (A). The opposite occurs near the shroud wall (B) where the meridional velocity decreases under the influence of the concave wall. This results in an increased blade loading to keep the flow tangent to the blade.

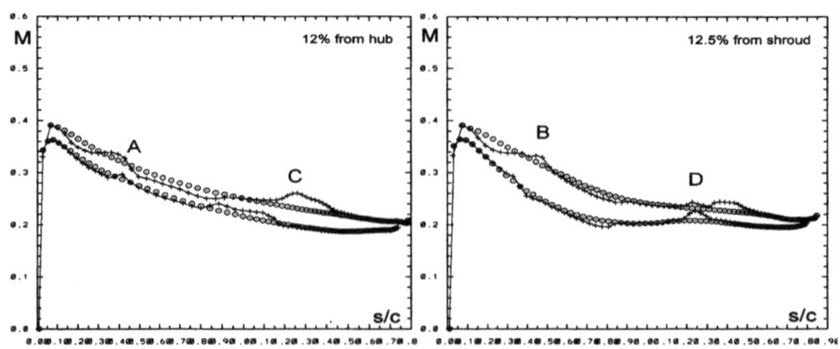

Fig. 8. Mach number distribution on the initial blade at 12% blade height from hub (left) and shroud

The second critical zone is at the end of the cross-over where the impact on the curvature disappears and the meridional velocity becomes again uniform over the passage width. The meridional velocity component suddenly decreases on the hub contour (end of convex wall) and the flow turns to more tangential. This results in a local increase of the blade loading (C). A sudden decrease of the blade loading is observed on the shroud wall (D) when the meridional flow suddenly accelerates when going from the concave to the straight wall.

The main purpose of the vane redesign by means of the inverse design method is to eliminate these local flow accelerations and decelerations and as such, avoid flow separation in the cross-over and return channel.

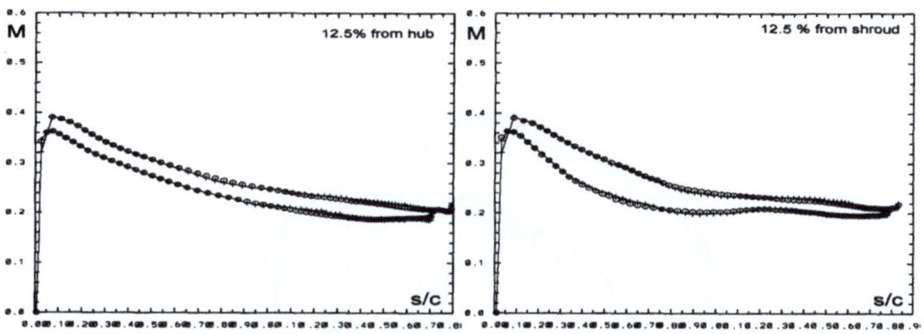

Fig. 9. Imposed (o) and calculated (+) Mach number distribution on the redesigned vane at hub and shroud

The smooth Mach number distribution imposed on the suction and pressure side at hub and shroud during the inverse design are shown by circles on Fig. 8. As the meridional contour is unchanged during this design process one can not expect big changes of the average velocity. The irregular blade loading however is replaced by an almost constant one. Attention is given to keep the integral of the blade loading: unchanged in order not to modify the outlet flow angle.

$$\int_{LE}^{TE} \left(W_{ss} - W_{ps}\right) \, ds \tag{6}$$

The inverse solver was fully converged after 550 time steps and the imposed and calculated Mach number distribution agree very well as shown in Fig. 9.

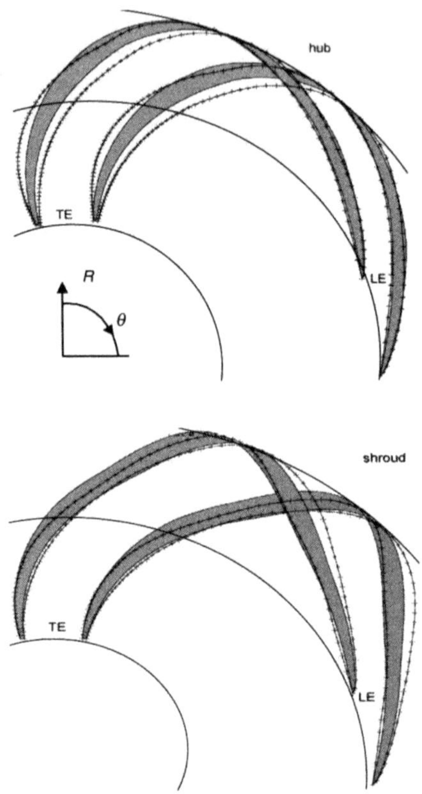

Fig. 10. Comparison between original (grey) and redesigned (+++) vanes

The original (grey) and redesigned (+++ lines) vanes are compared on Fig. 10. The main modifications of the blades are at the cross-over and downstream of it. They are more clearly visible on the m-$R\theta$ plots on Fig. 11.

The vane thickness near the shroud is increased at the start of the cross-over. This increased blockage compensates for the deceleration that normally takes place at the start of the concave side of the cross-over and pushes the flow towards the hub side. The blade thickness is decreased at the hub to compensate for the acceleration that takes place at the start of the convex side of the cross-over. The shroud vane thickness decreases at the end of the cross-over to limit the flow acceleration at the end of the concave wall. A rapid increase of the vane thickness is observed at the end of the cross-over at the hub to limit the deceleration at the end of the convex wall.

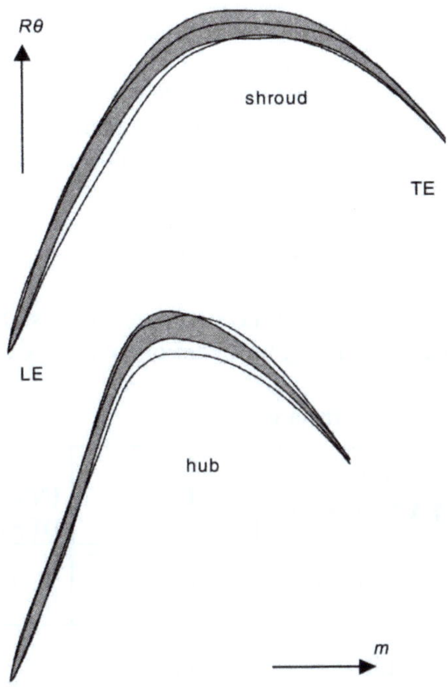

Fig. 11. Comparison between original (grey) and redesigned (—) vanes

5. Overall Performance

The main parameters to characterise the performance are the static pressure recovery:

$$C_p = \left(\overline{P}_3^s - P_2^s\right) / \left(P_2^o - P_2^s\right) \tag{7}$$

and total pressure loss coefficient:

$$\omega = \left(P_2^o - \overline{P}_3^o\right) / \left(P_2^o - P_2^s\right) \tag{8}$$

The latter one is of less importance in diffusers and return channels but, as shown by the following relation:

$$C_p + \omega = 1 - \left(\overline{P}_3^o - \overline{P}_3^s\right) / \left(P_2^o - P_2^s\right) \approx C_{p_{is}} \tag{9}$$

indicates, for a given geometry, how much of the possible static pressure rise is not achieved because of total pressure losses.

All calculations are made with the Navier Stokes solver for the inlet flow conditions specified in Table 1. An attempt was made to keep the mass flow constant.

However, as one imposes the outlet static pressure, the mass flow will be influenced by any change in pressure rise. One has therefore iteratively modified the outlet static pressure until the mass flow was within some tolerances around the target value. Anyway, the absolute inlet flow angle is kept constant at 67.6° so that a change in mass flow will not modify the vane incidence angle. Mach numbers are very low so that one can assume that the small change, corresponding to the difference in mass flow, will not significantly influence the non-dimensional coefficients C_p and ω. The averaging of the exit pressures is based on the conservation of mass, momentum and energy. Values are listed in Table 1 together with the corresponding performance parameters.

The results show a large increase of the static pressure rise for the CBL (Controlled Blade Loading) blade extending upstream of the cross-over.

Table 1. Comparison of overall performance

Parameters	**NON-EXTENDED**	**CBL**	**CBL + INVERSE DESIGN**
P_2^o [Pa]	299700.	299700.	299700.
P_2^s [Pa]	158760.	182298.	174800.
P_3^o [Pa]	231601.	262248.	262854.
P_3^s [Pa]	225713	257524.	256594.
ω	**0.483**	**0.319**	**0.295**
C_p	**0.475**	**0.641**	**0.655**
$\dot{m}$ [kg/s]	4.71	4.00	4.49

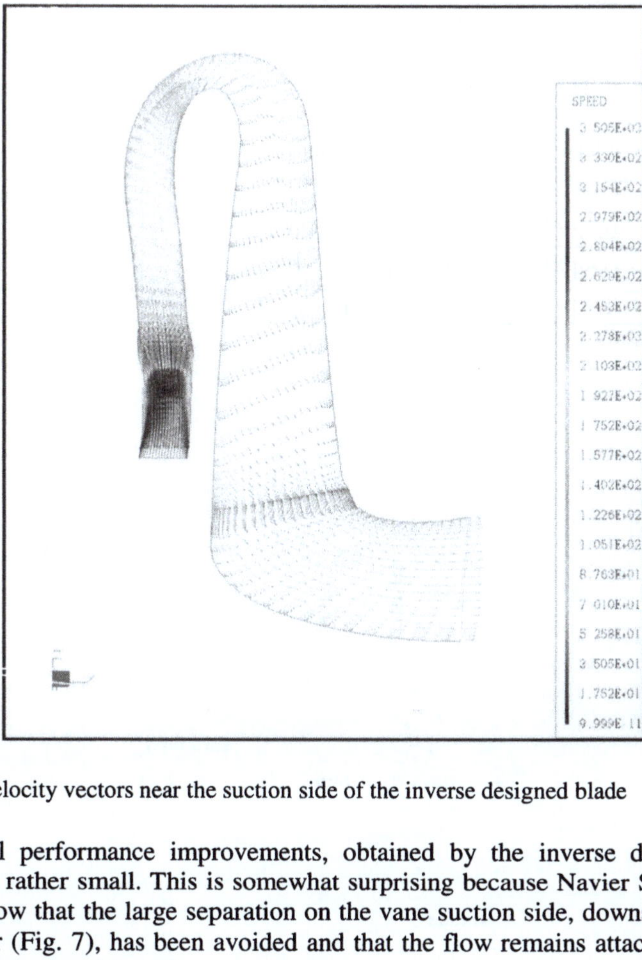

Fig. 12. Velocity vectors near the suction side of the inverse designed blade

Additional performance improvements, obtained by the inverse design of the blade, are rather small. This is somewhat surprising because Navier Stokes calculations show that the large separation on the vane suction side, downstream of the cross-over (Fig. 7), has been avoided and that the flow remains attached until the vane trailing edge (Fig. 12).

6. Influence of Lean

It can be expected that these long flow channels of small height will generate large secondary flows. The previous inverse design, being made by means of an Euler solver, can not account for this and following evaluation about the impact of lean on performance has been made with the Navier Stokes solver.

A common way to control secondary flows is by means of lean and sweep. It is unlikely that sweep may have any impact on these long vanes because of the small height and only the impact of lean has been investigated. It is introduced by rotating all points on the blade shroud over $\Delta\theta$=4.0° around the symmetry axis into the direction of the suction side (negative lean) and towards the pressure side (positive lean). The hub section is unchanged and the vanes are reconstructed by straight lines from hub to shroud. This circumferential shift of the shroud results in a maximum lean of 27.0° inside the cross-over (at the point of maximum radius).

Table 2. Impact of lean on return vane performance

	CBL	No lean	Negative lean	Positive lean
ω	0.3277	0.2785	0.2635	0.2802
C_p	0.6478	0.6691	0.6817	0.6380
β_3	4.25	-5.37	-2.99	10.75

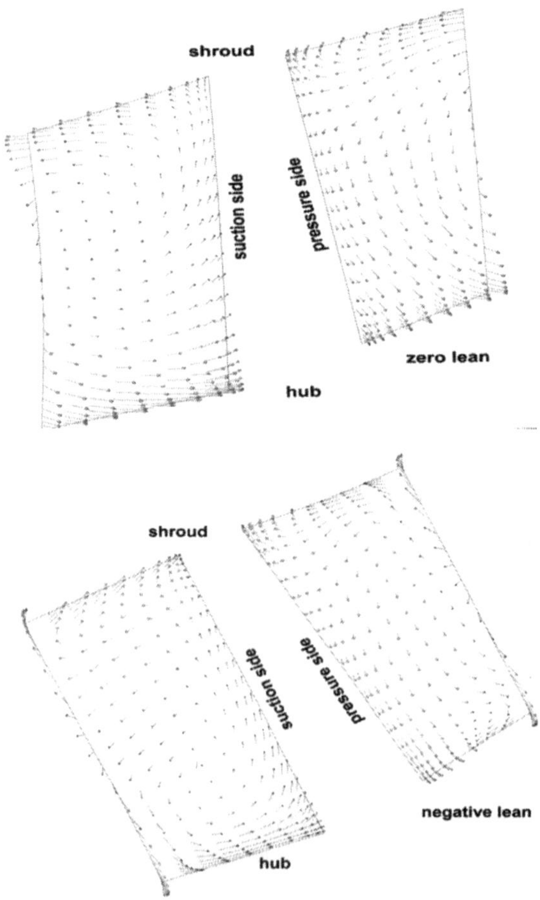

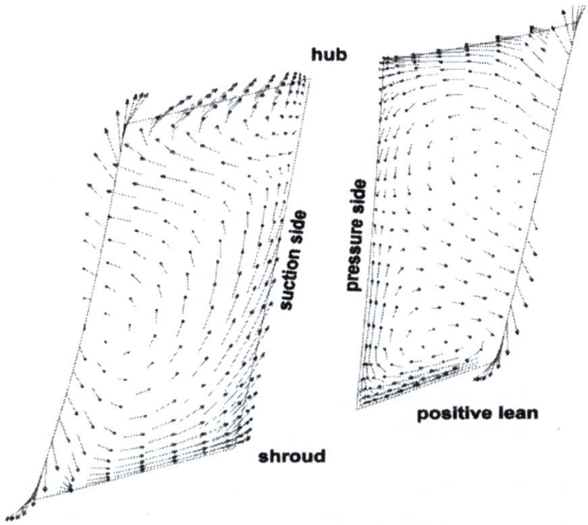

Fig. 13. Velocity vectors projected on a cross section near the vane trailing edge

The impact of lean on performance is listed in Table 2. One observes an additional increase of the pressure rise with negative lean and a 5% decrease with positive lean.

The impact of lean on secondary flows and vane outflow conditions is illustrated by the projection of the velocity vectors in a grid surface shortly upstream of the vane trailing edge (Fig. 13). Shown is the flow in one half pitch on each side of the vane. The left part of each figure should be placed next to the right one to obtain an idea of the complete passage vortex. The non negligible counter-clockwise vortex at 0.0° lean (Fig. 13) does not only become weaker but is even locally inverted when introducing negative lean. The higher losses at positive lean are due to the much stronger counter-clockwise vortex near the shroud surface and an additional clockwise vortex near the hub pressure side corner.

7. Conclusions

Large separation bubbles are likely to occur on the shroud side of the cross-over and in the hub suction side corner of classical return channels. The first one is caused by a sudden deceleration at the inlet of the cross-over. The latter one is caused by the sudden deceleration of the flow at the exit of the cross-over in combination with incidence problems resulting from the strong flow variation from hub to shroud.

Extending the vanes upstream of the cross-over may result in a considerable improvement of the performance. It requires however a careful design of the blade, in which the blade loading is controlled. It provides the potential for a reduction in overall dimension and cost without loss in performance.

Avoiding the flow separation on the vane suction side, by redesigning it by means of an inverse method, provides some additional gain in performance.

A correct leaning of the vanes also contributes to better performance by reducing the vorticity resulting from secondary flows.

Optimizing the cross-over curvature radius and divergence of the return channel may result in further improvements of performance but have not been investigated.

References

[1] Demeulenaere A., Van den Braembussche R.A. (1998), Three-dimensional Inverse Method for Turbomachinery Blading Design, ASME Journal of Turbomachinery, Vol. 120.

[2] Demeulenaere A. (1997), Conception et development d'une methode inverse pour la generation d'aubes de turbomachines, Ph.D Thesis, von Karman Institute, Belgium.

[3] Lenke L.J., Simon H. (1998), Numerical Simulation of the Flow through the Return Channel of Multistage Centrifugal Compressors. ASME 98-GT-255.

[4] Meng S.Y., Jackson E.D. (1983), The Continuous Diffusion Crossover System Design. ASME FED-Vol3. Return Passages of Multistage Turbomachinery.

[5] Rothstein E. (1984), Experimentelle und Theoretische Untersuchung der Strömungsvorgänge in Rükfürkanälen von Radialverdichterstufen, insbesondere solchen mit geringen kanalbreiten. PhD. Thesis, Aachen, Germany.

[6] Simon H., Rothstein E. (1983), On the development of return passages of multistage centrifugal compressors. ASME FED-Vol3. Return Passages of Multistage Turbomachinery.

[7] Thygesen R. (2000), Optimization of Return Channel Blades for Radial Compressors. VKI PR 2000-21.

[8] Van den Braembussche R.A. (1990), Design and Optimisation of Centrifugal Compressor, VKI Course Notes 141.

[9] Veress A. (2001), Inverse Design on Return Flow Channel for Multistage Radial Compressor, VKI-PR 2001-27

The Flow and Head Distribution within the Volute of a Centrifugal Pump in Comparison with the Characteristics of the Impeller without Casing

P. HERGT, Dipl.-Ing.
67059 Ludwigshafen
Germany

S. MESCHKAT, Dipl.-Ing., B. STOFFEL, B., Prof. Dr.-Ing.
Chair of Turbomachinery and Fluid Power, Darmstadt University
Magdalenenstr. 4, 64289 Darmstadt
Germany

Abstract. Measurements of the unsteady velocity, pressure and flow angle have been carried out at the impeller outlet of a centrifugal pump with and without volute casing at 5 operating points using the hotwire technology and a fast response single hole cylindrical probe. The test fluid was air.

While the velocities and pressures depend only on the axial coordinate and are rotationally symmetrical, if there is no casing around the impeller, the influence of the volute on the circumferential distribution of these quantities increases with the deviation of the operating point from the design point.

In respect to the local throughflow distribution, this influence is much more pronounced in comparison to the pressure distribution.
Key Words: spiral casing Pump, rotor stator interaction, experimental investigation

Nomenclature

b	$[mm]$	width of the vaneless radial diffuser, inlet width of the volute (42 mm)
c_m	$[m/s]$	meridional velocity at r_M
r_M	$[mm]$	radial position of the measuring section (208.5mm)
u_2	$[m/s]$	peripheral velocity at the impeller outlet
x	$[mm]$	axial distance from the front shroud at r_M
H	$[m]$	total head
Q	$[m^3/s]$	flow rate
Q_{des}	$[m^3/s]$	flow rate at design point

ε	[°]	circumferential angle from the tongue in the sense of rotation
φ	[-]	flow coefficient $= c_m / u_2$
φ_{des}	[-]	flow coefficient at design point
$\overline{\varphi}$	[-]	local mean value of φ $\left(= \int_0^1 \varphi \, d\left(\frac{x}{b}\right) \right)$
$\overline{\overline{\varphi}}$	[-]	local mean value of φ $\left(= \frac{1}{2\pi} \int_0^{2\pi} \overline{\varphi} \, d\varepsilon \right)$
ψ	[-]	pressure coefficient $= \frac{2gH}{u_2^2}$

1. Introduction

Single stage centrifugal pumps are mostly designed and produced with a volute casing. It is well known that – in this case – the flow at the impeller outlet can only be nearly independent from the circumferential position in the volute if the volute geometry is well designed and the pump is operated at its design point. For off-design operation of volute casing pumps, the volute casing geometry doesn't "fit" the outflow from the impeller. This leads to a non-uniform distribution of (time averaged) pressure and velocity along the circumference of the volute.

Different approaches exist for modeling this type of rotor-stator-interactions in turbomachines (mixing plane, frozen rotor, transient), see e.g. [2, 5]. To compare the capability of the respective modelizations in taking into account and describing important features of off-design flow in volute casing pumps, detailed flow field measurements and evaluations are needed.

A research project was carried out at the Laboratory for Turbomachinery and Fluid Power of the Darmstadt University of Technology which aimed at gaining more insight into the aspects of rotor-stator interactions in volute casing pumps and – additionally – creates a data basis for the validation of CFD simulations. Likewise experimental Investigations on rotor stator interactions can be found in [1, 2, 4-8].

2. Test rig

The test rig and its instrumentation which was described in more detail in [4] are suited for measurements of transient velocity and pressure distribution for different circumferential positions inside the volute near to the impeller outlet. For this purpose and to enable a continuous variation of the measuring positions along the whole circumference of the volute, the volute casing of the test pump can be turned by an electrical motor around the pump axis while the mechanism for traversing the probes is mounted to a fixed inner part of the casing and remains at a constant location. Table 1 shows the technical data of the pump.

Table 1. Technical specifications

Impeller outlet diameter	:	405	mm
Impeller outlet width	:	38	mm
Impeller inlet diameter	:	240	mm
Blade number	:	7	
Outlet angle, shroud	:	25	°
Outlet angle, hub	:	29	°
Specific speed	:	35	min^{-1}
Rotational speed	:	3000	min^{-1}
Design flow rate	:	1380	m^3/h

As can be seen in Fig. 1, the inner part of the casing is mounted on two vertical uprights on front side and rear side while the turnable volute casing is supported on the inner parts (see Fig. 2) by ball bearings. For the sealing between both parts of the casing, sheets of felt are used.

Fig. 1. Test rig configuration without volute

Fig. 2. Test rig with mounted volute

The configuration with mounted volute is shown in Fig. 2

In the case of the test rig configuration without the volute part of the casing (Fig. 1), flow measurements are possible for free impeller discharge where the pressure at impeller outlet is constant (= atmospheric pressure) along the impeller outlet.

The analysis of flow was performed by the use of hotwire anemometry and a fast response single hole cylindrical probe [4] which were located 6 mm radially outward of the impeller outlet. Using this single hole cylindrical probe with an integrated high frequency response pressure transducer gives information on the unsteady static and total pressure of the flow from which also the flow velocity can be calculated. But from the experience gained, more reliable information on the unsteady velocity and flow angle is found from a single-wired hotwire probe.

Both probes are traversed across the width of the impeller outlet. As the probes are turned stepwise around their longitudinal axis at each axial position while the flow direction is nearly perpendicular to the longitudinal axis, the flow angle can be determined by calculating the position of the maximum of measurement data for all time steps (as an ensembled average value in dependence of the angular impeller position and as a time-averaged value, as well).

The circumferential distribution of pressure on the surface of the single hole cylindrical probe is detected through a 0.3mm diameter hole. A calibration for different constant velocities allows to find the exact maximum pressure position and to determine a circumferential position on the cylinder surface where the measured pressure equals the local static pressure in the flow. The angle of the position where the measured pressure indicates the static pressure of the flow remains constant for a wide range of velocities as described in Schlichting [3]. Because transient effects in unsteady flow influence the circumferential pressure distribution on the probe surface and thus especially the right position where the surface pressure equals the local static pressure of the flow, the determination of the flow direction and magnitude of velocity is done additionally by the hotwire technique.

3. Results and Discussion

In the following, distributions of time-averaged measurement data will be discussed.

Impeller without casing, throughflow:

The distribution of the throughflow component φ is plotted versus the dimensionless axial coordinate in Fig. 3 for various values of the normalized flow rate Q/Q_{des}. The character of the curves is well known from other publications and shows that the main throughflow takes place near the rear shroud.

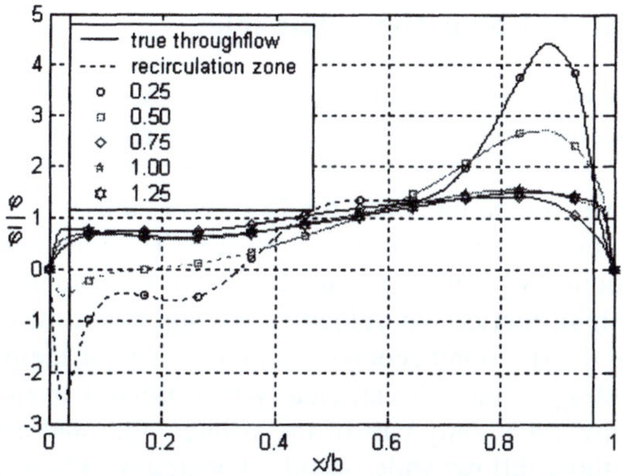

Fig. 3. Distribution of the meridional velocity, without casing

At strongly reduced flow rates (Q/Q_{des} = 0.50 and 0.25), backflow occurs at the measuring location near to the front shroud. Because of the flow recirculation inside and outside the impeller, part of the outflow from the impeller serves to compensate the backflow in the balance of mass flow. Therefore, only the remaining part of the outflow is representative for the mass transport through the pump and is called "true throughflow". In Fig. 3, the broken part of the curves indicates the width of the recirculation zone within which the part of the outflow only compensates the backflow. All distributions are rotationally symmetrical.

Impeller without casing, head:

The distribution of the local head is shown in Fig. 4. While the head is nearly independent of the axial coordinate at high flow rates, higher heads are measured near the rear shroud at low flow rates. This is somewhat surprising because this is also the area with the higher meridional velocities, i.e. local througflow (see Fig. 3

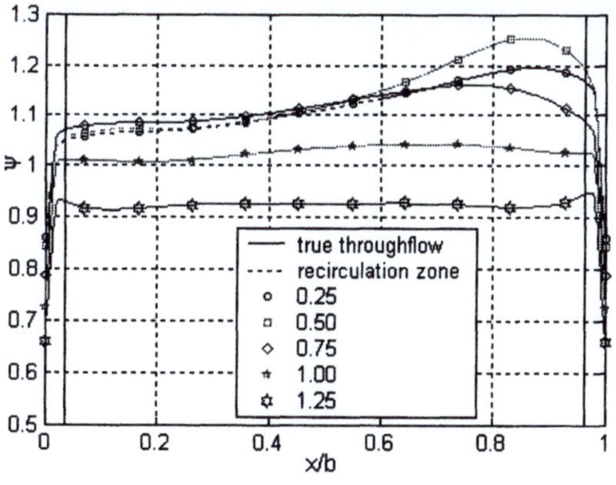

Fig. 4. Distribution of local head, without casing

Throughflow distribution at the volute inlet:

The meridional component characterizing the local throughflow in the measuring section (that is now the volute inlet) is shown for 5 values of the normalized flow rate in Fig. 5.

The broken curves on the diagrams for 75%, 50% and 25% of the design flow rate represent the limit of the true throughflow area. The remaining area characterizes the recirculation zone with a negative and the corresponding positive flow rate. Whereas the high meridional velocities are still found near the rear shroud, the circumferential distribution becomes the more irregular the more the flow rate deviates from the design value. It can clearly be observed from the distributions, that at part load flow rates, only part of the volute width and circumference contribute to the true throughflow while outside this area backflow and forward flow compensate each other and no net mass flow results. The borderline between the backflow area and the forward flow area can be identified from the isocontour having the value 0 of the dimensionless meridional component.

At 125% of the design flow rate, the highest throughflow component has been measured approximately 30° upstream of the tongue, what means that the local flow rate increases along the circumference as is found also by theoretical considerations.

In contrary to this, the local flow rate decreases along the circumference if the pump is operated at low flow rates which is also in correspondence to the theory.

At 25% of design flow rate, less than half of the volute inlet cross section contributes to the throughflow. The rest consists of a strong recirculation especially just upstream of the tongue where no true throughflow exists. Referred to the impeller frame of reference, it becomes evident that blade channels passing this area of the volute don´t contribute to the net impeller mass flow and are exposed to strong backflow once per each impeller revolution.

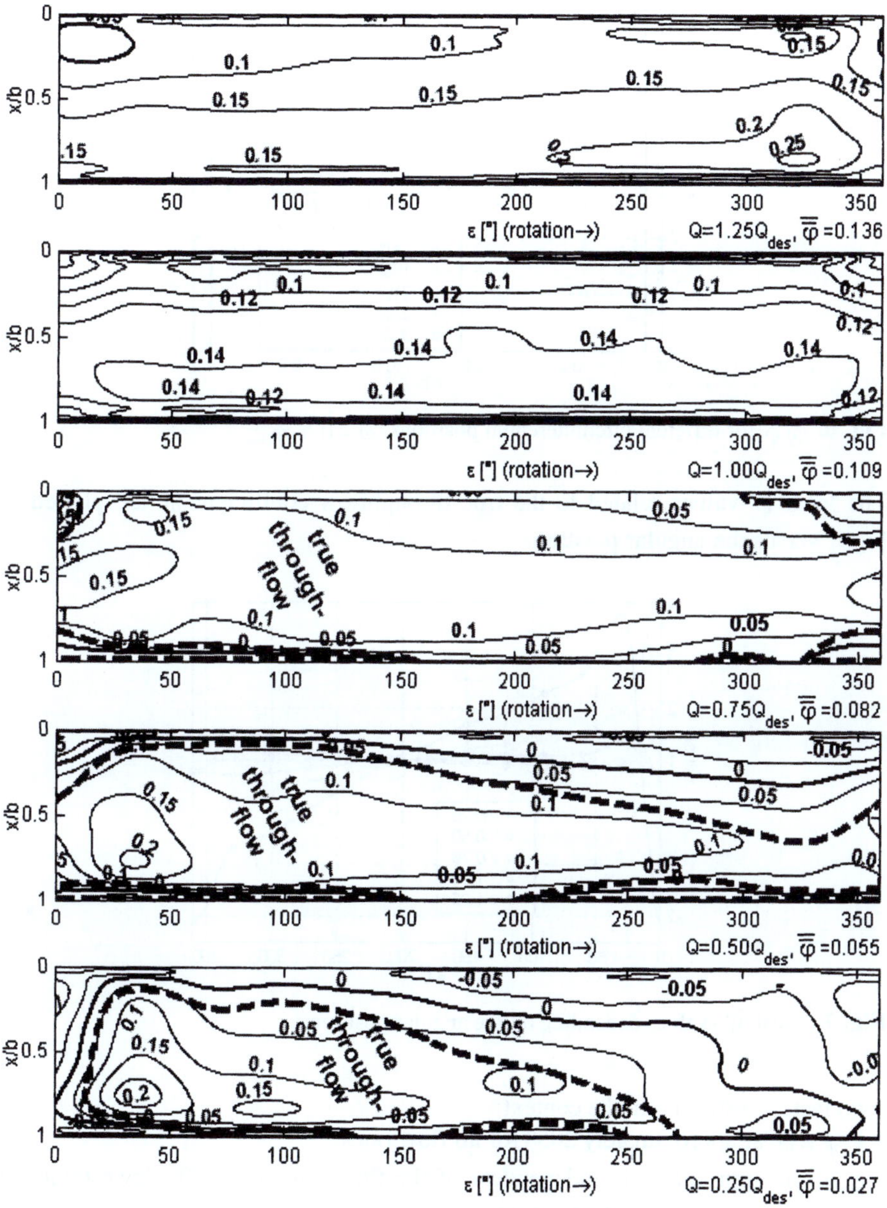

Fig. 5. Distribution of meridional component at the volute inlet

The curves in Fig. 6 underline the findings for 50% flow rate and some selected circumferential positions.

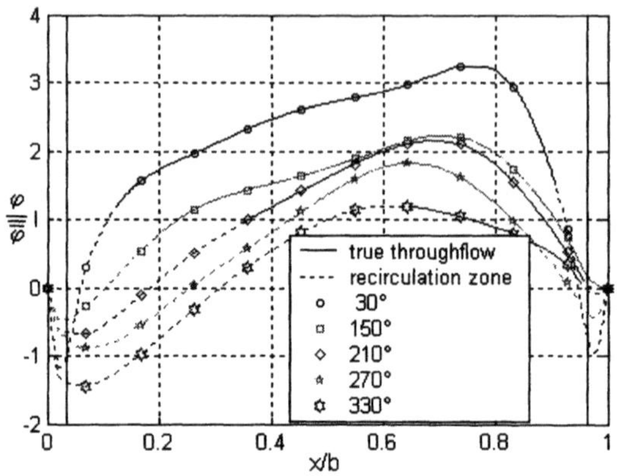

Fig. 6. $\varphi/\overline{\overline{\varphi}}$ for different circumferential positions. $\varphi = 0.5 \cdot \varphi_{des}$

The local $\overline{\varphi}$-values related to the true throughflow coefficient $\overline{\overline{\varphi}}$ are plotted in Fig. 7 versus the angular position.

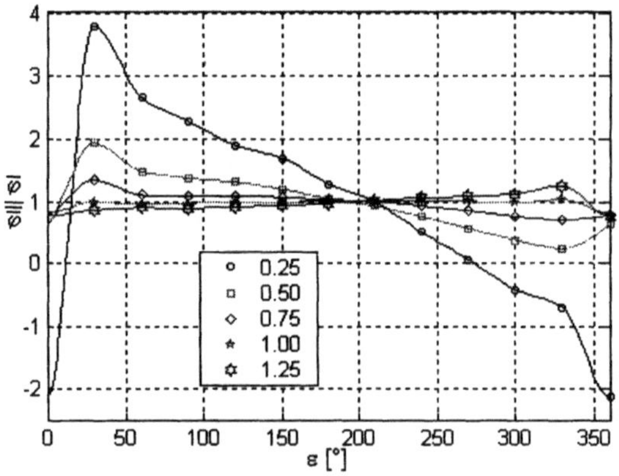

Fig. 7. Local $\overline{\varphi}$ -values depending on the angular position

Two facts are striking in the context:

$\overline{\varphi} = \overline{\overline{\varphi}}$ at approximately 205° for all flow rates.

$\overline{\varphi}$ is approximately independent of the flow rate at ca. 30° downstream of the tongue (3.8x0.25 ≈ 1.95x0.5 ... etc.)

This seems to be an interesting feature of the impeller-volute interaction but can not yet be explained presently.

Integrating the local flow rates along the circumference leads to Fig. 8.

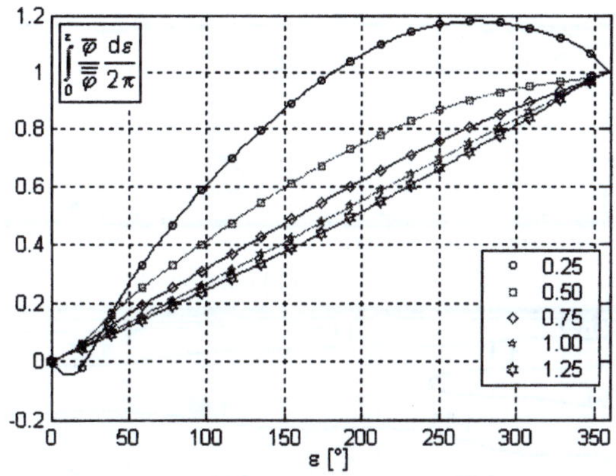

Fig. 8. Flow rate integrated along the circumference

As it could be expected already from Fig. 5 there is a linear dependence between the flow rate and the angular position at $\varphi = \varphi_{des}$, while the local flow rate increases up to $1.2\text{x}\overline{\overline{\varphi}}$ ($\varepsilon \approx 270°$) and falls down to $1.0\text{x}\overline{\overline{\varphi}}$ in the following sections due to the strong backflow in this area (see Fig. 5).

Head distribution at the volute inlet:

In addition to the throughflow distribution of Fig. 5, the distribution of head is shown in Fig. 9. Again, the broken lines characterize the true throughflow areas.

In comparison to the distribution of the meridional velocity, the distribution of head is more regular along the width of the measurement section for overload and design point. For part load, the shape along the width deviates from a flat distribution to a distribution with increased head near the rear shroud as this can also be found for the head distribution of the impeller without casing in Fig. 4.

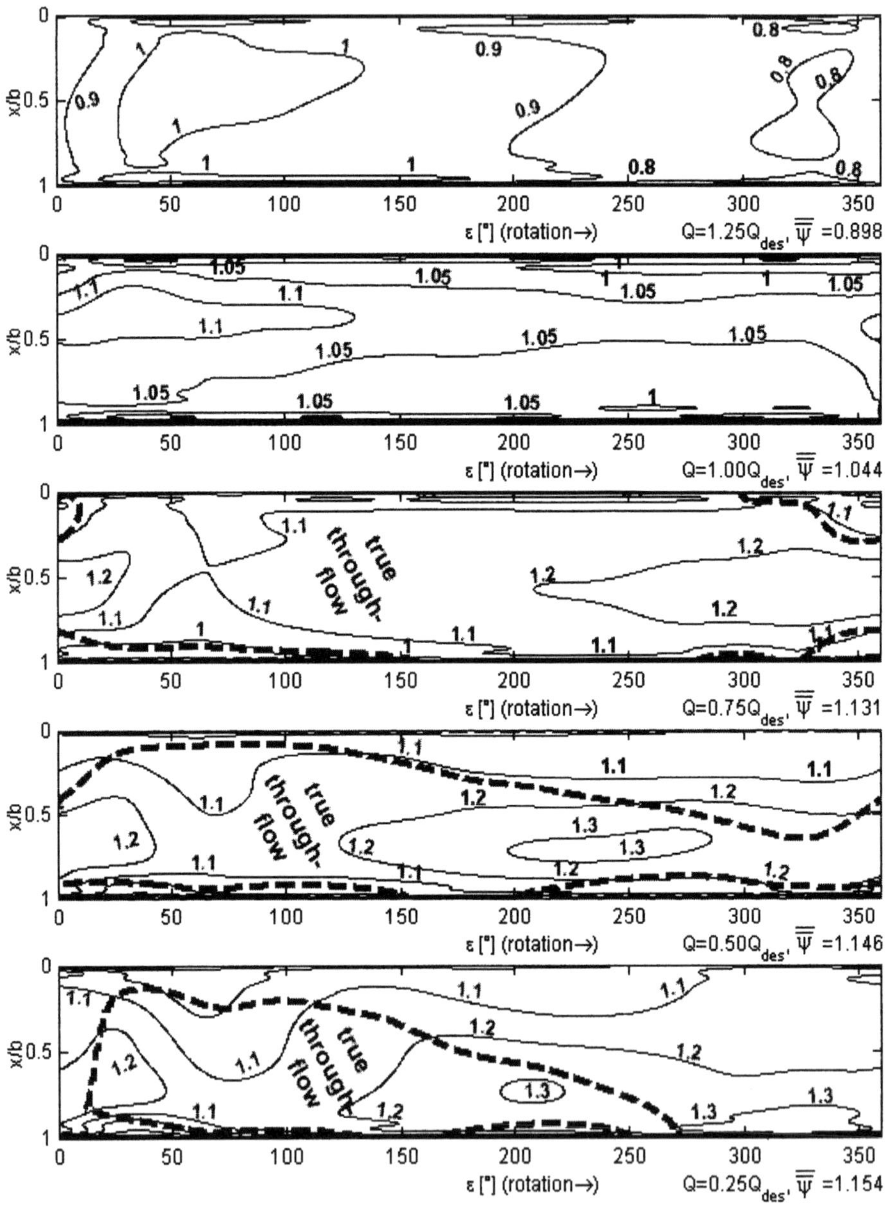

Fig. 9 Distribution of local head at the volute inlet

As expected, the circumferential distribution becomes more irregular with increasing deviation from the design flow rate. This could at least partially be caused by the variation of the circumferential velocity components.

The lowest pressure coefficients have been measured in the backflow zones ($\varphi < 0$) where a part of the energy has been lost by wall friction and mixing losses.

A detailed explanation of the relations between flow and head distribution is rather difficult due to the lack of information about what is caused by separation and recirculation within the impeller on the one side and within the volute only on the other side.

To get this information, measurements within the impeller would be necessary in addition.

4. SUMMARY

Distributions of flow and head at the impeller outlet in a centrifugal pump with volute casing were shown. Some typical distributions for these kinds of pumps were explained and discussed. Most of the results are strictly related to theory.. Thus the findings and the qualitative shape of the illustrated graphs and distributions can serve as an orientation for computational fluid dynamics and may encourage to do likewise analyses in the frame of postprocessing of such calculations.

Acknowledgments

Thank has to be expressed to the KSB-foundation whose financial and technical support made this work possible

References

[1] Flörkemeier KH (1977) Experimentelle Untersuchungen zur Optimierung von Spiralgehäusepumpen mit tangentialem und radialem Druckstutzen. PhD-Thesis, TU Braunschweig

[2] Fritz J (1999) Strömungswechselwirkungen in hydraulischen Maschinen. PhD-Thesis, TU München

[3] Schlichting H, Gersten K (1999) Boundary Layer Theory. Springer-Verlag, Berlin, New York

[4] Stoffel B, Meschkat S (2002) The local specific head at different circumferential positions in a volute casing centrifugal pump in comparison to the characteristic curve of the single rotor. Proceedings of the Hydraulic Machinery and Systems 21st IAHR Symposium, Lausanne.

[5.] Treutz G (2002) Numerische Simulation der instatonären Strömung in einer Kreiselpumpe. PhD-Thesis, TU Darmstadt

[6] Weiß K (1995) Experimentelle Untersuchungen zur Teillastströmung bei Kreiselpumpen. PhD-Thesis, TU Darmstadt
[7] Ubaldi M (1996) An Experimental Investigation of Stator Induced Unsteadiness on Centrifugal Impeller Outflow. Journal of Turbomachinery, Vol.118
[8] Ubaldi M (1993) Relative Flow and Turbulence Measurements Downstream of a Backward Centrifugal Impeller. Journal of Turbomachinery, Vol.113

Investigation of Unsteady Flow Phenomena in a Side Channel Vacuum Pump by PIV

Michael SCHROLL
Research Engineer
Sales & Application Manager
Intelligent Laser Application GmbH, Germany
Tel.: +49.2461.690431
Tel.: +49.2461.690439
e-mail: schroll@ila.de
internet: **www.ila.de**

Abstract The unsteady turbulent flow phenomena in a Side Channel Vacuum Pump and the mechanism of work transfer are still unknown in detail. The analysis of the flow based on the laser optical Particle-Image Velocimetry (PIV) is subject of this fluidic investigation [3].

The Side Channel Vacuum Pump is adapted to the special requirements for flow investigations with laser optical measuring methods. The unsteady turbulent flow field is investigated in 16 axial and 5 radial measurement layers each for 4 operating points. A more detailed insight into the flow phenomena and the mechanism of work transfer of Side Channel Vacuum Pump is given by the shown distributions of velocity, pressure, energy and turbulent characteristic parameters.

Nomenclature

a	$[m/s]$	sonic velocity
A	$[m^2]$	cross sectional area
K	$[m^2/s^2]$	Bernoulli constant
Ma	[-]	Mach number
n	$[min^{-1}]$	rotational speed
p	$[mbar]$	pressure
Q	$[m^3/s]$	volume flow
r	$[m]$	radius, radial coordinate
Re	[-]	Reynolds number
T	$[K]$	temperature
Tu	[%]	turbulence
u	$[m/s]$	circumferential velocity

v	$[m/s]$	absolute velocity
x	$[m]$	impeller based coordinate
y	$[m]$	impeller based coordinate
Y	$[kJ/kg]$	polytropic available work
z	$[m]$	impeller based coordinate
α	$[°]$	impeller rotation angle
φ	[-]	flow rate
κ	[-]	kappa number
Π	[-]	pressure ratio
ρ	$[kg/m^3]$	air density
ω	$[s^{-1}]$	angular velocity
Ψ	[-]	pressure number

Subscripts and Superscripts

a	outside (German: aussen)
i	inside (German: innen)
i	isotropic
K	channel (German: Kanal)
L	air (German: Luft)
N	nominal
opt	optimum
r	radial
S	suction side
u	circumferential (German: Umfang)
v	velocity
x	impeller based coordinate
y	impeller based coordinate
zul	admissible (German: zulässig)
0,1,2	gate valve (ambient), suction-, pressure side
Δ	difference value

1. Introduction

Motivated by the keen industrial demand for Side Channel Compressor or Side Channel Vacuum Pump (Fig. 1) its investigation comes increasingly in the field of focus for German industry and universities.

All gained cognitions after date about the unsteady turbulent flow phenomena in a side channel compressor and the mechanism of work transfer are predominantly out of results of pressure measurements. But they still disallow to merge the both partly antithetic models of flow.

The analysis of the enforced turbulent flow in the side channel based on experimental results is subject of this fluidic investigation and reached with the laser optical PIV [3].

The installed test facility for the Side Channel Vacuum Pump must be adapted to the special requirements for flow investigations with laser optical measuring methods in the complex geometry of the side channel. The unsteady turbulent flow field should be investigated with PIV in 16 axial and 5 radial measurement layers each for 4 operating points. This extensive information achieved by PIV provides a detailed and quantitative insight into the flow velocity fields in the range of 180° rotation angle.

Flow fields acquired within the side channel can be correlated with the dynamic pressure. These pressure measurements were done at the University of Applied Science Merseburg, Co-Institute Fluid- and Turbomachinery on an identical Side Channel Vacuum Pump. This correlation should enhance the hypothetical thesis for flow phenomena models in the investigated Side Channel Vacuum Pump.

The Bernoulli constant can be calculated for the generated meridional slice out of the acquired velocities and pressures. To characterise the turbulent flow the isotropic turbulence ($Tu_{i.v.K}$) can be calculated out of the unsteady oscillating motions.

2. Experimental Apparatus and Procedure

2.1. Test Facility

Because of their multi-finned impeller, Side Channel Vacuum Pump generate very low pulse suction air which make them extremely suitable for a large number of industrial applications needing precisely defined functions, like:

- vacuum technology for printing, chemical and procedural industry,
- cooling solvent for high power laser used in the manufacturing industry,
- aeration in sewage plants and water preparation systems and

Internally the Side Channel Vacuum Pump consists of a large number of circular hollow rings, half of which is in the upper and lower side of the pump housing; the corresponding other half is cast into either side of the impeller which has scoops on both sides.

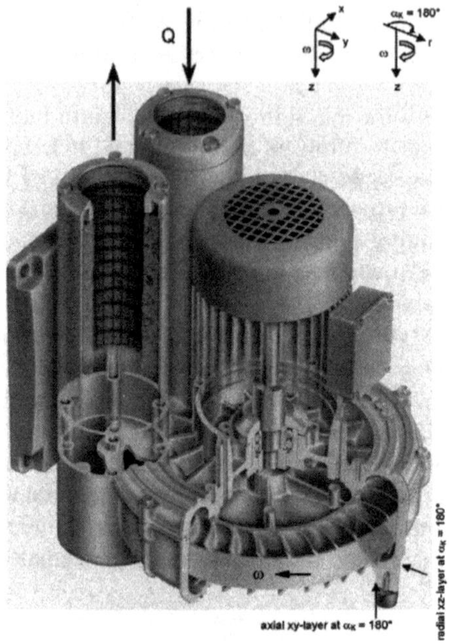

Fig. 1. Side Channel Vacuum Pump SV 5.690/1 in double suction flow configuration, impeller with scoops on both sides [1]

When the impeller turns, air trapped in each hollow ring or chamber is thrown outwards and then compressed by centrifugal force, running into the side channels before entering the next chamber in a screw-type motion. This is repeated several times and compresses the air in several stages throughout the turn of the impeller.

Figure 2 shows the experimental set up around the Side Channel Vacuum Pump installed at the Institute of Hydraulic Turbomachinery and Fluid Mechanics, Technical University of Berlin, Germany. The Side Channel Vacuum Pump is adapted to the special requirements of laser optical measurements techniques like PIV, particularly with regard to the casing cover and its optical access.

The Side Channel Vacuum Pump is RPM-regulated (100...3,600 min^{-1}) by a frequency changer. The four investigated operating points (Table 1) can be achieved by using the gate valve. Gate valve and particle input are located within the suction pipe. The particle input consists out of a mixing reservoir, a magnetic rod and a rotating mixer magnet. The mix out of ethyl alcohol and polyamide powder is sucked in by the vacuum through a perforated tube in the suction branch. After evaporation of the ethyl alcohol the polyamide particles (mean diameter 5μm) are consistent distributed over the whole side channel. Due to particles diameter, which is respirable, the exhaust air must be filtered at the outlet, using a filter class H13.

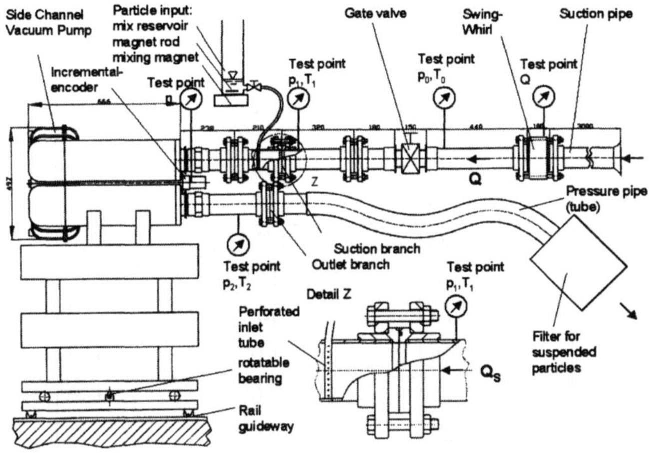

Fig. 2. Experimental set up: Side Channel Vacuum Pump with test points for characteristic curve

For necessary modifications of the casing cover to realize optical access for PIV (Fig. 3), the huge dimensions of the Side Channel Vacuum Pump SV 5.690/1 (Fig. 1) are helpful. Its impeller has 48 vanes on both sides, given radius are: outside $r_a = 0.230$ m, side channel $r_K = 0.034$ m and inside $r_i = 0.162$ m. With the frequency changer the Side Channel Vacuum Pump is regulated to $n = 3{,}000\ \text{min}^{-1}$, this produces circumferential velocities of $u_a = 72.22$ m/s und $u_i = 50.9$ m/s at impellers outside and inside radius.

The minimal differential pressure of the Side Channel Vacuum Pump is reached in over-load operating with $\Delta p = 60$ mbar at its maximum admissible volume flow of $Q_{S.zul.max} = 0.206\ \text{m}^3/\text{s}$. The last operating point, named extreme part-load, with the maximum differential pressure is chosen to $\Delta p = 320$ mbar due to the thermal limitation of acrylic glass. At extreme part-load the Side Channel Vacuum Pump reaches the minimum admissible volume flow of $Q_{S.zul.min} = 0.083\ \text{m}^3/\text{s}$. The velocity inside the channel v_K follows dependent upon volume flow to $v_{K.min} = 23.0$ and $v_{K.max} = 56.6$ m/s. Reference point for the volume flow should be the suction branch. The volume flow measured with the Swing Whirl before the gate valve must be converted to:

$$Q_S = Q\,(p_0/p_1)\,(T_1/T_0)\,. \tag{1}$$

The Best Efficiency Point (BEP) at $Q_{S.N} = Q_{S.opt} = 0.163\ \text{m}^3/\text{s}$ was acquired during measurements with a torque gauge at the University of Applied Science Merseburg, Co-Institute Fluid- and Turbomachinery on an identical Side Channel Vacuum Pump but a casing cover with original geometries. The chosen nominal-loaded operating point is very close to the BEP. Table 1 gives an overview about the most important measurement categories at the investigated operating points.

Table 1. Overview about the most important measurement categories

Operating point	Over-load	Nominal-load	Part-load	Extreme part-load
Δp in mbar	65	130	260	320
T_2 in K	309	320	353	383
n in min^{-1}	3,000	3,000	3,000	3,000
Q_S in m^3/s	0.206	0.175	0.120	0.083
Y in kJ/kg	5.8	11.9	28.1	35.7
v_K in m/s	56.6	49.3	33.0	23.0
$Q_S/Q_{S.opt}$	1.26	1.07	0.74	0.51
φ	0.78	0.68	0.46	0.32
ψ	2.22	4.54	10.75	13.68
Π	1.07	1.14	1.35	1.44

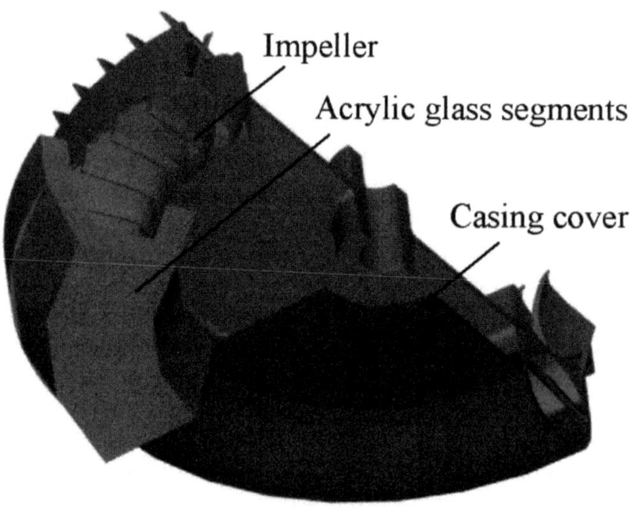

Fig. 3. Casing cover with acrylic glass window

2.2. Casing Cover Modifications

Modifications to the complex geometry of the casing cover are necessary to enable PIV measurements inside the Side Channel Vacuum Pump. Perpendicular optical accesses for laser light sheet and camera are realized in coplanar acrylic glass segments polygonal close to the toroid original geometry. The remaining trapezoid side channel area inside the acrylic glass window stays equal to the original toroid geometry.

Both needed optical accesses are combined in the acrylic glass window, consisting of eight segments. It is located opposite to the interrupter at the rotation angle of $\alpha_K = 180°$ and covers a range of 60° rotation angle.

2.3. PIV System and Procedure

The used PIV system at this time was partly a loan of ILA GmbH and due to this close to a commercial standard system consisting of:

- double pulsed Nd:YAG laser with 25 mJ at 532 nm (Continuum Minilight)
- an articulated mirror arm
- a light sheet optic with coated lenses
- 1x1k double shutter CCD camera (PCO SensiCam)
- PIV Syncronizer (ILA GmbH)
- VidPIV4.x Software (ILA GmbH)

As seeding particles the white and high reflective 5 µm polyamide powder Orgasol® is used, given into the flow as a mixture with ethyl alcohol (114 g Orgasol® in 1 l ethyl alcohol, Seeding/Air: $1{:}1.7{*}10^{-5}$, Concentr.: $19{*}10^{9}$ m^{-3}) as written above. The flow is measured correlated to the rotation angle of the impeller in 16 axial and 5 radial layers as shown in Figure 4.

For measurements in axial layers the light sheet is radial adapted to the Side Channel Vacuum Pump and the camera is looking parallel to the drive shaft. The 16 axial layers starts at $z = -15$ mm, looking inside channel between two vanes, rising up to the casing cover at $z = 36$ mm with an increment of 3 mm. Between 2…8 mm optical distortions caused by a glue splice makes measurements impossible. For measurements in radial layers the camera is looking radial and the light sheet is adapted parallel to the drive shaft. The first of the 5 radial layers is located at $r = 172$ mm, close to the inner radius r_i, the last one is located at $r = 212$, close to the outside radius r_a.

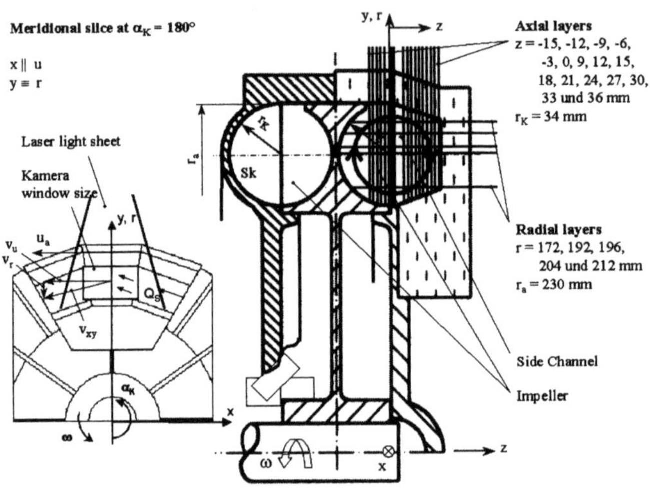

Fig. 4. Position of the 16 axial and 5 radial measurements layers

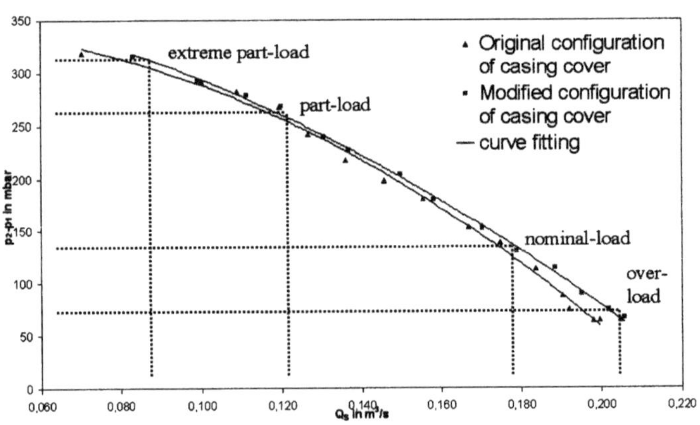

Fig. 5. Characteristic curves in comparison

The acquired absolute v_x-v_y-flow fields are converted from the not fixed Cartesian coordinate system of the moveable CCD camera to the polar coordinate system of the fixed Side Channel Vacuum Pump and shifted to their absolute position in a 3D data volume by several transformations using MATLAB® routines. Out of the 3D volume data set meridional slices at $\alpha_K = 180°$ are generated. These slices are similar to the investigated pressure measurement plane at the University of Applied Science Merseburg, Co-Institute Fluid- and Turbomachinery on an identical Side Channel Vacuum Pump [5], allowing comparing velocity and pressure directly as well as combining them in the Bernoulli constant. Furthermore the polar velocity components v_r and v_u and their statistical variation over in minimum 30 up to 1,000 single measurements are calculated in MATLAB® and prepared for combined contour/vector plots.

To ensure that pressure and velocity can be correlated, the characteristic curves of both machines in Berlin and Merseburg are compared in Figure **5** out of which four significant operating points are chosen for this investigation (Table 1).

3. Unsteady and Compressible Flow

The flow inside a side channel compressor is described as a forced unsteady turbulent vortex flow [6]. This can be ascertained as well in the investigated machine for the velocity and the pressure [6,2] because every rotation the impeller is passing through the complete pressure range and the interrupter with its expansive flow. This causes highly dissipation losses as heat radiation [7] and the work process must be considered as polytrophic. But investigations of the casing temperature shows that the temperature is nearly not rising below rotation angles of $\alpha_K = 240°$ and close to the ambient temperature. This allows to consider the work process as approximately adiabatic and irreversible at the investigated rotation angle of $\alpha_K = 180°$ with the reached pressure and temperature ratios of $p_2/p_1 = 1.07...1.44$ and $T_2/T_1 = 1.05...1.30$. This means additional the complete specific dissipation energy remains inside the compressor [4].

The specified adiabatic flow can be regulated to quasi-steady operating points with reasonable accuracy, allowing to calculate an average of several rotation correlated PIV measurements, but loosing the unsteady information like turbulence in the flow. The turbulence is in close relation to the Reynolds number

$$Re = v_K \; D_{hyd} / \upsilon . \tag{2}$$

With the existing geometry the channel velocity is calculated to:

$$v_K = Q_S / A_K = Q_S / \pi \, r_K^2 , \tag{3}$$

giving Reynolds numbers of $Re = 2.5 \times 10^5$ and 1.0×10^5 with $v_K = 56.6$ and $v_K = 23.0$ m/s for over-load and extreme part-load. Both Reynolds numbers indicates a turbulent flow high above the critical Reynolds numbers of $Re = 0.023 \times 10^5$. Turbulent flows can be quantified with the turbulence

$$Tu = \left(1/\bar{v}\right) \sqrt{\left(\overline{v_x'^2} + \overline{v_y'^2} + \overline{v_z'^2}\right)/3} \; . \tag{4}$$

Having isotropic turbulence Eq.(4) reduces to

$$Tu_i = \left(1/\bar{v}\right) \sqrt{\overline{v_x'^2}} \; . \tag{5}$$

Due to its low image capture frequency acquiring the fluctuation of v_x directly with PIV is not possible Therefore the variance s^2 of v_x and v_y over a large number of single PIV measurements (up to 1000) is used to calculate the turbulence

$$Tu_{i.v.K} = \left(1/\overline{v_K}\right) \sqrt{\overline{v_x'^2}/2} \tag{6}$$

Thermal and caloric ideal gases can be handled as incompressible at Ma≤0.4, accepting an error of max. 8% for air density variation [4]. With the maximum admissible volume flow of $Q_{S.zul.max} = 0.206\ m^3/s$, giving the maximum channel velocity $v_{K.max} = 56.6$ m/s, the sonic velocity

$$a_L = \sqrt{\kappa \left(p_l / \rho \right)} \tag{7}$$

results to $a_L = 349.1$ m/s and the maximum Mach number

$$Ma = v_K / a_L \tag{8}$$

to $Ma_{max} = 0.162$. At this Mach number inside the Side Channel Vacuum Pump at $\alpha_K = 180°$ considering

$$\rho_0/\rho_0 = \left[1 + \left((\kappa - 1)/2 \right) Ma^2 \right]^{\frac{1}{\kappa-1}} \tag{9}$$

the accepted error in air density variation reduces to negligible

$$100 \left(\left(\rho_0 - \rho \right)/\rho \right) \leq 1.4\% \ , \tag{10}$$

permitting to discus the investigated flow fields approximately as incompressible. This is specially to consider calculating the Bernoulli constant.

For the meridional slice pressure and velocity is acquired, both on a measuring grid with increments of 1 mm in y- and 3 mm in z-direction. For the overlapping areas the incompressible Bernoulli constant K can be calculated out of this both measurement categories for quasi-steady operating points disregarding all field forces to:

$$K = \left(v_{xy}^2 / 2 \right) + \left(p/\rho \right). \tag{11}$$

Due to collision risk with the impeller and missing mechanical guidance close to the casing cover the pressure measurements are limited between z = 3…27 mm. A glue splice in the acrylic glass window causes optical distortions between 2…8 mm making PIV measurements impossible. The remaining overlapping accessible area for the Bernoulli constant results in z = 9…27 mm, still the bigger part of the meridional slice.

4. PIV Results and Discussion

4.1. Velocity Fields

The PIV acquired velocity fields are shown in vector plots combined with coloured contour plots of the absolute velocity projected onto the measurement layer xy or xz. The results are standardised with the circumferential velocity at the outside radius $u_a = 72.22$ m/s. Low velocities down to zero are blue coloured. With rising velocities the colour changes over cyan, green, yellow and orange to the red coloured maximum standardised velocity of v_{xy}/u_a or $v_{xz}/u_a = 1.2$. Please excuse the missing grey colour bar for printed paper. This is due to the large amount of information. If needed please asked for a coloured digital copy of this paper at schroll@ila.de. Right sided on the picture you find information about the position of the laser light sheet (measurement layer) relative to the impeller and casing cover, about the coordinates, rotational direction, camera cut-out and operating point. Figures 6 and 7 show the absolute velocity v_{xy}/u_a in the axial layer z = 0 mm for over-load and extreme part-load in comparison and to give an example which pictures are used in Figure 8. Fig. 8 is an overview of the absolute velocities and the flow field variations inside the Side Channel Vacuum Pump for uniformly distributed selected axial xy-measurement layers with respect to the operating point. The deep blue regions for negative z inside the impeller (2 left sided rows) are caused by vanes solid material or shadows considered as zero.

On the other hand Figure 9 gives an overview about the absolute velocities for radial xz-measurement layers with respect to the operating point. Figure 10 shows the meridional slices with respect to operating point generated out of the 3D volume data set of velocities.

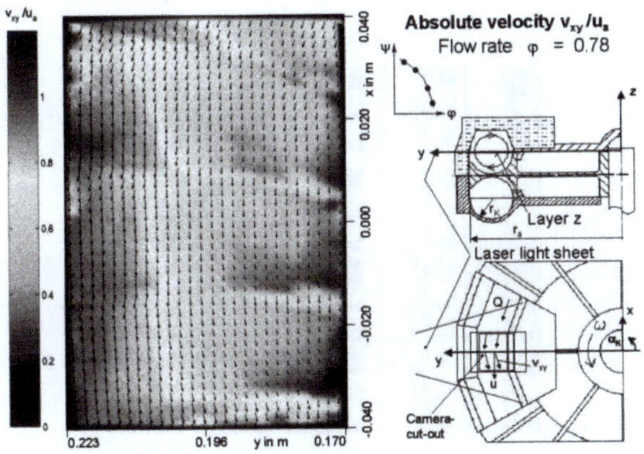

Fig. 6. Absolute velocity v_{xy}/u_a in the axial layer z = 0 mm for over-load

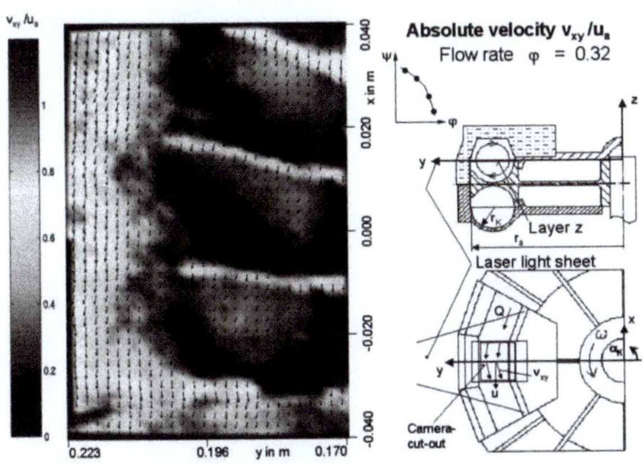

Fig. 7. Absolute velocity v_{xy}/u_a in the axial layer z = 0 mm for extreme part-load

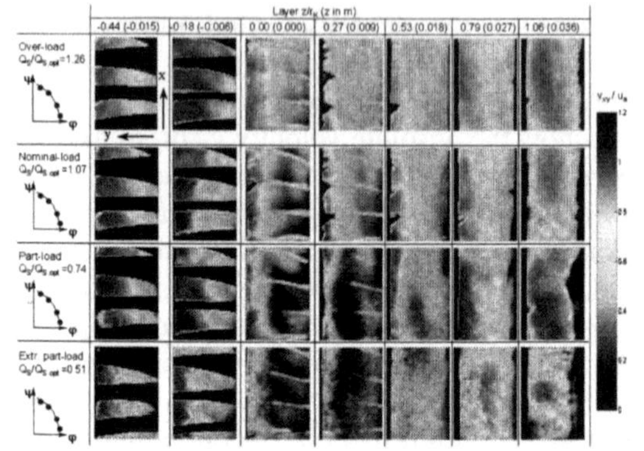

Fig. 8. Selected overview of absolute velocities v_{xy}/u_a in axial layers z = -15...36 mm with respect to operating point

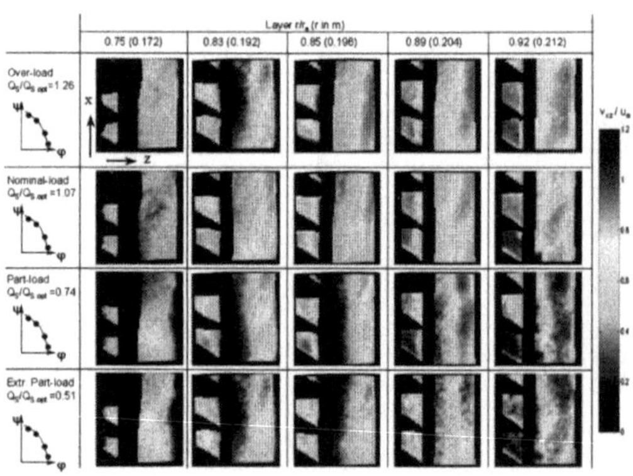

Fig. 9. Overview of absolute velocities v_{xz}/u_a in radial layers r = 172...212 mm with respect to operating point

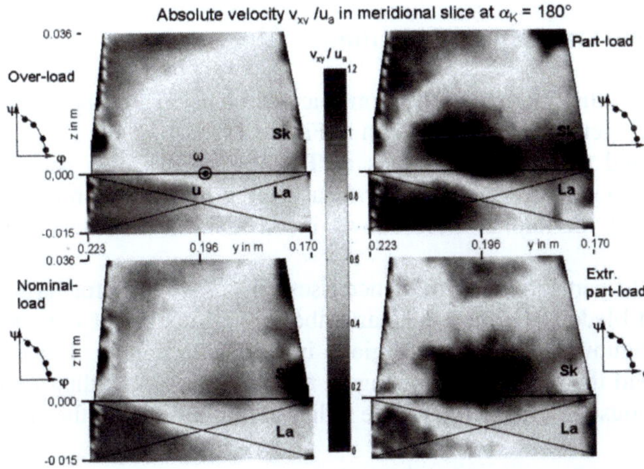

Fig. 10. Overview of absolute velocities v_{xz}/u_a in meridional slice with respect to operating point

The flow field for overload inside the impeller is homogenous and equal to the local circumferential speed. Changing from over-load to part-load areas with velocities v<u increases from the inside radius to the whole channel and reaching down to the blade bottom z = -15 mm. Areas of low velocities are recognizable at the trailing edges of the blade at z = 0 mm, increasing in size changing to part-load and reaching already the next blade for extreme part-load. At z = 9 mm this separated areas accumulate to the low speed centre of circular flow, best seen in the meridional slices. The size of this vortex and velocities gradient increases again throttling the volume flow. At z = 18 mm the recognizable influence of interaction from the impeller disappears.

The meridional slice for over-load shows rising velocities inside the impeller from inside to outside radius equal to the local circumferential speed too. Additional visible is the crossover of this high velocities from inside the impeller into the free side channel at the outside radius, ending in a large region under the casing cover at z = 36 mm. This structure stays stabile for all operating points, but more distinctive and with higher gradients. Locally velocities of v>u are reached. From the inside radius close above the impeller a large area of very low velocities is formed for less volume flow. Together with the axial and radial vector plots this structure indicates a circular flow in clockwise direction, leaving the impeller at outside radius and re-enter at inside radius. Notable is the decreasing of this structure for extreme part-load. At this extremely low flow rate the flow structure seems to collapse.

4.2. Unsteady Oscillating Velocities

The meridional distribution of the isotropic turbulence $Tu_{i.v.K}$ between 0...60% in respect to the operating point is shown in Figure 11.

At over- and nominal-load values of $Tu_{i.v.K}$>20% are only sporadic and predominantly at boundaries with higher measuring errors due to reflexion. The bigger part of the meridional slice shows inconsiderable turbulence values under 10%.

Changing to part-load the turbulence rises up to 30...50% for larger areas, specially behind blade's trailing edges and above the impeller at outside radius, but without contiguous homogenous regions in the side channel. Proceeding to extreme part-load the turbulence rises again to $Tu_{i.v.K}$>50%, building now a contiguous homogenous regions at the outside radius at half altitude of the side channel.

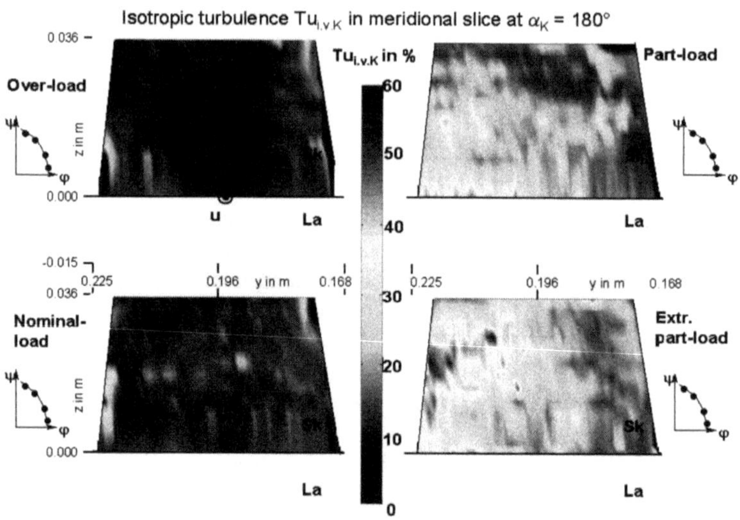

Fig. 11. Overview of isotropic turbulence in meridional slice with respect to operating point

4.3. Bernoulli Constant

The contour plots of the Bernoulli constant in Figure 12 uses different colour legends to distinguish the single distributions, regardless to the large range over all four operating points.

Pressure [5] and velocity distributions are in opposite directions along the altitude of the side channel and along the blade to higher radius. This allows to suppose a nearly continuous Bernoulli constant, meaning an iso-energetic flow. But there is no complete compensation over the whole side channel slice plane, indicating again the unsteady flow. In the region of high y and z values are high velocities together with high pressures. Energy conversion occurs principally in this regions.

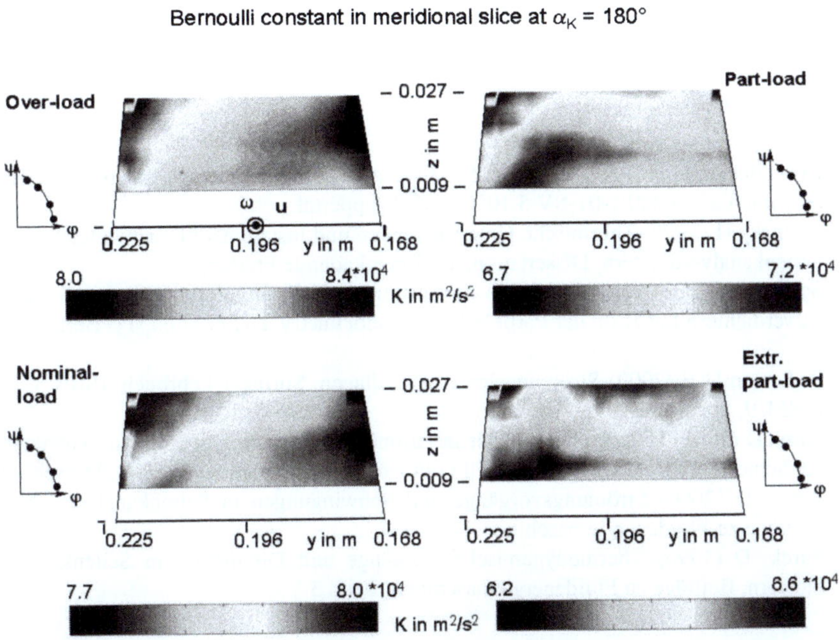

Fig. 12. Overview of Bernoulli constant in meridional slice with respect to operating point

5. Summary

The primary objective, an advanced analysis of the enforced turbulent flow in the side channel based on experimental PIV results, is reached. Acquired velocity fields are correlated to parallel measured pressure fields. The results giving a detailed insight into the working mechanism and the specific energy distribution at the rotational angle of $\alpha_K = 180°$.

Notable in the velocity fields, turbulence and Bernoulli distributions is the obvious interaction of the rotating cascade to the flow for higher radius with increased power pulse current transports in opposite to the inside radius area with nearly no noticeable work transfer to the fluid.

In over-loaded and nominal-loaded operation the flow follows closely to the tangential circumferential velocity. The work transfer must be done predominantly by the enforced unsteady and turbulent flows of the rotating cascade, inducing a power pulse current transport sufficient to generate the low differential pressure at high volume flows.

Changing to part-load operation the flow gets more and more in the typical circular flow, leaving the impeller with high velocities up to v>u at the outside radius and re-entering the impeller on a very low velocity level at the inside radius. The loss in velocity is transferred into pressure energy. This explains the Bernoulli constant distribution with high values near the casing cover, couldn't be caused by interaction of the rotating cascade.

References

[1] Gebr. Becker GmbH & Co. (1998) Seitenkanal-Vakuumpumpe und –Verdichter, Baureihenheft SV 5.300/1-01-SV 5.1050/1-01, Wuppertal

[2] Galinsky H (1999) Thermische Untersuchungen und instationäre Druckverteilungen in Seitenkanalverdichtern, Dissertation, TU Bergakademie Freiberg

[3] Schroll M (2003) Untersuchungen der instationären Strömungsvorgänge in Seitenkanalverdichtern mit Hilfe der Particle-Image Velocimetry, Dissertation, TU Berlin

[4] Siekmann H E (2000) Strömungslehre, Grundlagen, Springer-Lehrbuch, ISBN 3-540-66851-9, Springer-Verlag, Berlin

[5] Surek D (2001) Untersuchungen der instationären Strömungsvorgänge in Seitenkanalverdichtern, Zwischenbericht zum DFG-Forschungsvorhaben Su 163/3-1, Merseburg

[6] Surek D (2000) Strömungsvorgänge und Schwingungen in Seitenkanalverdichtern, Beiträge zu Fluidenergiemaschinen, Band 5

[7] Surek D (1996) Thermodynamische Vorgänge und Dissipation in Seitenkanalverdichtern, Beiträge zu Fluidenergiemaschinen, Band 3

MIX
Papier aus verantwortungsvollen Quellen
Paper from responsible sources
FSC® C105338

If you have any concerns about our products,
you can contact us on
ProductSafety@springernature.com

In case Publisher is established outside the EU,
the EU authorized representative is:
Springer Nature Customer Service Center GmbH
Europaplatz 3, 69115 Heidelberg, Germany

Printed by Libri Plureos GmbH
in Hamburg, Germany